Aquatic Protected Areas as Fisheries Management Tools

Funding for this book was provided by

National Oceanic and Atmospheric Administration
Office of Oceanic and Atmospheric Research
National Sea Grant College Program

Fisheries and Oceans Canada

Great Lakes Fishery Commission

National Oceanic and Atmospheric Administration National Ocean Service

National Oceanic and Atmospheric Administration Fisheries

National Oceanic and Atmospheric Administration
Undersea Research Program

U.S. Geological Survey Biological Resources Discipline

U.S. Minerals Management Service

Aquatic Protected Areas as Fisheries Management Tools

Edited by

J. Brooke Shipley

American Fisheries Society, 5410 Grosvenor Lane, Suite 110,
Bethesda, Maryland 20814, USA

American Fisheries Society Symposium 42

Proceedings of the American Fisheries Society/Sea Grant Symposium
Aquatic Protected Areas as Fisheries Management Tools
Held in Quebec City, Quebec, Canada
11–12 August 2003

American Fisheries Society
Bethesda, Maryland
2004

Suggested citation formats are:

Entire book

Shipley, J. B., editor. 2004. Aquatic protected areas as fisheries management tools. American Fisheries Society, Symposium 42, Bethesda, Maryland.

Chapter in book

Houde, E. D., and S. J. Roberts. 2004. Marine protected areas: an old tool for new circumstances. Pages 23–35 *in* J. B. Shipley, editor. Aquatic protected areas as fisheries management tools. American Fisheries Society, Symposium 42, Bethesda, Maryland.

Cover image by David McClellan, National Oceanic and Atmospheric Administration (NOAA) Fisheries

Printed in the United States of America on acid-free paper.

Library of Congress Control Number 2004108653
ISBN 1-888569-62-X
ISSN 0892-2284

American Fisheries Society web site: *www.fisheries.org*

American Fisheries Society
5410 Grosvenor Lane, Suite 110
Bethesda, Maryland 20814-2199, USA

For my parents,
'cause they knock my flip-flops off.

Contents

Preface

The American Fisheries Society (AFS) and the National Sea Grant College program continued their past collaborative initiatives to provide in-depth discussion of timely issues by hosting the featured symposium, "Aquatic Protected Areas as Fisheries Management Tools," held during the 2003 AFS Annual Meeting on August 11–12, 2003, in Quebec City, Quebec, Canada. This third collaboration follows previous AFS/Sea Grant symposia, the 1998 "Fish Habitat: Essential Fish Habitat and Rehabilitation" and the 2001 "Fisheries in a Changing Climate."

The aquatic protected areas (APA) symposium focused upon the recent shift in management attitudes to change the single-species assessments to ecosystem-based management. Such protected areas are being looked upon as means to accomplish a variety of objectives, including (but not restricted to) protection of fish stocks, control of fishing effort, protection of critical fish habitat, creation of spawning and recruitment refugia, and enhancement of species biodiversity. However, the principal focus for this project was limited primarily to a critical examination of the role of protected areas as tools for fisheries management. Some of the topics covered included the design and selection of APAs, evaluation of effectiveness, management issues, and economic and legal aspects, as well as the presentation of various case studies. Internationally recognized scholars were invited to present and discuss the concepts and science behind aquatic protected areas and their use as management tools.

I would like to extend a thank you to our keynote speakers and panelists. Daniel Pauly, Trevor Ward, and Ed Houde presented to a packed room and were very much appreciated for imparting cutting-edge information. In addition to the presenters, four panelists from each sector of fisheries debated the topic of reserves for management use. Bill Amaru presented the viewpoint of commercial fishermen; Tundi Agardy represented the non-governmental organizations; Ryck Lydecker gave the recreational fisherman's viewpoint; and Rebecca Lent contributed the governmental perspective. Thank you for engaging the attendees in such lively debate and for enlightening all present about each sector's views.

Ron Baird and Gus Rassam of Sea Grant and AFS, respectively, are the principal organizers and leaders of the joint AFS/Sea Grant Fellowship, which enables a graduate student to interact with the some of the best leaders and researchers in the fisheries field. It was my honor to receive this fellowship, and I would like to extend my deepest heartfelt thanks to them for this wonderful opportunity.

With this opportunity, however, came the very large responsibility for assisting the steering committee to organize, schedule, and manage the symposium itself. The initial meeting of all parties was during the 2002 AFS Annual Meeting in Baltimore, Maryland, where I received much-needed advice on the work that was ahead of us. I would like to thank each member of the steering committee; without you, we would still be developing a logo: Emory Anderson, Robert Brock, Elizabeth Burkhard, Colleen Charles, Barry Costa-Pierce, Carlos Fetterolf, Kenneth Frank, Churchill Grimes, Bill Haglan, Thomas Hourigan, Chuck Krueger, Scott McKinley, Barbara Moore, Gus Rassam, Robert Stickney, James Tilmant, Charles Wahle, and Susan White. While every one of these steering committee members was always available to answer questions or provide advice, there are those I would especially like to acknowledge for the assistance they gave me above and beyond the call of duty: Emory, Robert, Gus, and Carlos, thank you.

Quebec City was a picturesque location, and I would like to thank those responsible for the organizing and planning of this conference site. I would like to acknowledge Martin Castonguay and Christine Fletcher, who spent countless hours helping to get our symposium off the ground during the planning phase.

All members of the National Sea Grant Office and the American Fisheries Society headquarters provided support whenever it was needed. In particular, I would like to thank Lee Benaka, the original AFS/Sea Grant fellow, who met with me on my first day and imparted much-used advice, as well as Betsy Fritz, Jennifer Gallivan, Gail Goldberg, Jeremy Corff, Mark Ehrick, and Tracy Klein, who continually assisted me throughout this process to assure a quality AFS symposium and these subsequent proceedings.

This symposium would not have been possible without the support of sponsors, to which I am indebted. Special thanks to the American Fisheries Society, Fisheries and Oceans Canada, Great Lakes Fishery Commission, National Ocean Service, National Oceanic and Atmospheric Administration (NOAA), NOAA Fisheries, National Park Service, National Sea Grant Office, Undersea Research Program, U.S. Geological Service–Biological Resources Discipline, U.S. Fish and Wildlife Service, and U.S. Minerals Management Service.

Finally, I would like to thank my family and friends for the support they have given me over the past 18 months. I would also like to thank my dissertation committee for granting me a leave from my doctoral program in order to accept the AFS/Sea Grant fellowship.

It has been my pleasure to be involved in this enterprise because of the people I have worked with, the knowledge I have gained, and the opportunities that have arisen. I hope you enjoy reading these proceedings and, as I trust that you will as much as I do, appreciate the reviewers who evaluated and ensured the quality.

J. Brooke Shipley
Editor

Foreword

RONALD C. BAIRD

National Sea Grant College Program,
National Oceanic and Atmospheric Administration,
1315 East-West Highway, SSMC-3, Eleventh Floor,
Silver Spring, Maryland 20910, USA

The year 1998 marked the establishment of a fellowship sponsored jointly by the National Oceanic and Atmospheric Administration's (NOAA) Sea Grant Program and the American Fisheries Society (AFS). The recipient was to help edit a proceedings volume of a special symposium at the AFS Annual Meeting. The symposium was to be organized around a "management critical" theme in fisheries science. The idea was to create a forum that helps frame the critical policy issues from the perspectives of major stakeholders, delivers a synopsis of the "state of the science" on the subject, and provides guidance as to future policy needs and research direction.

Our first symposium, in 1998, led to the publication of a well-received volume on essential fish habitat. A sequel, in 2001, on climate and fisheries proved the paradigm sound. With encouragement from Rolland Schmitten and subsequent directors of the National Marine Fisheries Service (NMFS) and AFS, we continue the series with this third symposium, entitled "Aquatic Protected Areas as Fisheries Management Tools." Again, we address a topic of great concern for contemporary fisheries management both in terms of policy and science supporting that policy. The issue of protected areas is not only timely but will assume increasing importance in a fisheries management context worldwide.

Recognition and thanks are in order to the many symposium cosponsors and to AFS for their direction and hard work. A special word of praise to the Sea Grant Fisheries Theme Team, the cosponsors steering committee, and especially Emory Anderson, the NMFS liaison to the National Sea Grant Office. Finally, we all owe a debt of gratitude to Brooke Shipley, AFS/Sea Grant Fellow, and to the many contributors who have made this third volume a great sequel to those that have gone before.

The intent of these symposia is to challenge us to develop new ideas and approaches to the very serious and difficult business of practical fisheries management. As our first two volumes and a voluminous literature make clear, declining yields of commercial species, climate change, and degradation of ecologically important habitats point to the difficulty and shortcomings of current fisheries management practices.

With continuing human population growth and associated impacts to fisheries resources, there is heightened interest in tools, such as aquatic protected areas, that more effectively conserve stocks that are under increasing stress. Likewise, social values and changing economic factors impinge on the fisheries management equation. Examples include protection of unique environments (e.g., coral reefs), the growth of interest in recreational fishing, and marine aquaculture. In the case of aquatic protected areas, as with other management innovations, significant controversy can arise regarding the implementation of such tools, their purpose, and the science base.

This volume is a sequel to the broader 2001 study by the National Research Council on marine protected areas. This volume focuses specifically on the scientific basis for the use of aquatic protected areas as a tool in fisheries management and on the questions of governance in that context.

Several themes are evident here, including that holistic (ecosystem-based) management approaches are increasingly viewed as essential to sustainable fish production; that site-based, geographically explicit knowledge on relevant time and space scales is critical to effective management; and that multiple-use reconciliation and adaptive management practices that include protected areas as one of many tools are integral to practical, successful resource management.

This volume, sponsored by AFS, NOAA's Sea Grant Program, NMFS, and other cosponsors, represents an important milestone in developing the basis for the practical application of an increasingly important management option in the effective conservation of invaluable natural resources.

A Summary and Overview of the Symposium

J. Brooke Shipley

American Fisheries Society/Sea Grant Fellow, American Fisheries Society,
5410 Grosvenor Lane, Suite 110, Bethesda, Maryland 20814, USA

The third special symposium to originate from the American Fisheries Society (AFS) and National Sea Grant College collaboration, "Aquatic Protected Areas as Fisheries Management Tools," was held at the AFS Annual Meeting in Quebec City, Quebec, Canada, on August 11–12, 2003. This groundbreaking symposium resulted in a collection of papers that will serve as scientific road maps for future action.

The highly discussed, studied, and debated topic of marine protected areas (MPAs) has been expanded and more broadly defined, at least for the purposes of this project, to include freshwater concerns and engage government agencies whose research and management mandates include freshwater fisheries resources. While it may be clear that protected areas are an appropriate way to protect habitat, cultural resources, and biodiversity, the current debate revolves around whether they are appropriate as management tools. Therefore, there was agreement that the symposium should focus on the use of no-take aquatic protected areas as fisheries management tools. The focus was on the application of this management approach rather than on arguing whether it was a needed or valid tool.

The symposium opened with a panel of constituent speakers, followed by a short question-and-answer session. Following the panelists, keynote speakers introduced the symposium by providing a general overview of protected areas in terms of fisheries management, ecosystem management, take versus no-take, biodiversity, cultural resources, habitat, and so on. The remaining presentations fell into broad categories of design and selection, evaluation and effectiveness, case studies, and economic and legal issues. At the conclusion of the symposium, the constituent panel reconvened for brief summations and a final question-and-answer session.

The constituent stakeholders panel was comprised of a representative from each of four major fisheries sectors: government management, commercial fishing, recreational fishing, and environmental organizations. After each of the representatives completed their presentations, the floor was opened for questions. The common theme was the importance of understanding and appreciating the need to protect our living resources but doing so with caution. The need for scientific research and involvement by all stakeholders in creating a given protected area was emphasized. A general fear was expressed that regions would be set aside without proper scientific reasoning and without the input from those who historically use the areas, thereby resulting in fishermen, both recreational and commercial, being excluded, with no scientific basis. While environmental groups generally support the idea of areas closed to human impacts, without public support to close an area to fishing and without the supporting scientific reasoning, there would be no point in pursuing the closure.

After the constituent panel made its presentations and fielded questions from the audience, the keynote speakers gave in-depth perspectives on protected areas and their use. Ed Houde (Chesapeake Biological Laboratory, Solomons, Maryland) presented a comprehensive review of protected areas showing that they had been used historically but were now being employed in different ways (i.e., as management tools). He concluded "that MPAs, combined with other management tools, can protect habitats and nursery grounds, reduce bycatch, lower fishing mortality on pre-recruits, promote 'spillover' of recruited fish to unprotected areas, and provide a buffer against uncertainty. Poor knowledge of stage-specific dispersal patterns of fish and of fishermen behavior (i.e., distribution of effort) beyond reserve boundaries constrain endorsements of MPAs by some scientists and managers."

The second keynote speaker, Trevor Ward (University of Western Australia, Crawley), stated, "Four main types of potential benefit can be identified for MPAs in fisheries—biological support for the exploited stock, socioeconomic benefits for the fishers, biodiversity benefits for the fished ecosystems, and increased fishery stability. The extent to which any MPA will achieve these benefits depends on its design, its integration with the broader fishery management system, and the effectiveness of the existing stock management."

Daniel Pauly (University of British Columbia, Vancouver) completed the keynote presentations by assuring the audience that marine reserves are "not a new invention of arm-chair ecologists, designed to torment hardworking, overregulating fishers. Rather, they would only reestablish the natural structures which have enabled earlier fisheries to maintain themselves for centuries. Getting this message across to a wide public is urgent, given the hardening stance of the fishing industry against marine reserves and the fact that, presently, they cover an infinitesimally small fraction of the world ocean."

After the constituent panel and the keynote speakers introduced the basics of protected areas and the reasoning both for and against them, the scientific papers began, with the first group covering the design and selection of protected areas. This group included papers from Louis Botsford (University of California, Davis), Robert Brock (National Oceanic and Atmospheric Administration [NOAA] Fisheries, Silver Spring, Maryland), Mary Yoklavich (NOAA Fisheries, Santa Cruz, California), and Lisa Wooninck (NOAA Fisheries, Santa Cruz, California). While Brock reported on one of the few marine reserves in the United States (Dry Tortugas National Park, Florida) and how it was created and Yoklavich spoke on the 2001 creation and use of another reserve near the Pacific Coast (Cowcod Conservation Areas), Botsford described the use of models to help in the design of marine reserves and their potential effect on existing fisheries. Wooninck finished this section by describing a recent study that evaluated the current protected areas within the NOAA Fisheries realm through an "assessment cover[ing] four broad categories of information: (1) basic attributes: where, who, when, and how of site establishment; (2) level of protection and restrictions; (3) management and support; and (4) goals and monitoring."

Robert Shipp (University of South Alabama, Mobile), Andre Punt (University of Washington, Seattle), and Patrick Christie (University of Washington, Seattle) gave presentations on how to evaluate the usage of protected areas. Shipp presented one of the few dissenting scientific voices, stating that "increased harvest from fish stocks is often included among the many benefits attributable to marine reserves. Examination of the most targeted finfish stocks from the coasts of the United States finds that this is rarely the case. This is because the stocks are highly mobile or are not overfished, and overfishing is not occurring. Traditional fishery management tools, such as seasonal closures, quotas, and bag limits, are proving effective in management of most marine stocks and are preferred over marine reserves for purposes of maintaining or increasing finfish yield." Punt pointed out that "MPAs provide a new opportunity to estimate quantities, such as natural mortality and growth, needed when defining management reference points. However, MPAs, if effective, will further distort any spatial variation in the density and age structure of the population, thereby further increasing the extent to which the homogeneous distribution assumption underlying most stock assessments is violated." Christie concluded this group of papers by describing the work he had been part of for many years in Southeast Asia. In that part of the world, it is evident that while a protected area may benefit the biological species, the social implications resulting from the protection could classify the reserve as a failure. He said, "In general, the 'too little, too late' scenario results in a poor understanding of frequently contentious social interactions operating on multiple levels (local, national, international, gender, class, ethnicity), unintended negative consequences, missed opportunities for positive change and reallocation of resources, and an incomplete scientific record. Marine protected areas are destined to fall short of biological and social goals unless social sciences are effectively integrated into the design and evaluation process."

Ten case study papers were presented by Steven Murawski (NOAA Fisheries, Woods Hole, Massachusetts), James Bohnsack (NOAA Fisheries, Miami), Rod Bertelsen (Florida Fish and Wildlife Commission, Marathon), Janet Ley (Australian Maritime College, Tasmania), Malcolm Clark (National Institute of Water and Atmospheric Research, New Zealand), Ken Frank (Fisheries and Oceans Canada, Dartmouth, Nova Scotia), David Reid (Ontario Ministry of Natural Resources, Owen Sound), Jim Taggart (U.S. Geological Survey, Juneau, Alaska), and Jonathan Fisher (University of Pennsylvania, Philadelphia). Topics spanned many fields and geographical regions, such as trawl closures on sea mounts in New Zealand to protect orange roughy *Hoplostethus atlanticus*, protection of reef fish and spiny lobster *Panulirus argus* in the Florida Keys area, groundfish closures in the Gulf of Maine and southern New England as well as off Nova Scotia, lake trout protection in Lake Huron, and gill-net closures for a game fish in tropical Australia. In virtually all cases, the fish stock(s) in question benefited in terms of increased numbers, size, age composition, etc. from the protection afforded them by the prohibition of harvesting in specific areas.

The final set of papers dealt with economic and legal issues. Rita Curtis (NOAA Fisheries, Silver Spring, Maryland) discussed the economic issues revolving around and due to protected areas. Kristen Fletcher (University of Mississippi Sea Grant Law Center) gave a memorable presentation about the legal aspects of protected areas, saying, "Preserving public lands through restrictive uses is not a new concept in resource management; it has been recognized in U.S. law for over 100 years. Generally, however, these efforts have been directed to areas above the high water mark. With the increase in use, questions regarding the legal underpinnings of no-take zones—the authority necessary to create and manage no-take zones—as well as other potential legal challenges have arisen."

The symposium concluded with the reconvening of the constituent panel. The representatives for government management, environmental organizations, commercial fishing, and recreational fishing summarized their thoughts and views and, again, fielded questions from the audience. While many people support the idea of protected areas, there is a desire for intelligent fisheries management. Stakeholders want managers to use science as well as public opinion to determine the structure of protected area networks, particularly since many fisheries management programs are beginning to develop "ecosystem-based management" practices. The general consensus was that protected areas are useful tools for fisheries management but that scientific research must be the background for the implementation and development of reserves. President Franklin D. Roosevelt once said, "There is nothing so American as our national parks . . . The fundamental idea behind the parks . . . is that the country belongs to the people, that it is in process of making for the enrichment of the lives of all of us." He was right.

Stakeholder Perspectives on Aquatic Protected Areas

Ronald C. Baird

National Sea Grant College Program,
National Oceanic and Atmospheric Administration,
1315 East-West Highway, SSMC-3, Eleventh Floor,
Silver Spring, Maryland 20910, USA

Essential fish habitat (EFH) provisions of the Sustainable Fisheries Act of 1996 marked a major milestone for fisheries management. The law explicitly requires that fisheries managers consider habitat quality in management plans. Thus was established the beginning of a legal framework for ecosystem or holistic approaches to fisheries management. Essential fish habitat clearly targets habitat degradation and associated causes of loss of fisheries resources including fishing activities themselves. Current practices have not led to sustainable fisheries, and habitats critical to fish production continue to degrade. Aquatic protected areas (APAs) have emerged from holistic management ideas and are designated, discrete geographic areas, with the purpose of conservation or enhancement of fisheries production. However, while straightforward in concept, APAs pose many practical problems as management tools. Foremost is the issue of multiple-use reconciliation. Public policy is a politically mediated process. Stakeholder interests will, therefore, ultimately determine the policy dimensions of APAs in management contexts. Critical first steps in helping develop APAs as practical management tools are to both engage stakeholder concerns and provide scientific information as a foundation for informed public policy.

To initiate those first steps, various stakeholders made formal presentations at the beginning session and provided remarks in response to the ensuing scientific presentations at a discussion forum ending the symposium. What follows is a summarization of those sessions, including audience input.

Stakeholder perspectives were presented from the following sectors: fisheries management, conservationist non-governmental organizations (NGOs), commercial fishing, recreational fishing, and federal science management. Input was also provided by a diverse audience of users and scientists. There was general agreement that aquatic protected areas, as defined here, have been commonly used in fisheries management practice for some time. Moreover, the traditional reasons for implementing such protected zones have been well understood, and, for the most part, there is agreement on traditional objectives of such regulations. That said, concern was expressed about the proliferation of APAs, new and novel applications, and the lack of both a sound, scientific foundation and meaningful engagement of stakeholder interests by management agencies. Most acknowledged that the context for use must be adequately defined and that APAs do not solve all management problems.

More specifically, fisheries managers are bound by legislation to prevent overfishing, rebuild stocks, and manage living resources on an ecosystem basis. Already in law are mandates for Federal Fisheries Management Zones, habitats of particular concern (EFH), and rare or sensitive habitats. Hence, the current policy milieu specifically authorizes APAs as a tool in fisheries management practice.

The NGOs view APAs in a broader conservation context and generally embrace APAs as an effective tool in the transition of society to sustainability. They are viewed as versatile, having many objectives while acting as both a forum for multiple-use reconciliation among users and valuable in evaluating the efficacy of management.

For commercial fishermen, closed areas are generally well accepted as long as the benefits are well articulated and the needs of all users are taken into consideration. However, proliferation is seen as an issue. Cost, benefits, risks, and ecosystem management in the context of all users must be part of an open democratic process.

The recreational fishing community, a large and growing stakeholder group, is especially concerned about proliferation and extremism in the application of APAs. An anti-fishing mentality and a view of APAs as a blanket solution to environmental problems without identifying the primary causes of the problem are major issues. A sound scientific rationale for any APA designation is crucial.

Comments from the constituent's forum at the end of the symposium further reinforced many of the initial perspectives while highlighting a number of others based on the many presentations made over the 2-day period. Among the issues that were reinforced was the criticality of public support and the current high level of confusion about objectives of APAs. Scientific research is seen as essential for the long term. In addition, specific indicators of performance consistent with those objectives need be articulated. Finally, a comprehensive monitoring program adequate to measure those indicators over the lifetime of the designated APA should be integral to the designation process.

It is acknowledged that, in a great many current APAs, increases in local biomass and individual mean size are evident in observed stocks. Likewise, evidence is accumulating in catch data that indicates increased yield from areas adjacent to APAs. The multiplicity of regulatory jurisdictions, policy objectives, and legal mandates has generated considerable confusion and misunderstanding. Greater clarity of purpose and a simplified legal and regulatory framework are needed.

Successful application of APAs must have a strong regional basis and consensus. Place-based considerations and the importance of local engagement were emphasized. Aquatic protected areas are often of multi-year duration. Consequently, there was concern that adequate planning, performance measures, and a long-term resource base be identified before implementation. Decision rules and periodic cost–benefit analyses are important throughout the life of the project.

In summary, current management practices have clearly fallen short of sustainability of harvestable fisheries resources. As we transition to ecosystem approaches to fisheries management practices, APAs can be valuable tools in that transition. The application of APAs, however, must be consistent, with clear, well-understood, and accepted objectives based on sound science; realistic, demonstrable performance measures; and stakeholder engagement.

Symbols and Abbreviations

The following symbols and abbreviations may be found in this book without definition. Also undefined are standard mathematical and statistical symbols given in most dictionaries.

A	ampere		in	inch (2.54 cm)
AC	alternating current		Inc.	Incorporated
Bq	becquerel		i.e.	(id est) that is
C	coulomb		IU	international unit
°C	degrees Celsius		J	joule
cal	calorie		K	Kelvin (degrees above absolute zero)
cd	candela		k	kilo (10^3, as a prefix)
cm	centimeter		kg	kilogram
Co.	Company		km	kilometer
Corp.	Corporation		l	levorotatory
cov	covariance		L	levo (as a prefix)
DC	direct current		L	liter (0.264 gal, 1.06 qt)
D	dextro (as a prefix)		lb	pound (0.454 kg, 454g)
d	day		lm	lumen
d	dextrorotatory		log	logarithm
df	degrees of freedom		Ltd.	Limited
dL	deciliter		lx	lux
E	east		M	mega (10^6, as a prefix); molar (as a
E	expected value			suffix or by itself)
e	base of natural logarithm (2.71828…)		m	meter (as a suffix or by itself); milli
e.g.	(exempli gratia) for example			(10^{-3}, as a prefix)
eq	equivalent		mi	mile (1.61 km)
et al.	(et alii) and others		min	minute
etc.	et cetera		mm	millimeter
eV	electron volt		mol	mole
F	filial generation; Farad		N	normal (for chemistry); north (for
°F	degrees Fahrenheit			geography); newton
fc	footcandle (0.0929 lx)		N	sample size
ft	foot (30.5 cm)		NS	not significant
ft³/s	cubic feet per second (0.0283 m³/s)		n	ploidy; nanno (10^{-9}, as a prefix)
g	gram		o	ortho (as a chemical prefix)
G	giga (10^9, as a prefix)		oz	ounce (28.4 g)
gal	gallon (3.79 L)		P	probability
Gy	gray		p	para (as a chemical prefix)
h	hour		p	pico (10^{-12}, as a prefix)
ha	hectare (2.47 acres)		Pa	pascal
hp	horsepower (746 W)		pH	negative log of hydrogen ion
Hz	hertz			activity

ppm	parts per million	V	volt
ppt	parts per thousand	V, Var	variance (population)
qt	quart (0.946 L)	var	variance (sample)
R	multiple correlation or regression coefficient	W	watt (for power); west (for geography)
r	simple correlation or regression coefficient	Wb	weber
		yd	yard (0.914 m, 91.4 cm)
rad	radian	α	probability of type I error (false rejection of null hypothesis)
S	siemens (for electrical conductance); south (for geography)	β	probability of type II error (false acceptance of null hypothesis)
SD	standard deviation		
SE	standard error	Ω	ohm
s	second	μ	micro (10^{-6}, as a prefix)
T	tesla	$'$	minute (angular)
tris	tris(hydroxymethyl)-aminomethane (a buffer)	$''$	second (angular)
		$^{\circ}$	degree (temperature as a prefix, angular as a suffix)
UK	United Kingdom		
U.S.	United States (adjective)	%	per cent (per hundred)
USA	United States of America (noun)	‰	per mille (per thousand)

Part I:
Panelists

American Fisheries Society Symposium 42:3–4, 2004

Extended Abstract

Area-Based Management and Sustainable Fisheries under the Purview of the National Marine Fisheries Service (NOAA Fisheries)

WILLIAM T. HOGARTH, REBECCA LENT, AND ROBERT J. BROCK[1]

National Oceanic and Atmospheric Administration, National Marine Fisheries Service,
1315 East-West Highway, Silver Spring, Maryland 20910-3282, USA

The National Oceanic and Atmospheric Administration's National Marine Fisheries Service (NOAA Fisheries) is entrusted with the management of the living marine resources (LMRs) of the United States. Since the United States possesses the world's largest exclusive economic zone (EEZ), managing LMRs within this broad expanse has proven difficult and, at times, highly contentious. There are multiple uses of the EEZ (e.g., extraction, tourism, conservation) that are sometimes very much at odds with one another. Mandates call for managing sustainable fisheries (extraction) while at the same time protecting marine mammals, endangered species, and non-target species (conservation).

Traditional fisheries management in the United States is carried on today in a way that is very similar to that used by the U.S. Commission of Fish and Fisheries (the precursor to NOAA Fisheries) since 1871. The intent of fisheries management has always been to control fishing mortality rates. This has been accomplished by regulating the type and size of the gear used, the times of the year where fishing is open to extraction, and the locations where fishing is allowed. Through regulation of mesh net size, juvenile fish pass through the net and are protected. Regulating fishing gear such as bottom trawls seeks to protect both bottom habitat as well as demersal species. In recent times, the idea of limited entry (e.g., number of vessels permitted to fish), individual fishing quotas that contain total allowable catches, and the number of days at sea that are allowed for fishing are further attempts to lower fishing mortality rates and change the behavior of the fishers. Regulations on locations and times of the year that fishing is allowed have always attempted to protect species when they are most vulnerable to harvest (aggregating in order to spawn).

Fisheries are managed in an arena of uncertainty because of an incomplete understanding of complex trophic-level dynamics that drive fish populations as well as changing abiotic factors (NRC 1999). Realizing that single-species management does not adequately address issues such as trophic-level interactions, essential fish habitat, bycatch and discards, and abiotic forcing (NMFS 2001), NOAA Fisheries has implemented area-based management to deal with these uncertainties, and councils have done ecosystem-based management.

Throughout the EEZ, NOAA Fisheries has many area-based management units that offer additional protection to finfish, habitat, threatened and endangered species, and the habitat upon which they depend. The 23 NOAA Fisheries-administered federal fishery management zones found throughout the EEZ were created to better protect focal fish species and assemblages. For example, the Northeast Multispecies Closures on Georges Bank in the Gulf of Maine prohibit all bottom trawling for groundfish. This prohibition protects commercially important demersal species such as Atlantic cod *Gadus morhua*, haddock *Melanogrammus aeglefinus*, and winter flounder *Pseudopleuronectes americanus*. The 13 federal threatened and endangered

---

[1] Corresponding author: Robert.Brock@noaa.gov

critical habitat and species protected areas restrict types of fishing gear being deployed in specific geographic areas in order to protect endangered species such as the northern right whale *Eubalaena glacialis*. The Oculina Banks Habitat Area of Particular Concern off the Atlantic coast of Florida is one of the oldest protected areas (1984) administered by NOAA Fisheries and prohibits bottom trawling and other activities in order to protect the rare ivory tree coral *Oculina varicosa*. An excellent inventory of area-based protected sites can be found on NOAA's National Marine Protected Area Center web site (*http://mpa.gov*).

Although the term "marine protected area" seems to be relatively recent, area-based management has been implemented by NOAA Fisheries for decades. By protecting not only focal species (e.g., Atlantic cod) but also such things as predator–prey interactions and habitat, it is hoped that area-based management will be an improvement over the shortcomings of the single-species management of the past (NRC 2001). These improvements will be realized in rebuilt fish stocks and sustainable fisheries for the future.

References

NRC (National Research Council). 1999. Sustaining marine fisheries. National Academy Press, Washington, D.C.

NRC (National Research Council). 2001. Marine protected areas: tools for sustaining ocean ecosystems. National Academy Press, Washington, D.C.

NMFS (National Marine Fisheries Service). 2001. Marine fisheries stock assessment improvement plan. Report of the National Marine Fisheries Service National Task Force for Improving Fish Stock Assessments, 2nd edition, revised. NOAA Technical Memorandum NMFS-F/SPO-56.

American Fisheries Society Symposium 42:5–7, 2004

Synergy in Aquatic Protection:
A Pragmatic Approach to Protection and
Enhancement of Marine Ecosystems

WILLIAM AMARU

25 Portanimicut Road, South Orleans, Massachusetts 02662, USA

Abstract.—The protection of certain fishing grounds during spawning has traditionally played a role in fisheries management in New England and is a major part of the present-day stock rebuilding program. The closures of four large areas now closed year-round to gear capable of taking groundfish have resulted in the recovery of a number of stocks of important groundfish species in the Georges Bank and Gulf of Maine area and have been understood and accepted by many in the commercial fishing industry. Although successful, these and similar areas elsewhere do not constitute a needed ecosystem approach toward improved protection of fish and their habitat. The terrestrial example of the National Park system is suggested as the approach to be applied in the marine environment for the creation of marine reserves. Conservation organizations and regulatory agencies need to argue effectively to promote the advantages of protecting marine environments for the benefit of the fish species, their habitat, and the citizens that use those resources. The selection and establishment of protected areas must be based on clearly stated objectives and must address the needs of all users. While specific federal agencies are clearly mandated by law to take the lead in providing long-term protection for the marine environment, public involvement and opinion must be included.

"Closed Area" Fisheries Management in New England

My career as a commercial fisherman spans the past 30 years. When I started fishing as a long-liner from Cape Cod, Massachusetts, the industry was very much as it had been since early in the last century. A group of people from tight-knit communities and extended families, we fished much like our fathers before us. The worldwide depletions of commercially valuable fish were well underway but had not had a significant impact nor been fully appreciated or understood in the United States outside the small community of scientists who studied long-term and short-term trends in abundance. The need to describe what led up to the ultimate collapse and now ongoing rebuilding of many fish stocks is beyond the scope of this brief paper. It is necessary to point out, however, that protecting certain grounds where fish were known to congregate during spawning played a role in New England's fisheries management program since the 1970s; select-area protection remains a major aspect of rebuilding

to this day. In fact, some of the very first steps taken to curb overfishing were to seasonally close large areas of spawning grounds on Georges Bank and adjoining waters. While this was done originally without concern for protecting anything except the target fishes, the net result has been the creation of temporary de facto habitat protection zones. If these actions had been accompanied by meaningful reductions in fishing pressure, there is a very good chance that the horrific depletions we have seen in New England may not have occurred. Unfortunately, we will never know.

In the context of present management for New England groundfish protection, four large areas are now closed year-round to gear capable of taking groundfish. Other areas adjacent to the coast are closed from several to 6 months a year and form a matrix of "rolling closures." The purpose is to give spawning Atlantic cod *Gadus morhua* and other species an opportunity to mate unencumbered, protected from the perturbations of fishing gear as the spring spawning cycle "rolls" up the coast with the advance of the weather patterns. While none of these examples provides a zone of protection aimed to enhance and protect at the ecosystem level, they are tools in common use and are understood and accepted by many in the commercial fishing industry nationally.

The key to their being accepted in New England is the positive and ongoing effect they have had on rebuilding fish populations. Where stock rebuilding has occurred, there has been an observable connection to the large, closed areas adjacent to and within their ranges.

What seems so obvious today was not so obvious some years ago. Battles fought on the management level (by the New England Fishery Management Council) to select the areas, which now include over 11,112 km² (6,000 square nautical mi), were never easy. Over the period of time these areas have been closed, certain fish stocks within them have shown strong growth. It would appear the closures have, in fact, given these species the opportunity to spawn, grow, and, in most cases, help resupply surrounding waters with an increased number of new entrants to the fishery. Research indicates that complex bottom communities left "fallow" improve and offer increased protection to many food and prey species. Empirical data provided by fishing fleets working the areas surrounding the closed zones show catch per unit of effort has increased. Another data set showing improvement in these populations is the product of the National Oceanic and Atmospheric Administration (NOAA) Fisheries spring and fall surveys of groundfish of Georges Bank and the Gulf of Maine. The data currently show reemerging populations of yellowtail flounder *Limanda ferruginea*, haddock *Melanogrammus aeglefinus*, witch flounder *Glyptocephalus cynoglossus*, winter flounder *Pseudopleuronectes americanus*, and others, all at densities above the average swept-area coefficients for similar habitats outside the closures. Although not originally intended to be a specific beneficiary of these closed areas, sea scallops *Placopecten magellanicus* also increased greatly in abundance, and several years ago, when these areas were reopened to tightly controlled scalloping, tremendous yields were realized.

While never easy, the choice to provide special protection to selected grounds has proven itself. The evidence indicates that predatory apex bottom fish populations like the gadoids are benefiting. In the complex world of often politically charged resource allocation and management, any success story is welcome. This is an example of such a story. What remains missing from these examples is an ecosystem-wide approach toward greater protection from any introduced disturbance. That kind of protection would afford all species present an environment in which to be conceived, born, grow, and die naturally, anathema to some resource users, a long-awaited measure of protection to others.

Protection of Marine Ecosystems

There are some examples of synergy in habitat protection that we can draw on. The following best describes the way we might go forward.

Early in the 20th century, the decision to protect unique terrestrial habitats in North America was made by a giant in the fledgling American conservation movement. Theodore Roosevelt, whose passions included big game hunting and fishing, established the first national program to set aside unique, large land areas for the preservation and enhancement of the flora and fauna found there. The National Park program is a product of this forward-thinking president who understood the value of giving special protection to selected sights and who possessed the political acumen to execute them.

A century later, the same rationale applies to the marine environment. Efforts have been undertaken to identify and offer special protection to sensitive or rare biological niches around the coasts of the United States. Perhaps the best national example of this type of protection can be found in the National Marine Sanctuaries Program.

This program is an effort to start a national platform of underwater parks and reserves whose designation will afford them special protection from development and exploitation while allowing continued access to users. The designation of a marine reserve in the purest sense of the word, however, is conspicuously absent. Except for a few areas encompassing sensitive and rare coral sights in the southeastern United States and on the West Coast off Southern California and Alaska, there are few true marine "sanctuaries." The use of political acumen of the kind used by Teddy Roosevelt is missing in its pursuit. The highly charged world of resource allocation is at the center of this failure. Wherever one stands on the issue of designating marine areas for protection, modern conservation organizations and governmental agencies have not argued effectively to promote the advantages of protecting marine environments. Why is this so? What can be done to change it?

In the middle of the last century, writings by environmental leaders like Aldo Leopold and Rachel

Carlson created a national movement. An awareness of pollution brought about by rapid and often uncontrolled development caught on with the public. Following this revelation, state and local agencies entrusted to protect the health of freshwater and saltwater environments wrote scores of new regulations. The broader issue of protection for the marine environment through marine reserves has since received much in the way of discussion but little in the way of real protection. Examples of conservation organizations promoting closures of fishing grounds for the purpose of protecting habitat are common. What is far less common is connecting habitat protection to stock health and increased benefits for those who lose access to traditional fishing grounds. The means by which Teddy Roosevelt created a national movement to identify and protect habitats for the preservation of wild lands and animals was by associating the benefits that protection would bring to all users. Creatures and their habitats would be protected, but benefits would accrue to all citizens. Access would be increased but limited. People would become more aware of the diversity of habitats while not negatively impacting them.

Conclusion

We can see that habitat protection can result from a force of personality and political conviction, as in the power of persuasion Teddy Roosevelt used so effectively in the last century. This example, while important, does not provide an answer for addressing the slow progress made in setting aside marine environments for the greater good of the aquatic whole. What

lies at the center of implementing stronger protection for marine environments?

For me, the most fundamental explanation remains the same: address the needs of the marine environment in the context of all its users. Selection of protected sites must be based on a combination of the needs of all players. There needs to be accompanying justification and a real effort to provide equal benefits. Objectives need to be very clearly stated. Areas should be as large as needed but never larger than required to meet their specific objectives. The surrounding areas should fit together as a matrix of interconnecting habitat corridors where additional benefits can be distributed to adjoining ecosystems. Marine protection should not be seen as accruing only to the dynamic conditions within a site. They should be seen in their bigger context as affecting the flora and fauna of all the creatures associated with the protection.

The means to accomplish this goal are well known. They exist in the democratic, open, public process. The agencies mandated by law to carry out the work are well established, and their constituent commercial and recreational fishing communities and others are well versed in the process. In conjunction with the regional fishery management councils and the National Ocean Service Sanctuary Program, NOAA Fisheries should work with the public to move forward. The public should, and must be, considered equals in their knowledge, and their needs and opinions should never be underestimated. When these conditions are met, the real groundwork for providing long-term protection for the marine environment and for the people who live by it will be in place.

American Fisheries Society Symposium 42:9–13, 2004

Nongovernmental Organization Views on Advantages and Disadvantages of Aquatic Protected Areas

TUNDI AGARDY[1]

Sound Seas, 6620 Broad Street, Bethesda, Maryland 20816-2604, USA

Abstract.—Nongovernmental conservation and environmental organizations have generally embraced aquatic protected areas (APAs) as effective tools in the arsenal available to protect biodiversity, safeguard ecological integrity, and allow sustainable use of marine and coastal ecosystems. Aquatic protected areas are extremely versatile and can be applied to achieve a whole range of environmental and social outcomes, including enhanced fisheries production; greater involvement of users in management of an area; protection of cultural, historical, or ecologically important sites; and creating venues for scientific research, both basic and applied. Among the many advantages of APAs are two important benefits that other management and conservation measures rarely provide: (1) they present a small-scale arena in which government, the private sector, and the public at large can adopt cooperative or complementary management measures to benefit all parties, and (2) they allow manipulative experiments to be carried out to measure the impacts that humans have on ecosystems as well as the efficacy of management. In fact, in many parts of the world, we would not have the means to gauge management, discern trends in environmental condition, or start the process toward integrating management of coastal and marine areas without the presence of APAs. However, APAs sometimes present problems for environmental or conservation organizations—for instance, when they are not designed to be self financing and thus drain organizational resources. Another problem with APAs—and this applies to other sectors besides just nongovernmental organizations—is that they can create the illusion that marine management problems are being dealt with sufficiently when, in fact, they are not. Any organization or institution working to promote APA establishment should be wary of this potential pitfall and work to ensure that the contexts in which APAs sit are adequately protected as well.

Introduction

Environmental groups or nongovernmental organizations (NGOs) function as purveyors of information, as translators of scientific and management language to the vernacular, as honest brokers between competing interests, as advocates and lobbyists for certain types of reform or regulatory measures, and as adversaries to management agencies and industry by practicing environmental litigation. In many of these roles, environmental groups have been seen as the antithesis to development, to business interests, and to the needs of many user groups. Yet today, environmental groups play an increasingly important nonadversarial role in demonstrating how conservation and sustainable use can be accomplished through on-the-ground conservation projects that benefit users, community groups, business, and national interests.

Due to the ballooning scope of both global coastal degradation and fisheries conflicts, environmental groups have recently begun to get more involved in fisheries management and conflict resolution. In tackling fisheries issues, most organizations attempt to base their projects and advocacy on the best available scientific information. These groups sometimes undertake in-house scientific research, predictive modeling, and meta-analysis. However, in most cases, the NGOs are recipients of scientific information and act as liaisons between the scientific community, decision makers, and the public. Key scientific information that forms the underpinnings of campaigns and field

---

[1] E-mail: tundiagardy@earthlink.net

projects address three facets of sustainability: (1) the levels of resource removal that can be realized without adverse impact on the ecosystem given the particular environmental condition of the ecosystem at time of harvest, (2) the least invasive means by which that harvest can be undertaken at desired levels of harvest, such that habitat impacts and bycatch are minimized, and (3) the most appropriate stocks for large scale harvest, those being stocks that are not the sole representatives of genetically unique organisms and those for which the ecological role of the species is neither critically important nor redundant.

Environmental groups increasingly influence the direction of resource management and habitat protection in many parts of the world, addressing marine as well as the more traditional terrestrial issues of conservation. Commonly held roles for such groups include synthesizing understanding about marine issues, highlighting findings in a way that can be communicated to the general public, and advocating for policy reform and incentives that will change human behavior to make resource use more sustainable. In addition, and perhaps most importantly, environmental groups play a crucial role in leading by example, demonstrating how environmentally sound and socially beneficial resource use can be achieved through field projects and interventions. One of the most powerful tools NGOs use in this regard is protected areas: marine protected areas, which are usually established as in-water designations, and aquatic protected areas (APAs) that include marine protected areas but also include coastal protected areas such as estuarine reserves, wetlands protected areas, etc.

Nongovernmental Organizations and Aquatic Protected Areas

If a common environmentalist attitude toward APAs can be said to exist (and this is a dangerous assumption given the diversity of groups and their approaches), it is to synthesize existing information, communicate it, and advocate for change in policy and regulations (Agardy 2000). In addition, some groups advocate shifting the burden of proof when evaluating fishing impacts on ecosystems. Shifting the burden of proof has received recent attention in the fisheries management community (Dayton 1998), but there remain misconceptions (Agardy 2000). Much of the conservation community advocates shifting this burden of proof in evaluating the prospective impacts of new fisheries, expanded fisheries, or new technolo-

gies and suggests that regulators permit such fisheries development only when proof of no likely impact exists.

Many environmental NGOs also advocate establishing strictly protected marine reserves to further our understanding and protect species, habitats, and ecological processes (Sumaila et al. 2000). Such reserves are implemented in a variety of fashions: as components within larger, multiple-use protected areas that seek to accommodate a wide array of users, as single elements in scientifically designed reserve networks, and as one tool of many used in corridor approaches, coastal management, and regional planning.

Environmental groups have been demanding better information for some years on the true, ecosystem-wide impacts of fisheries activity, particularly in cases where new fisheries are being launched, where major gear modifications are taking place, or where major expansion of fishing effort is occurring. Most groups also advocate greater use of marine protected areas and fisheries reserves as a tool to strengthen management and to provide control sites to further scientific understanding and promote adaptive management. Finally, environmental groups have played a key role in developing case studies where government bodies work in concert with user groups and communities to demonstrate how comanagement can be achieved. In pushing for these approaches to marine conservation, environmentalists underscore the need for a holistic perspective on conservation problems and a holistic approach to their solution (Agardy 2003).

Environmentalists have often been labeled "nature-centric," a disparaging term used to contrast the value systems of those who would put biodiversity conservation above considerations of human needs. While it is true that many environmental groups see their niche in extremist positions that provide counterbalance to the wise-use movement and other equally extreme anti-environmental philosophies, most groups practice a human-centric conservation that recognizes human needs, especially those of marginalized coastal communities and traditional fishers. These groups work to understand the social costs of massive economic development, the tension between artisanal and commercial fisheries, and the social impacts of large-scale fisheries and aquaculture operations that act to exclude local peoples and steer profits away to large multinational corporations (Kurien 1978; Kurien and Achari 1994). In fact, when developing small-scale models of integrated coastal management, environmental groups only seem to succeed when these more local concerns have been adequately appraised and addressed.

When humans are not considered bona fide elements of ecosystems and human needs are ignored in either the rush to develop or the defensive move to protect the environment, social conflicts worsen. The resulting social effects include social disruption (Acheson 1987; Berkes 1987), migration and resulting interference with traditional sustainable patterns of resource use in areas of in-migration, environmental refugee movements that put pressures on scarce resources or vulnerable ecosystems, undermined national security, and resource-use conflicts that sometimes escalate into resource wars (Poggie and Polnac 1988; McGoodwin 1990). In places where artisanal and commercial fisheries clash with intensity, such as much of coastal West Africa, for instance, documenting and empirically assessing the problem is key to being able to deal with it. Further, resource management authorities must look to ensure equitable sharing of benefits whenever possible (de Fontaubert et al. 1996) and explore ways to address more local needs in addition to national economic interests. Comanagement has much potential in this regard and may prove to be the way forward in many settings (Stroud 1994). Comanagement that recognizes the legitimacy of traditional management regimes, such as those demonstrated by marine tenure or community-imposed limited access (Ruddle and Johannes 1985; Ruddle 1988; Ruddle et al. 1992), have especially good chances of succeeding (Dyer and McGoodwin 1994). Successful models of fisheries management that have addressed social issues well seem to be those that clearly define roles and responsibilities of local communities and government authorities and recognize the benefits of participatory planning and management (Pinkerton 1987; Jentoft and McCay 1995).

Given the current situation, the roles that environmentalists can play in brokering information and communicating it in ways that the public can grasp are increasingly critical. In adopting a holistic approach to describing human impacts on marine environments, we need objectivity and scientific rigor, but we also must present the big picture (Caddy 1993). Recognizing linkages is a prerequisite, and hard as it may be, we need to weed through all human impacts that affect marine systems and fisheries potential simultaneously. It is critically important to recognize and prioritize true threats to ecosystem health and function, identify the underlying drivers behind these threatening human activities, and design marine protected areas that address these threats. Though the latter may seem obvious, it is clear that imposition of "generic"

APAs through a one-size-fits-all approach has led to many situations where uses of an area are curtailed without lessening the imminent threats to the resource base or marine biodiversity.

Environmental NGOs play a key role in highlighting what works and what does not. Such groups take on field projects and interventions that demonstrate how these principles can be effectively applied. Aquatic protected areas are key here because they are most often the venues for such demonstration projects. In addition, marine reserves and other protected areas are crucial in serving as benchmarks and baselines in furthering our understanding of how ecosystems function and how humans affect such functioning. These two crucial roles of APAs are often overlooked, yet in some areas, they may be more important than the habitat protection afforded by the APA regulations. In such cases, APAs can steer us in the direction of effective integrated coastal management and a better understanding of marine systems and our impacts on them.

Caveats and Conclusions

Aquatic protected areas provide versatile and powerful tools for combating environmental degradation, overexploitation of resources, and ignorance. However, APAs are not a panacea and can only do so much. APAs are, after all, islands of protection in seas that are threatened by many simultaneous and cumulative threats. Even the best-designed and best-implemented protected area will not amount to much if the context in which that protected area exists—namely the wider region—is allowed to deteriorate.

Aquatic protected areas also often confer heavy burdens of cost on environmental NGOs that push for their establishment. This is because many APAs are idealistically designed to address real and proximate threats and, in so doing, curb certain profit-making uses of ocean space and resources. Not only are alternative livelihoods for displaced users often not considered in the planning process but often neither are self-financing mechanisms. The result is that environmental NGOs often have to "prop up" protected area management, significantly draining resources from other projects and programs. And when these financial resources are removed and the umbilical cord is cut, many protected areas deteriorate and disappear. The consequences of these failures are significant and can threaten future projects when the confidence and trust of communities and decision makers are threatened.

Perhaps the greatest shortcomings of APAs is that they can—and often do—create a false sense of security that marine conservation issues are being adequately addressed, even when the APA is designed to tackle only a single issue or satisfy only a single user group. In a recent paper, I and other coauthors discuss this problem in relation to 20% targets for no-take areas within APAs (Agardy et al. 2003). We present the following hypothetical situation: a mythical country's new government regulations require an agency with jurisdiction over a highly used, degraded coastal area to set 20% of the area aside as no take. The managing authority rushes to fulfill the target, imposing no-take restrictions in a single reserve in the most remote part of the area, where extractive activities are essentially nonexistent anyway (explaining why a quick response was possible in the first place). Having achieved the target, the decision maker can pat himself or herself on the back, congratulate the managing agency, and walk away from the real and persistent problems that remain: uncontrolled use in areas outside the 20% restricted area, use conflicts and animosity toward regulators, point and nonpoint source pollution, environmentally damaging coastal development and construction, etc. Consequently, we are left with a situation in which the 20% target has indeed been reached, and yet 80% of the ecosystem remains as threatened—or even worse off—than before the management measure was instituted.

The condition of the marine environment, with its implications for nature and human well-being, warrants quick responses. In the rush to come to the ocean's rescue, our tendency is to search for and apply simplistic models of management. If we have come to learn anything from our relatively brief flirtation with marine management, however, it is that generic models do not work well. Aquatic protected areas must be designed to match specific circumstances of ecology, of threat, and of human need that all vary from place to place. They must never be implemented as the sole solution to the complex set of problems that marine conservation presents but must be developed in the context of effective, regional-scale coastal and ocean management. Environmental groups can move us in this direction but only if they work closely with governmental agencies, user groups, and donors. A truly integrated response is what the world oceans need.

References

Acheson, J. M. 1987. The lobster fiefs revisited: economic and ecological effects of territoriality in the Maine lobster industry. Pages 37–65 *in* B. J. McCay and J. M. Acheson, editors. The question of commons: the culture and ecology of communal resources. University of Arizona Press, Tuscon.

Agardy, T. 2000. Effects of fisheries on marine ecosystems: a conservationist's perspective. ICES Journal of Marine Science 57:761–765.

Agardy, T. 2003. An environmentalist's perspective on responsible fisheries: the need for holistic approaches. *In* M. Sinclair and G. Valdimarsson, editors. Responsible fisheries in the marine ecosystem. CABI, Wallingford, UK.

Agardy, T., P. Bridgewater, M. P. Crosby, J. Day, P. K. Dayton, R. Kenchington, D. Laffoley, P. McConney, P. A. Murray, J. E. Parks, and L. Peau. 2003. Dangerous targets: differing perspectives, unresolved issues, and ideological clashes regarding marine protected areas. Aquatic Conservation: Marine and Freshwater Ecosystems 13:1–15.

Berkes, F. 1987. Common property resource management and Cree Indian fisheries in Subarctic Canada. Pages 66–91 *in* B. J. McCAy and J. Acheson, editors. The question of the commons: the culture and ecology of communal resources. University of Arizona Press, Tuscon.

Caddy, J. F. 1993. Towards a comparative evaluation of human impacts on fishery ecosystems of enclosed and semi-enclosed seas. Reviews in Fisheries Science 1(1):57–95.

Dayton, P. K. 1998. Reversal of the burden of proof in fisheries management. Science 279:821–822.

de Fontaubert, C., D. Downes, and T. S. Agardy. 1996. Biodiversity in the seas: protecting marine and coastal biodiversity and living resources under the convention on biological diversity. Island Press, Washington, D.C.

Dyer, C. L., and J. R. McGoodwin. 1994. Introduction. Pages 1–15 *in* C. L. Dyer and J. R. McGoodwin, editors. Folk management in the world's fisheries: lessons for modern fisheries management. University of Colorado Press, Niwot.

Jentoft, S., and B. McCay. 1995. User participation in fisheries management, lessons drawn from international experiences. Marine Policy 19(3):227–246.

Kurien, J. 1978. Entry of big business into fishing, its impact on fish economy. Economic and Political Weekly 13(36):1557–1565.

Kurien, J., and T. R. Achari. 1994. Overfishing the coastal commons: causes and consequences. Pages 218–244 *in* R. Guha, editor. Social ecology. Oxford University Press, New Dehli.

McGoodwin, J. R. 1990. Crisis in the world's fisheries: people, problems, and policies. Stanford University Press, Stanford, California.

Pinkerton, E. 1987. Intercepting the state, dramatic processes in the assertion of local management rights. Pages 344–369 *in* B. J. McCay and J. M. Acheson, editors. The question of the commons: the culture and ecology of communal resources. University of Arizona Press, Tuscon.

Poggie, J. J., Jr., and R. B. Polnac. 1988. Danger and rituals of avoidance among New England fishermen. Maritime Anthropological Studies (MAST) 1(1):66–78.

Ruddle, K. 1988. Social principles underlying traditional inshore fishery management systems in the Pacific Basin. Marine Resource Economics 5:351–363.

Ruddle, K., E. Hviding, and R. E. Johannes. 1992. Marine resource management in the context of customary tenure. Marine Resource Economics 7:249–273.

Ruddle, K., and R. E. Johannes. 1985. The traditional knowledge and management of coastal ecosystems in Asia and the Pacific. UNESCO, Jakarta, Indonesia.

Stroud, R. H., editor. 1994. Conserving America's fisheries. National Coalition for Marine Conservation, Marine Recreational Fisheries 15, Savannah, Georgia.

Sumaila, U. R., S. Guenette, J. Adler, and R. Chuenpagdee. 2000. Addressing ecosystem effects of fishing using marine protected areas. ICES Journal of Marine Science 57:752–760.

American Fisheries Society Symposium 42:15–19, 2004

How the Organized Recreational Fishing Community Views Aquatic Protected Areas

RYCK LYDECKER[1]

BoatU.S. (Boat Owners Association of The United States),
880 South Pickett Street, Alexandria, Virginia 22304, USA

Abstract.—The focus of this paper is saltwater angling in U.S. waters, and the term marine protected areas (MPAs) is used throughout the paper because of its common acceptance among most fishing and nonfishing stakeholders. The organized recreational fishing community— composed of angler groups, for-hire fishing providers, and the tackle industry—views MPAs with great skepticism at best and outright opposition and hostility at the extreme. That is because this stakeholder community equates the term "MPA" with "no-fishing zone." Executive Order 13158 defines the term "marine protected area" only in broad terms as "any area of the marine environment that has been reserved by [law] or regulations to provide lasting protection for part or all of the natural or cultural resources therein." Yet public discourse in marine resource management, public policy, and nonfishing stakeholder circles defines, or at least implies, the very discrete (and to anglers, very inflexible and threatening) definition of "no-fishing" or "no-take" zone. A review of the official positions on MPAs as voiced by organizations within the organized recreational fishing community is presented. Much of the information is derived from a meeting of U.S. recreational fishing leaders convened in February 2003 to examine a variety of issues related to marine fish management, particularly at the federal level, and angler participation in the process. Marine protected areas were identified as a major topic for discussion among representatives to the Sportfishing Leadership Conference, which had federal fishery managers in attendance. The author's contention that no single issue has so galvanized the U.S. sportfishing community is supported using excerpts from position papers, newsletters, Internet web sites, and the sportfishing consumer and industry press.

Introduction

It has been said that the problem with marine protected areas (MPAs) is that the term holds different meanings for different people (NOAA 2000). I submit that the real problem with MPAs is that the term means *precisely the same thing* to different people. More specifically, the term MPA generally conveys identical meaning to members of the two stakeholder groups that really count in the public policy process that is now unfolding in regards to marine resource management.

The two groups are identified here as "environmentalists" and "anglers," for want of better terminology, although the terms are by no means mutually exclusive (nor are they intended to be in this context).

Use the term marine protected area within earshot of an environmentalist and a recreational angler and both of them will hear "no-take zone" or "no-fishing zone." Of course, this may or may not be the speaker's intent, and therein lies the problem, at least in public discourse. Certainly, a similar communications problem does not exist within the academic and resource management communities represented here today where the issue of marine protected areas is not nearly so cut-and-dried. Or is it?

Misunderstood Meanings

Referring to the graphic used on the website announcing this symposium (Figure 1), the casual observer would equate the term—in this case "aquatic protected area"—with no-fishing zone. If that was not the intent of the organizers, it is not obvious here. In fact, the words "fully protected areas" appeared

---

[1] E-mail: RLydecker@BoatUS.com

Figure 1.—Graphic from the Aquatic Protected Areas as Fisheries Management Tools Symposium web site (AFS 2003).

in the symposium subtitle. Even without that, the impression is clear—no fishing—and that, unfortunately, is the recreational fishing community's perception of the agenda behind the rush to impose MPAs.

From what I have been able to glean from the MPA debate, it appears that two definitions for the term "marine protected area" actually have standing in the public policy arena, the first being the language published by the National Academy of Sciences, which defines MPA as a "geographic area with discrete boundaries that has been designated to enhance conservation of marine resources" (NRC 2001).

The second is contained in Executive Order 13158 (U.S. Office of the Federal Register 65:105 [31 May 2000]:34909), which defines marine protected areas as "any area of the marine environment that has been reserved by [law] or regulations to provide lasting protection for part or all of the natural or cultural resources therein."

No problems there. By either definition, the United States has had MPAs for 30 years or more, and recreational anglers will argue that they have been in the forefront of creating managed areas that fit these definitions. The signing of Executive Order 13158 in May 2000 really put the issue squarely in front of the organized recreational fishing community. Since that time, the concern—the fear—most often voiced by anglers and the organizations that represent them derives from the meaning attached to the term MPA by certain nongovernmental organizations (NGOs) that, for convenience, we have labeled "environmentalists."

According to Michael Nussman, president of the American Sportfishing Association (ASA), a fishing tackle trade group, the impetus for creating no-take zones in the United States started in 1996 when Congress passed the Sustainable Fisheries Act, which called for the restoration of some 100 severely depleted fish populations.

"People in the environmental community are frustrated that some fish stocks are not being restored in what they view as an appropriate time period," Nussman reports, "so now they are pushing no-take reserves as the solution" (personal communication cited *in* Lydecker 2002).

Some time after Executive Order 13158 was signed, the National Oceanic and Atmospheric Administration (NOAA) invited me to attend a stakeholder meeting on MPAs. The group included many representatives of the various environmental organizations, and part of our charge was to work in small groups to develop a definition for the term "marine protected area." This proved a delicious assignment for some in attendance, and the discussions in our small groups leaned repeatedly and continually toward a "keep out" definition. My only ally for common sense in the room, it seemed, was a commercial fisherman from the Florida Keys named Tony Iarocci, a member of the South Atlantic Fishery Management Council and, like me, a very short-lived appointee to the Marine Protected Area Federal Advisory Council. As I recall the discussions, most of the representatives of the environmentalist NGOs not only wanted to apply a "no extractive uses" definition to the term, they wanted it codified in federal law.

It is just that kind of talk that has generated skepticism, if not downright opposition, from the recreational fishing community, and it was about that same time that I really began to hear from BoatU.S. members regarding this issue. Since about 280,000 of our 550,000 boat-owner members identify themselves as anglers, this is an issue of great concern to us. (To be fair, however, it should be noted that our individual members have varying views on MPAs, depending, of course, on their understanding of what is meant by the term. Some do, in fact, support no-take reserves as a means to rebuild stocks.)

Organized Sportfishing's Position

Early in 2001, the American Sportfishing Association developed an opinion/editorial article warning of a growing threat to sportfishing from "extremist environmental groups" that intended to use MPAs as tools to accomplish anti-fishing goals. It reads, in part:

These groups, in an ever-expanding campaign to increase membership and advance their misguided agenda, have said to heck with America's 12 million saltwater anglers, nearly all of who [*sic*] also own a boat. The Ocean Conservancy for one proclaims that large swaths of ocean

should be turned into "ocean wilderness areas" where no one but a privileged few should tread.

There's nothing wrong with the concept of protecting ocean areas, but they should be modeled after our terrestrial system of national parks where public recreation such as fishing and boating is allowed, but commercial exploitation is not. Some of the nation's best freshwater fishing occurs in our national parks, wildlife refuges and national forests. (Nussman 2002)

The article was distributed to the recreational boating and sportfishing press and reprinted widely in the United States and abroad. It may have been the single greatest factor to focus the attention of individual anglers on the MPA issue. The article also directed anglers to a "Freedom to Fish" web site (*www.freedomtofish.org*) intended to build a grassroots defense against misguided MPA proposals.

In cooperation with the Coastal Conservation Association (CCA), a 75,000-member sportfishing group with 15 state chapters in the Gulf and Atlantic states, ASA worked with congressional leaders to draft the Freedom to Fish Act. That bill, first introduced in the U.S. Senate on August 2, 2001, and in the House of Representatives on October 11, 2001, would prevent implementation of, in the words of Fred Miller, CCA government relations chairman, "blanket no-fishing zones [that] represent a quick and unfair approach, managing people rather than resources. The Freedom to Fish Act will help protect recreational anglers' freedom to fish without jeopardizing the proper conservation of the resource" (CCA 2001).

The official CCA policy on MPAs states that CCA will fight to protect access for recreational fishermen to all public fishing areas unless

• there is a clear indication that recreational fishermen are the cause of a specific conservation problem and that less severe conservation measures, such as gear restrictions, possession limits, size restrictions, quotas, or closed seasons will not adequately address the targeted conservation problem;

• the closed-area regulation includes specific, measurable criteria to determine the conservation benefit of the closed area on the affected stocks of fish and provides a timetable for periodic review of the continued need for the closed area at least once every 3 years;

- the closed area is no larger than that which is supported by the best available science; and

- provision is made to reopen the closed area to recreational fishing whenever the targeted conservation problem no longer exists (CCA 2003).

A review of the publications, web sites, and conversations with the principals of half a dozen large sportfishing organizations indicates a similar policy position on the issue. The Recreational Fishing Alliance (RFA), which claims 80,000 affiliated members with representation in all coastal states, has a similar, though slightly more expansive, policy statement that adds:

- The proposal [for an MPA] also should allow for other types of recreational fishing, such as trolling for pelagic species that would not have an impact on demersal stocks of concern, as an example; and

- The fishery management measures [within an MPA] are part of a fishery management plan as required by the Magnuson-Stevens Act as amended by the proposed Freedom to Fish Act (RFA 2003a).

An RFA position paper further describes the general challenges to recreational fishing:

We have consistently been under-represented in the regulatory process and received an unfair allocation of quotas as a result. And, if that wasn't enough, we are now firmly in the sights of a consortium of radical environmental groups that want to close vast areas of the ocean under the guise of "marine protected areas" and further restrict recreational fishing access to fisheries we were instrumental in helping rebuild. (RFA 2003b)

In February of this year, 80 representatives of the saltwater fishing and boating community met with leaders from NOAA for an expansive dialogue on marine conservation and recreation issues. The group "laid the groundwork for collaborative approaches to marine policy making in order to better represent and engage the agency's recreational fishing and boating constituency" (ASA 2003).

One of three general recommendations to NOAA and its National Marine Fisheries Service to come out of the Sportfishing Leadership Conference requested that the agency adopt a "common sense

approach on MPAs" and included this statement on the issue:

In the spirit of conservation, [NOAA/Department of Commerce should] create clearer definitions and specific science-based guidelines for the establishment and management of marine protected areas, including public involvement, conservation goals, provisions for enforcement, and evaluation criteria. (ASA 2003)

Summary

Given the foregoing information, it would appear that no single issue in recent memory has so galvanized the recreational fishing community. Perhaps this statement from David Cummins, president of the CCA, sums up the position of most angler organizations:

Time and area closures can be effective management tools when based on good scientific data, but arbitrary restriction of recreational anglers merely displaces fishing effort, increases regulatory confusion, increases user group conflicts and casts doubt on the entire fishery management process. It is a disservice to all U.S. citizens. (CCA 2001)

Thus, it would appear that only when the term "marine protected area" is used and understood by *all* stakeholders in its broadest sense will the organized sportfishing community be effectively engaged in discussions regarding management prescriptions within specific sites, existing or future, whatever they may be labeled.

References

AFS (American Fisheries Society). 2003. Aquatic protected areas as fisheries management tools. AFS. Available: *www.fisheries.org/apa_symposium/homepage.htm* (November 2003).
ASA (American Sportfishing Association). 2003. Saltwater community unites to tackle tough issues on marine conservation and recreation. American Sportfishing (ASA Newsletter) 7(1):9.
CCA (Coastal Conservation Association). 2001. Congress examines recreational angler's freedom to fish. CCA. Available: *www.joincca.org/press/2001/FTF1.htm* (December 2003).
CCA (Coastal Conservation Association). 2003. CCA's position on marine protected areas/no fishing zones. CCA. Available: *www.joincca.org* (November 2003).

Lydecker, R. 2002. Anglers de-bone no-take zone. BoatU.S. Magazine 7:18–19.

NOAA (National Oceanic and Atmospheric Administration). 2000. Marine protected areas of the United States. NOAA, Marine Protected Areas Center, November update, Silver Spring, Maryland.

NRC (National Research Council). 2001. Marine protected areas: tools for sustaining ocean ecosystems. National Academy Press, Washington, D.C.

Nussman, M. 2002. Environmental groups threaten to ban sportfishing. Guest editorial. Saltwater Sportsman 63(6):10.

RFA (Recreational Fishing Alliance). 2003a. Marine protected areas. RFA. Available: *www.savefish.com* (November 2003).

RFA (Recreational Fishing Alliance). 2003b. RFA position paper: marine protected areas (MPAs). RFA. Available: *www.savefish.com/mpaprop.html* (November 2003).

Part II:
Keynote Speakers

American Fisheries Society Symposium 42:23–35, 2004

Marine Protected Areas: An Old Tool for New Circumstances

EDWARD D. HOUDE[1]

University of Maryland Center for Environmental Science, Chesapeake Biological Laboratory,
Post Office Box 38, Solomons, Maryland 20688, USA

SUSAN J. ROBERTS[2]

Ocean Studies Board, National Research Council, The National Academies,
500 5th Street Northwest, Washington, D.C. 20001, USA

Abstract.—Reserves and sanctuaries are age-old tools for conservation of fishery resources, but they are underutilized in managing heavily fished stocks in stressed marine ecosystems. Conventional fisheries management, which emphasizes controlling catch and fishing effort, can be effective but is insufficient under some circumstances. A new emphasis on ecosystem-based fisheries management argues that spatial controls, including marine protected areas (MPAs), are desirable components of integrated ocean management plans. A National Research Council (NRC) study and report highlighted the importance of spatial heterogeneity and the need to incorporate it in fisheries management plans. The NRC study concluded that, under many circumstances, MPAs combined with conventional management can: protect habitats and nursery grounds, reduce bycatch, lower fishing mortality on prerecruits, promote "spillover" of recruited fish to unprotected areas, and provide a buffer against uncertainty. Poor understanding of stage-specific dispersal patterns of fish and behavior of fishers (i.e., distribution of effort) outside MPA boundaries constrains acceptance of MPAs by some scientists and managers. If implemented, MPAs must be enforced, monitored, and evaluated to ensure that innovative, incremental, and adaptive implementation of MPAs contributes substantively to precautionary fisheries management.

Introduction

To casual observers, the sea appears homogeneous and monotonous. Oceanographers, fisheries scientists, and especially fishers and fishery managers, are keenly aware that marine ecosystems consist of patchworks of habitats that support biological communities and fished stocks. This structural heterogeneity and associated variable productivity present challenges to fisheries management. Historically, coastal fisheries have been pursued throughout their ranges of occurrence, with few regulations to control access. Broad management zones, including exclusive economic zones, now are widely recognized (e.g., 0–4.8 km [0–3 mi], 4.8–321.9 km [3–200 mi], inland waters, and estuar-ies), but management at smaller scales, targeting specific habitats of fished stocks, is less frequent. Nevertheless, spatial management at many scales is not uncommon, and the concept of closed or reserved areas has a long history in marine and terrestrial ecosystems. The proportion of the sea in protected areas is far less than the proportion in terrestrial reserves, and the fraction classified as no-take reserves that prohibit fishing is miniscule (Carr et al. 2003).

Many of the world's fishery resources are over-exploited, some to a point of near collapse (Garcia and de Leiva Moreno 2003; Myers and Worm 2003). Conventional fisheries management, which emphasizes controlling fishing effort and catches, and thus fishing mortality rates, yields, yield per recruit, and spawning stock biomasses, has performed poorly or inconsistently in many instances. The world's aquatic ecosystems are now heavily fished, a circumstance that has emerged in the past 60 years. Conventional

[1] Corresponding author: ehoude@cbl.umces.edu
[2] E-mail: sroberts@nas.edu

fisheries management, despite its inconsistent record, has shown some recent successes and should not be abandoned. It is theoretically sound, but economic incentives to fish intensively and the unpredictable variability in aquatic ecosystems require innovative management that is prudent, precautionary, and averse to risk. Marine protected areas (MPAs) represent a relatively underutilized hierarchy of tools that can add desirable dimensions to conventional management.

The debate over potentials and efficacy of MPAs for fisheries management has intensified over the past decade. As a consequence, MPAs for resource management and conservation have been repeatedly reviewed (e.g., Murray et al. 1999; Fogarty et al. 2000; Jennings 2001; Jones 2002; Palumbi 2002; Polunin 2002). Expert panels have recommended that MPAs be included in the suite of tools available to fisheries managers (NMFS 1999; NRC 1999a, 2001). The case for expanded use of MPAs in fisheries is strengthened by these reports, but the debate continues because many economic, social, and biological issues remain unresolved. A National Academy of Sciences committee (NRC 2001) addressed these issues. Here, we summarize conclusions and recommendations of the committee, discussing them in the expanded context of the continuing debate.

The National Research Council (NRC) committee considered a broad scope of ecosystem issues in addition to fisheries (NRC 2001). Although the NRC report addresses these broader issues, our emphasis here is limited to MPAs and fisheries management. The consensus of the NRC committee was that MPAs adopted for marine fisheries and ecosystem management can contribute positively to spatially explicit management that recognizes the uniqueness of habitats and variability in their productive capacities. Adopting MPAs as a significant component of a suite of ecosystem-based approaches for fisheries management underscores the need to conserve the structure and productive capacity of ecosystems in addition to optimizing yields of individual stocks.

Area closures as a tool for fisheries management in the U.S. Exclusive Economic Zone are recognized explicitly by the Magnuson-Stevens Sustainable Fisheries Act (SFA; NMFS 1996), but the SFA contains little supporting language to encourage their innovative use or to conduct research on their potential. Marine protected areas for fisheries management, embedded in broader coastal zone management plans, constitute a vision for protecting and restoring habitat that also serves goals of sustainable fisheries management (NRC 2001). Congressional hearings in the United States on reauthorization of the SFA have debated the potential of MPAs as a fisheries management tool (H. R. 4749, May 2002). When the SFA is reauthorized, it is possible that spatial management tools such as MPAs may become more prominent in fishery management plans as an alternative or supplement to conventional management approaches.

National Research Council Report Conclusions

The NRC report (NRC 2001) concluded that, while not new in concept or application, spatial approaches are underutilized and potentially can improve management, especially to meet precautionary and risk-averse goals now designated by most resource management agencies. Properly designed MPAs can:

- protect nursery areas;

- protect or restore critical habitats;

- limit bycatch;

- protect threatened or endangered species;

- rebuild age and size structure of stocks (and increase fecundity);

- promote spillover and dispersal from protected to open fishing zones;

- reduce fishing mortality rates;

- reduce the burden of stock assessment science;

- provide insurance against the "uncertainties" of science and management; and

- promote education and research on marine ecosystems.

No single MPA can achieve all these goals, nor should it be expected that individual MPAs bring such broad benefits to fishery management.

The NRC committee reached its conclusions recognizing that performance of MPAs is itself subject to uncertainty. Design and implementation of MPAs for fisheries management cannot rely on a single prescribed formula because aquatic ecosystems are complex, and the needs for fisheries management vary widely. Many types of MPAs potentially could be adopted to support sustainable fishery management, ranging from small temporal closures that essentially "fine-tune" conventional, single-species management to complete closures of portions of ecosystems to fishing or any disturbance of living resources and habitats (see IUCN 1994).

The NRC committee concluded that MPAs, in many circumstances, can protect habitats, restore biodiversity, and rebuild or protect fishery resources. However, near-term benefits in yields, yields per recruit, or profits from MPAs to fishers are uncertain or may be negative if fishing effort is uncontrolled in areas open to fishing. In reviewing MPAs for fisheries management, Polunin (2002) also reaches this conclusion.

Marine protected areas can protect vulnerable habitats from destructive fishing practices and other human threats, and they may be particularly valuable for protecting nurseries that support prerecruit fish. Properly sited MPAs can reduce bycatch of fast-growing prerecruits of targeted species in addition to protecting substrates and structure of critical nursery habitats. Marine protected areas can reduce the unintentional catches of nontarget species, including endangered or threatened mammals, turtles, and birds, in fishing gears. And MPAs potentially can reduce excessive fishing mortalities on species such as the tropical groupers (family Serranidae, principally *Epinephelus* and *Mycteroperca* spp.) that form highly vulnerable spawning aggregations. Other management approaches, properly implemented, also can directly control effort and reduce fishing mortality. This does not negate the potential of MPAs as an alternative or supplementary approach in many instances. Finally, the argument that MPAs are insurance against the uncertainties of science and conventional management, as well as unpredictable climate and environmental shifts, has merit when considering MPAs as a management tool.

The main objective of MPA implementation for fisheries management will not always be increased yields or yield per recruit, as argued by Gerber et al. (2003). In heavily exploited stocks or those at risk of collapse, reductions in fishing mortality, stabilization of a declining stock, and rebuilding stock biomass are objectives for short-term and long-term management needs. In the long term, increases in equilibrium yields, yield per recruit, and profits to fisheries may be management goals, but in overexploited or threatened stocks, near-term goals are less burdened by such expectations.

Habitats and Spatial Management: A Role for Protected Areas

The National Marine Fisheries Service Fisheries Ecosystem Principles Advisory Panel (NMFS 1999) and the NRC committee (NRC 2001) strongly recommended incorporating protected areas and other spatially explicit approaches into ecosystem-based fisheries management. Habitat patchiness and quality are clearly recognized by fishers who concentrate effort in historically productive habitats where fish aggregate and are abundant. As a consequence, effects of fishing are most apparent in productive parts of coastal zones. Broader implementation of MPAs in fisheries management would (1) recognize explicitly the importance of marine habitats, (2) act to protect essential fish habitats, and (3) address the need to preserve the structure of marine ecosystems for sustainable fisheries.

Planning and Design

Successful MPAs result from careful planning and design, whether the overall goal is to manage fisheries or some combination of goals. Design of effective MPAs should proceed through four sequential stages (NRC 2001): (1) evaluate conservation needs at local and regional levels (i.e., identify the problem), (2) clearly define objectives and goals for establishing an MPA, (3) describe key biological and oceanographic features of the region, and (4) identify and choose site(s) that have highest potential for implementation. The last of these criteria recognizes political and social realities. Plans and designs that clearly address issues, conservation needs, probable costs, and benefits (Table 1) and which are supported by most stakeholders will have higher probability for implementation than those rife with contention and stakeholder mistrust.

The NRC report (NRC 2001) highlighted the importance of engaging stakeholders, including those geographically distant from a proposed site, in the planning and design phases of MPAs. Based on interviews, public hearings, and evaluations of existing MPAs, the report concluded that affected stakeholders, especially fishing interests, must truly participate in deliberations that could substantially impact their way of life and earning potential. In many cases, MPAs will have goals broader than fishery management (e.g., protection of biodiversity, rare species, habitats, and cultural sites), making it essential that the broad community of stakeholders be involved in the planning phase. This "bottom-up" strategy, while essential, cannot succeed without "top-down" guidance and leadership in the planning process by government agencies with jurisdictional

Table 1.—Costs and benefits associated with marine protected areas in fisheries management. Derived from Table 4-3 in NRC (2001).

Issue	Cost	Benefit
Yield	Decreased catch; negative impacts on other fisheries	Higher stock fecundity and recruitments; lower bycatch
Displacement	Increased fishing pressure in open areas	Reduced effort; protection for essential fish habitat
Management	New research and monitoring needs	Better estimates of population parameters
Economics	Disproportionate impact on local communities	Insurance against stock collapse
Non-market values	Loss of customary fishing areas and rights to access	Restored ecosystem, habitats, and species

authority and responsibilities. This "middle ground" approach (Jones 2002) may be critical for successful implementation of MPAs.

Selecting MPA sites is no simple task. Fortunately, site-selection criteria and algorithms (Airame et al. 2003; Leslie et al. 2003) are being developed to add objectivity to the process and several principles exist upon which to base MPA design (Botsford et al. 2003). Ecological criteria (Roberts et al. 2003) can be defined to provide a foundation for selection of MPA sites although, in our opinion, they should not be the sole basis for site selection. There are few case studies on MPAs in North American waters and little knowledge of long-term performance of sites protected for fisheries management. Two major and critical gaps in knowledge are (1) lack of information on movements of fish, and (2) poor understanding of behavior of fishers in response to area closures. Dispersal of fish eggs and larvae and migrations of older stages are critical factors in MPA site selection. The level of dispersal of early life stages and the so-called spillover of young fish from an MPA to areas open to fishing depend upon complex interactions between oceanographic factors and stage-specific behaviors of fish. Although knowledge is less than sufficient, modeling research demonstrates that dispersal, combined with behavior of fishers outside the MPA, is a major determinant of probable MPA success (Holland and Brazee 1996; Hastings and Botsford 1999; Sanchirico and Wilen 1999, 2001; Sladek-Nowlis and Roberts 1999; Fogarty et al. 2000; Gerber et al. 2003).

Marine protected area sites that serve as "sources" for dispersal of young fish or recruits to the fishery, rather than "sinks" receiving dispersed migrants, have the highest potential to make positive contributions to fisheries (Carr and Raimondi 1998; Crowder et al.

2000; Fogarty et al. 2000). Understanding dispersal potential of young fish from closed to open areas is a subject of intensive research, primarily based on models, to evaluate the possibility that spillover from reserves will seed fished areas in quantities high enough to benefit fisheries (Roberts et al. 2001; Gaines et al. 2003; Largier 2003).

Marine protected areas in productive areas that produce a surplus of young fish are more likely to serve as sources and produce the spillover that can replenish exploitable fish outside of MPAs. Siting of MPAs in areas that are productive nurseries for prerecruits can have clear benefits if it reduces their bycatch (Horwood et al. 1998), thus increasing production and recruitment to the stock. Designation of MPAs to protect prerecruit fish and their nursery habitats, when those habitats are well defined, was broadly supported in hearings and interviews conducted by the NRC committee (NRC 2001). Protection of nursery habitats also has broad support in the conservation and fishery management communities (Beck et al. 2001).

The Social Agenda

Use of the sea has increased dramatically in recent decades. Traditional fisheries resources are now essentially fully exploited (Pauly 1996; Garcia and Newton 1997). Globally and in North America, the major concern in fisheries management is overcapitalization (Garcia and Newton 1997; NRC 1999a; Garcia and de Leiva Moreno 2003). Surplus fishing capacity promotes excessive effort and fishing mortality. Marine protected areas may help indirectly to control effects of excess effort but will not solve the problem if effort displaced from MPAs is redirected and concentrated in open fishing areas (Holland and Brazee 1996; Sanchirico and Wilen 1999, 2001) with attendant excess landings, habitat damage, and increases in bycatch.

The open-access expectation of fishers, which has prevailed historically, is being challenged. "Privileged-access" approaches in fisheries management are being proposed and implemented worldwide. Selective licensing, individual fishing quotas, and other privileged-access approaches are hotly debated but in recent years have been increasingly adopted (NRC 1999b). Establishment of MPAs and no-take areas and the restriction to free access of traditionally-fished areas that is implied are often opposed by recreational and commercial fishers. Although individual fishers may support MPA concepts, majority support may only be forthcoming when stocks are severely overexploited or collapsed. In public meetings conducted by the NRC committee (NRC 2001), support by fishers for MPAs as a primary management tool often was lacking, except under conditions when stocks had collapsed. It is, however, important to note that many fishers did support designation of MPAs for reasons other than fisheries management.

As the public becomes more informed about values of marine ecosystems beyond direct economic returns and profits from extracted resources, the desire to protect and preserve these systems grows. So-called heritage and existence values of ecosystems now compete with traditional, extractive uses such as fisheries (NRC 2001). Creation of MPAs to serve essentially as parks or "fully-protected reserves" that preserve unique habitats, promote biodiversity, protect ecosystem functions and services, and conserve threatened or endangered species is a legitimate reason to establish MPAs (NRC 2001). Under some circumstances, such reserves may eventually benefit fisheries from spillover effects or restoration of habitat and biodiversity. The public and many fisheries managers now recognize that an ecosystem provides fundamental services that maintain structure and function, and so support the ecosystem's demand-derived services (e.g., fisheries) (Holmlund and Hammer 1999).

Size, Number, and Location

In our view, there is no well-defined rule specifying size, area, or numbers of MPAs to support fisheries management. It is undisputed that location, size, and number must be considered in developing an MPA strategy (NRC 2001). With respect to size, large MPAs, in some cases, may protect a larger fraction of enclosed spawning stock from fishing mortality and so increase spawning stock biomass (Mangel 2000) and other measures of reproductive output (e.g., eggs per recruit; Gerber et al. 2003). These potential benefits of MPAs for heavily fished and overexploited stocks were recognized long ago (Beverton and Holt 1957) and are being considered again in the context of emerging interest in MPAs (e.g., Guennette et al. 1998; Gerber et al. 2003). Permanent closure of 20% or more of fisheries ecosystems has been recommended to protect, on average, 20% of a spawning stock's biomass (NOAA 1990), a value once thought sufficient to stabilize and sustain high recruitments to exploited stocks. For long-lived, late-maturing, or moderately migratory species, 20% might be too low to achieve this goal, and a fishery reserve considerably larger, possibly in excess of 50% of the managed area, might be required if an MPA is the primary management tool (Lauck et al. 1998; Walters 2000).

Marine protected areas for fisheries management generally are expected to accomplish more than improving a stock's status within MPA boundaries. Their value may be judged by the quantity of fish spilled over the MPA boundaries into areas open to fishing. Numerous modeling studies have demonstrated that locations, sizes, numbers, and dimensions of MPAs, in combination with fish behavior and mobility, all can be critical in determining how effective an MPA is in protecting enclosed fish or in promoting spillover (Apostolaki et al. 2002; Gaines et al. 2003; Largier 2003; Shanks et al. 2003). As a stand-alone management approach, MPAs occupying as much as 30–70% of a management region may be required to optimize yield (Lauck et al. 1998; Walters 2000). In other situations, rather small MPAs, or networks (see below) of MPAs, if sited optimally, may play a significant role in protecting areas that support vulnerable spawning aggregations of reef fishes, eliminating bycatch of rare species, or conserving unique and productive habitats from destructive fishing.

Assigning simple geometries and configurations, with simulated organism behaviors, to hypothetical MPAs has provided insight into consequences for spillover or retention (Crowder et al. 2000) and their effects on potential yield, yield per recruit, or spawning potential (i.e., eggs per recruit; Gerber et al. 2003). In most situations, managers will be faced with selecting from several alternatives for MPAs with respect to sizes, geometries, and potential locations. Siting algorithms, such as "simulated annealing" (Airame et al. 2003; Leslie et al. 2003), which can identify and rank the quality of MPA sites or networks, may prove effective for recommending sizes and number of MPAs to meet management goals. This is especially true in complex, multispecies fisheries.

Networks and Zones

Aquatic ecosystems are fluid, three-dimensional spaces with a high degree of connectivity. Networks of MPAs, linked via ocean circulation and by life histories and behaviors of target organisms, potentially can provide the greatest benefits for strategic fisheries management and other MPA objectives. Networking implies that properly designed complexes of MPAs will be complementary. Arguments favoring networks emphasize connectivity among MPA sites (Jennings 2001; NRC 2001; Roberts et al. 2003) to optimize spillover effects. Effectively networked MPA systems require broad knowledge of oceanography, habitats, and the complex life histories of affected species. At present, such knowledge is limited or insufficient to design networks of MPAs that will increase productivity, stability, and profitability of fished stocks. Lack of knowledge does not negate the desirability of MPA networking. It emphasizes the need for careful research and planning before networks can be designated and implemented.

When the NRC committee was preparing its report (NRC 2001), Executive Order number 13158 was issued by President Clinton. It emphasized networking, calling for development and implementation of a coordinated network of MPAs in the U.S. coastal zone. Although the Executive Order's objectives reach beyond interests of fisheries management, its implications for fisheries are important. The Executive Order directed the National Oceanic and Atmospheric Administration (NOAA), in cooperation with the Department of the Interior, to establish an MPA Center and develop a framework for a national system of MPAs. The agencies have developed a web site to disseminate information about MPAs and agency programs (NOAA 2003a), initiated a national marine managed areas inventory, established the Institute for MPA Science and the Institute for MPA Training and Technical Assistance, and convened the MPA Advisory Committee to provide advice on a national system of MPAs.

The NOAA Marine Sanctuary Program has existed since 1972 and now includes 13 sanctuaries (NOAA 2003b). Many observers believe this program can serve as a framework and catalyst for development of MPA networks in the U.S. coastal zone. The NRC report (NRC 2001) and others have noted, however, that the NOAA sanctuaries have diverse goals, are broadly dispersed throughout the U.S. coastal zone, and seldom have fisheries management as a primary concern. There are exceptions; for example, the Florida Keys National Marine Sanctuary (FKNMS) includes fisheries in its management plans. Also, fishery management was a major impetus for establishing the Tortugas Ecological Reserve within the FKNMS to protect resident species and enhance spillover of larval fishes to Florida Bay and the Keys (NOAA 2003c). Similarly, the recent initiative to establish no-take reserves in the Channel Islands National Marine Sanctuary off California was fueled by concern over declines in abundance of many popular fish species. However, the NOAA sanctuaries program was not established to supplement fisheries management. If sanctuaries are to play a prominent role in establishing MPA networks to serve fisheries management, the sanctuaries program must assume additional authority over resource management decisions.

In the United States, the coastal zone and ocean fall under jurisdictions of several federal, state, and local authorities. Ultimately, effective resource management in coastal ecosystems will require zonation and cooperation, not only by users but also by management agencies with diverse responsibilities. The NRC committee recommended that MPAs for fisheries management ultimately be embedded in a broader strategic plan for coastal ocean management that addresses the full spectrum of human activities and needs for protection of ecosystem structure and function (NRC 2001).

Designating zones to alleviate conflicts and to partition coastal ecosystems into areas of acceptable uses will, in some instances, create MPAs for fisheries. Some zones will be established deliberately to meet fisheries management goals, but others may serve as de facto fishery reserves if fishing is incompatible with other prescribed zonal management. For example, zones abutting military facilities, high security areas, heavily used transportation routes, or critical breeding islands for marine mammals or birds may exclude fishing and, thus, effectively serve as no-take reserves. The NRC study (NRC 2001) recognized the broad spectrum of protected areas, reserves, and zoned activities that could be designated within the coastal ocean (Figure 1). Zones and networks of MPAs, some extending seaward from wetlands or terrestrial habitats to 370.4 km (200 nautical mi) offshore, could lie in overlapping jurisdictions of local, state, and federal authorities. In such situations, there is obvious need for close cooperation among fisheries and non-fisheries management agencies if zoning and MPA designations are to effectively support the respective management obligations.

In the case of fisheries, there are consequences of zoning over and above any benefits that may accrue from increased fisheries productivity in a zoned coastal ocean. Zones or networks of MPAs may force commercial and recreational fishers to operate in more distant areas, at increased costs. Effects on local communities also must be considered when MPAs are created, especially if they are embedded in larger zones that may more broadly restrict or ban traditional fishing gears and methods.

Economics (Costs and Benefits)

Economic benefits of MPA implementation are unlikely to be immediate, but if an MPA is effective, benefits may derive from long-term stabilization and recovery of depleted fish stocks in response to MPA implementation. There are economic costs and benefits associated with MPA-based management when compared to conventional fishery management approaches (Table 1) that should be considered at the outset in plans for MPAs. Except when stocks have collapsed, fishers may have little economic incentive to embrace MPAs. Near-term profitability from MPA implementation may, in fact, decline (Sanchirico and Wilen 2001). Discount rates are high; foregoing landings or profits today in the hope that MPAs will restore, rebuild, or improve landings and profitability on a decadal time scale is risky and often unacceptable from a fisher's point of view (Milon 2000). This concern is a compelling reason for fully

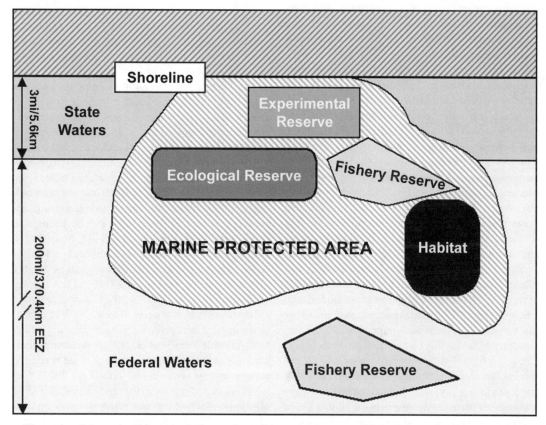

Figure 1.—Schematic of hypothetical zonation of the coastal ocean, designating various types of marine protected areas. The 0–3 and 3–200 nautical-mile state and federal jurisdictional zones (United States) are indicated (from NRC [2001] Figure 6-3). EEZ = exclusive economic zone.

including fishers in all deliberations on MPA planning and implementation.

Conventional Fisheries Management

It is difficult to foresee many situations in which MPAs are the sole management alternative in major fisheries. Marine protected areas are more likely to be one of a suite of measures to control fishing effort and improve status of stocks. Botsford et al. (2003) concluded that the effect of establishing an MPA on fishery yield will be similar to reducing fishing effort and that an MPA will have an effect on yield per recruit that is analogous to increasing age at first capture. Thus, MPAs, if accepted as a management tool, may eventually support increased yields and profits. Acceptance and implementation of MPAs to date have been most probable in overexploited fisheries that are near collapse. Under this circumstance, fishers often support MPAs as a supplement to conventional management, and a management agency is likely to significantly curtail fishing effort outside the MPA. This combination of measures is more likely than either measure alone to rebuild stocks to levels that are sustainable and profitable.

Conventional management will continue as the major approach to regulate fisheries because fishing in open areas beyond MPA boundaries will require controls to prevent excessive catches and effort. Evidence presented to the NRC committee (NRC 2001), and considered in most evaluations of MPAs (Halpern 2003), strongly supports the belief that no-take reserves promote increases in fish numbers, biomasses, age structure, and fecundities within reserve boundaries. This is especially true for site-attached organisms or those that do not migrate extensively (Polunin 2002). Evidence, based on models and arguments implying benefits of MPAs relative to conventional management for highly or seasonally migratory fishes is less compelling. Yet, some benefits of reduced fishing effort might be expected for such species if large MPAs were implemented or if smaller MPAs were sited on critical migratory routes (permanently or seasonally) where fish are both aggregated and highly vulnerable to fishing. As an example, closures of spawning tributaries are commonly employed in anadromous fish management. The large seasonal and permanent closures in the United States fishery for swordfish *Xiphias gladius* in the northwestern Atlantic designed to protect small swordfish and threatened sea turtles is an example of an MPA adapted for management of a highly migratory species (see NOAA 2000).

The NRC committee (NRC 2001) debated weaknesses and drawbacks of conventional fisheries management in the context of alternative MPA-based management. Quota and effort controls have not always met success, and many analysts believe that conventional approaches have failed to achieve sustainability (Botsford et al. 1997). In the United States during 2002, of 236 marine stocks with sufficient data, 34% were overfished, and of 274 stocks with sufficient data, 24% were experiencing overfishing (NMFS 2003). In several cases, conventional management is succeeding in restoring stocks to conditions that indicate sustainability, but in others (e.g., Pacific rockfishes *Sebastes* spp. and northwest Atlantic cod *Gadus morhua*), the likelihood of success is not apparent. When such circumstances prevail, a move toward ecosystem-based management (Brodziak and Link 2002) that includes MPAs as a significant component may be critical to restore a fishery to sustainability and profitability.

Quality of stock assessments is often uncertain. Appropriate stock-size or fishing-mortality reference points and targets are poorly known or imprecise for many fished stocks (NRC 1998). Uncertainties in assessments, combined with overcapacity (NRC 1999a), can lead to failed effort control, declining stocks, lower yields, and demands for better methods to stabilize stocks. Management agencies now are adopting more precautionary approaches in determining biological reference points (BRPs) and in selecting specific targets and thresholds in managed fisheries (Caddy and Mahon 1995). However, agreeing on appropriate stock assessment models, selecting BRPs (usually stock biomass or fishing mortality rates), and then setting appropriate targets and thresholds can be formidable tasks (NRC 1998; Restrepo et al. 1998; Mace 2001).

Can a combination of conventional and MPA-based management attain the elusive goals of stability and sustainability that we seek? Global fish catches are essentially static, and 25–30% of stocks are overfished (Garcia and de Leiva Moreno 2003). Conventional management potentially can return many stocks to sustainable levels, but present circumstances indicate that ecosystem-based approaches are needed to supplement the conventional tools (Link 2002). Can adoption of ecosystem-based approaches, including allocation of areas to no-take reserves, lower the risks of stock collapses and, thus, balance, to an extent, uncertainties in science and management that are mostly irreducible? The answer is not obvious since there also are risks and uncertainties attendant to MPA-based management. But when circumstances dictate (i.e., stocks are depleted and conventional management is ineffective), MPAs as a supplementary tool can add a degree of insurance and precautionary protection to stocks.

Performance Issues

Expectations

In a fisheries context, MPAs usually are designated with the expectation that benefits will be exported to a wider surrounding region that includes areas open to fishing. Meta-analysis of data collected on existing MPAs confirms that no-take reserves lead to increased abundance and sizes of protected species within a reserve's boundaries (Halpern 2003), but the benefits to surrounding areas are less predictable or, in many cases, unevaluated. Performances of MPAs with respect to spillover and enhancement of stocks will depend on the particular dispersal behaviors of organisms at each relevant life stage, which are poorly understood at present (Fogarty et al. 2000; Jennings 2001). Polunin (2002) indicates that benefits to fisheries through spillover of prerecruits are most often apparent in tropical reef fisheries for species that are relatively site attached during their recruited stages.

Evaluations

Expectations for MPA performance should be clearly expressed during the planning and design phase by reviewing the state of knowledge of the ecosystem, by conducting experiments when possible, and by modeling. Since any export of benefits beyond MPA boundaries depends on fisher behavior and on conventional management in the open fished area, these factors also must be considered at the outset. In some cases, there is strong evidence of benefits exported from MPAs to surrounding regions. One example is the documentation of many record-sized game fish species landed just beyond reserve boundaries of security areas adjacent to the Kennedy Space Center in Florida (Johnson et al. 1999; Roberts et al. 2001). In another example, vessel tracks of scallop vessels on Georges Bank (from vessel monitoring systems), off Massachusetts, are densest near the boundaries of three MPAs that restrict dredging and trawling (Fogarty et al. 2000), suggesting that spillover of

prerecruit sea scallops *Placopecten magellanicus* benefits the fishery. In these cases, it seems clear that spillover of large adult fish (Florida) and larval stages of scallops (Georges Bank) is enhancing the respective fisheries.

Absence of a rigorous experimental design prior to implementing MPAs often makes evaluation tenuous at best. Most MPA evaluations lack true controls or appropriate pretreatments and posttreatments (NRC 2001) to allow comparisons of fish stocks within and outside the reserves or to allow trends in yields, fishing mortality rates, recruitment rates, and economic benefits from a fishery to be compared with outcomes in the absence of MPAs. Overcoming design faults is not easily accomplished, especially when large areas of the ocean are proposed for MPAs. The usual replicated treatments and controls embodied in solid experimental design may not be feasible. Confounding of results from poor design may not be fatal to evaluation, but it may add years to the time required to reach conclusions about benefits or costs of MPA implementation.

Monitoring

Monitoring may be costly but is essential to judge performance of MPAs relative to expectations and goals established in the planning phase (NRC 2001). Effective monitoring should include regular assessments of habitat and of stock structure for fished and key unfished species both within and outside the boundaries of established MPAs. In addition, measures of fishery performance (biological and socioeconomic) must be routinely obtained. For fished species, indicators of age structure, spawning stock biomass, fishing mortality rates, and biological reference points associated with those indicators of stock status must be obtained and incorporated into assessment models. Monitoring results should be expeditiously reported and available to fishers.

Enforcement

Enforcement is required to meet management goals for either conventional or MPA-based management. Many of the thousands of protected areas in marine ecosystems throughout the world are thought to be ineffective because there is no enforcement of restrictions on activities, including fishing, within MPA boundaries. Reliable and regular enforcement is essential if objectives of an MPA are to be achieved or if potential benefits are to be evaluated.

Success of an MPA can lead to increased enforcement needs. For example, in the Georges Bank closures, the large increases in size and abundance of scallops made the benefits of illegally fishing within the closure (US$55,800 trip landed value; fishery management plan for Atlantic sea scallops; NEFMC 2000) large relative to the fine (as low as $5,000; M. McDonald, NOAA General Council, Northeast region, personal communication). Consequently, MPA design should include considerations of enforceability. Small MPAs will increase enforcement costs and difficulty; boundaries of an MPA should be clearly referenced by latitude and longitude coordinates. Exclusion of all fishing vessels is easier to enforce than prohibitions on certain gears, and use of vessel monitoring systems should be increased to simplify enforcement of fishing vessel activities.

Nonperformance

If monitoring and evaluations fail to demonstrate benefits of MPAs for fisheries or the ecosystem after an agreed-upon implementation period, then reevaluation is in order. Failure to meet management objectives should require design revision or termination of an MPA in favor of alternative management approaches. Spatially explicit management policies, including MPAs, should be implemented with the same adaptive flexibility as measures applied in conventional management.

Fisheries Ecosystem Plans

In the United States, the need to define essential fish habitat (EFH) and to manage fisheries to ensure their protection were highlighted in the amended Magnuson-Stevens Fisheries Conservation and Management Act (NMFS 1996). The Congressionally-mandated Ecosystems Principles Advisory Panel (NMFS 1999) emphasized the need for ecosystem approaches to improve fisheries management, including a major conceptual recommendation that each regional fisheries management council develop a Fishery Ecosystem Plan (FEP). These FEPs are envisioned to serve as umbrella plans for individual fisheries management plans (FMPs). The FEP will define important ecosystem-level concerns that must be addressed in FMPs. Habitat concerns and a role for MPAs were prominently considered in the panel's report (NMFS 1999). We believe that a reauthorized Magnuson-

Stevens Act could be effective in encouraging designation of MPAs as a tool to protect EFH and to promote spatially explicit resource management.

Concluding Remarks

Hundreds of MPAs are presently implemented in the United States coastal zone and thousands have been implemented globally, but a minuscule number has fisheries management as a primary goal, and even fewer are true no-take reserves. Debates on efficacy, legality, and possibilities for MPAs have multiplied in the past decade. It seems clear that MPAs will increase in importance in the United States during the next decade as ecosystem approaches for fisheries management and the needs to conserve biodiversity and habitats become more prominent. Regional fishery management councils in the United States presently are developing MPA plans within their respective jurisdictions. Careful planning, care in site selection, and consideration of possibilities for zoning and creation of MPA networks are needed now to assure that MPAs will meet their stated goals. Implementation of MPAs can help to satisfy research needs in fisheries by providing sites to obtain critical fishery management information (e.g., estimates of natural mortality, dispersal rates of fish, and role of habitat).

Marine protected areas are an old but underutilized tool that recognizes heterogeneity of aquatic habitats and the need to manage fisheries with that heterogeneity in mind. This approach is especially needed today in highly intensive fisheries where overcapacity and overfishing occur all too frequently. When those circumstances prevail, there is substantial evidence that MPAs can contribute positively to resource management. There also is evidence that significant socioeconomic disruptions to traditional commercial and recreational fisheries will result from adoption of MPAs as major components of management plans. In this regard, the need for stakeholder involvement in all phases of MPA planning and design is clear. The "middle ground" approach recommended by Jones (2002), in which agency-dictated, "top-down" direction is balanced with broad "bottom-up" stakeholder involvement is an appealing model for MPA planning.

We have emphasized how the pressure of new circumstances drives the need for additional tools such as MPAs in marine fisheries management. The frontiers of fishing are behind us. Few habitats now are unfishable, and therefore there are few natural refuges from fishing. Marine protected areas cannot be a stand-alone solution to fishery management problems. When circumstances dictate, they can be an effective component of management plans. Strengthening legislation to encourage consideration of MPAs and other spatially explicit management approaches is desirable. For example, language in reauthorization of the Magnuson-Stevens Act in the United States could be added to (1) address issues and identify probable benefits of MPAs, (2) develop criteria for MPA implementation, and (3) specify research needs.

Acceptance and implementation of MPAs should be based on criteria that allow managers to judge if MPAs are preferable to imposition of additional conventional management measures. Insurance against uncertainties of science and management is an acceptable, precautionary reason to recommend MPAs, especially in depleted fisheries that are threatened with collapse or are threatened by other human activities. There are many cases in which conventional management can address and correct overfishing. Even under that circumstance, MPAs may be a desirable element in a comprehensive management package if it can be demonstrated that the approach will be effective, economical, and enforceable. There are many additional reasons to implement MPAs beyond the goals for fisheries management. Creating parks, protecting unique habitats or historical sites, and protecting vulnerable species and biological communities (i.e., biodiversity) are valid reasons to create MPAs. Marine protected areas will not solve the problems of marine fisheries, but they can play a role in the solution as part of a balanced management portfolio.

Acknowledgments

A Committee of the Ocean Studies Board, National Research Council, National Academy of Sciences, conducted the study that we frequently cite (NRC 2001). Committee members included E. Houde (Chair), F. Coleman, P. Dayton, D. Fluharty, G. Kelleher, S. Palumbi, A. Parma, S. Pimm, C. Roberts, S. Smith, G. Somero, R. Stoffle, and J. Wilen. S. Roberts was the Study Director. We are grateful to committee members for their contributions to the NRC study. We freely acknowledge their important influence on the content of the present paper but note that extensions of ideas, references to recent literature on MPAs, and many opinions and conclusions are ours alone and may not represent views of all committee members.

References

Airame, S., J. E. Dugan, K. D. Lafferty, H. Leslie, D. A. McArdle, and R. R. Warner. 2003. Applying ecological criteria to marine reserve design: a case study from the California Channel Islands. Ecological Applications 13(Supplement):S170–S184.

Apostolaki, P., E. J. Milner-Gulland, M. K. McAllister, and G. P. Kirkwood. 2002. Modelling the effects of establishing a marine reserve for mobile fish species. Canadian Journal of Fisheries and Aquatic Sciences 59:405–415.

Beck, M. W., K. L. Heck, K. W. Able, D. L. Childers, D. B. Eggleston, B. M. Gillanders, B. Halpern, C. G. Hays, K. Hoshino, T. J. Minello, R. J. Orth, P. F. Sheridan, and M. P. Weinstein. 2001. The identification, conservation, and management of estuarine and marine nurseries for fish and invertebrates. BioScience 51:633–641.

Beverton, R. J. H., and S. J. Holt. 1957. On the dynamics of exploited fish populations. United Kingdom Ministry of Agriculture and Fisheries, Fisheries Investigation Series 2, Number 19, London.

Botsford, L. W., J. C. Castilla, and C. H. Peterson. 1997. The management of fisheries and marine ecosystems. Science 277:509–515.

Botsford, L. W., F. Micheli, and A. Hastings. 2003. Principles for the design of marine reserves. Ecological Applications 13(Supplement):S25–S31.

Brodziak, J., and J. Link. 2002. Ecosystem-based fishery management: what is it and how can we do it? Bulletin of Marine Science 70:589–611.

Caddy, J. F. and R. Mahon. 1995. Reference points for fisheries management. FAO Fisheries Technical Paper 347.

Carr, M. H., J. E. Neigel, J. A. Estes, A. Adelman, R. R. Warner, and J. J. Largier. 2003. Comparing marine and terrestrial ecosystems: implications for the design of coastal marine reserves. Ecological Applications 13(Supplement):S90–S107.

Carr, M. H., and P. T. Raimondi. 1998. Concepts relevant to the design and evaluation of fishery reserves. Pages 27–31 in M. M. Yoklavich, editor. Marine harvest refugia for West Coast rockfish: a workshop. NOAA Technical Memorandum NOAA-TM-NMFS-SWFSC-255.

Crowder, L. B., S. J. Lyman, W. F. Figueira, and J. Priddy. 2000. Source-sink population dynamics and the problem of siting marine reserves. Bulletin of Marine Science 66:799–820.

Fogarty, M. J., J. A. Bohnsack, and P. K. Dayton. 2000. Marine reserves and resource management. In C. Sheppard, editor. Seas at the millenium. Elsevier, London.

Gaines, S. D., B. Gaylord, and J. L. Largier. 2003. Avoiding current oversights in marine reserve design. Ecological Applications 13(Supplement):S32–S46.

Garcia, S., and C. Newton. 1997. Current situation, trends, and prospects in world capture fisheries. Pages 3–27 *in* E. L. Pikitch, D. D. Huppert, and M. P. Sissenwine, editors. Global trends: fisheries management. American Fisheries Society, Symposium 20, Bethesda, Maryland.

Garcia, S. M., and I. de Leiva Moreno. 2003. Global overview of marine fisheries. Pages 1–14 *in* M. Sinclair and G. Valdimarsson, editors. Responsible fisheries in the marine ecosystem. CABI Publishing, Wallingford, UK.

Gerber, L. R., L. W. Botsford, A. Hastings, H. P. Possingham, S. D. Gaines, S. R. Palumbi, and S. Andelman. 2003. Population models for marine reserve design: a retrospective and prospective synthesis. Ecological Applications 13(Supplement):S47–S64.

Guennette, S., T. Lauck, and C. Clark. 1998. Marine reserves: from Beverton and Holt to the present. Reviews in Fish Biology and Fisheries 8:251–272.

Halpern, B. 2003. The impact of marine reserves: do reserves work and does reserve size matter? Ecological Applications 13(Supplement):S117–S137.

Hastings, A., and L. Botsford. 1999. Equivalence in yield from marine reserves and traditional fisheries management. Science 284:1537–1538.

Holland, D. S., and R. J. Brazee. 1996. Marine reserves for fisheries management. Marine Resource Economics 11:157–171.

Holmlund, C. M., and M. Hammer. 1999. Ecosystem services generated by fish populations. Ecological Economics 29:253–268.

Horwood, J. W., J. H. Nichols, and S. Milligan. 1998. Evaluation of closed areas for fish stock conservation. Journal of Applied Ecology 35:893–903.

IUCN (The World Conservation Union). 1994. Guidelines for protected area management categories. International Union for the Conservation of Nature and Natural Resources, Gland, Switzerland, and Cambridge, UK.

Jennings, S. 2001. Patterns and prediction of population recovery in marine reserves. Reviews in Fish Biology and Fisheries 10:209–231.

Johnson, D. R., N. A. Funicelli, and J. A. Bohnsack. 1999. Effectiveness of an existing estuarine no-take fish sanctuary within the Kennedy Space Center, Florida. North American Journal of Fisheries Management 19:436–453.

Jones, P. J. S. 2002. Marine protected area strategies: issues, divergences and the search for middle ground. Reviews in Fish Biology and Fisheries 11:197–216.

Largier, J. L. 2003. Considerations in estimating larval dispersal distances from oceanographic data. Ecological Applications 13(Supplement):S71–S89.

Lauck, T. C., C. W. Clark, M. Mangel, and G. R. Munro. 1998. Implementing the precautionary principle in fisheries management through marine reserves. Ecological Applications 8(1):S72–S78.

Leslie, H., M. Ruckelshaus, I. R. Ball, S. Andelman, and H. P. Possingham. 2003. Using siting algorithms in the design of marine reserve networks. Ecological Applications 13(Supplement):S185–S198.

Link, J. S. 2002. Ecological considerations in fisheries management: when does it matter? Fisheries 27(4):10–17.

Mace, P. M. 2001. A new role for MSY I single-species and ecosystem approaches to fisheries stock assessment and management. Fish and Fisheries 2:2–32.

Mangel, M. 2000. Trade-offs between fish habitat and fishing mortality and the role of reserves. Bulletin of Marine Science 66:663–674.

Milon, J. W. 2000. Pastures, fences, tragedies and marine reserves. Bulletin of Marine Science 66:901–916.

Murray, S. N., R. F. Ambrose, J. A. Bohnsack, L. W. Botsford, M. H. Carr, G. E. Davis, P. K. Dayton, D. Gotshall, D. R. Gunderson, M. A. Hixon, J. Lubchenco, M. Mangel, A. MacCall, D. A. McArdle, J., C. Ogden, J. Roughgarden, R. M. Starr, M. J. Tegner, and M. M. Yoklavich. 1999. No-take reserve networks: sustaining fishery populations and marine ecosystems. Fisheries 24(11):11–25.

Myers, R. A., and B. Worm. 2003. Rapid worldwide depletion of predatory fish communities. Nature (London) 423:280–283.

NEFMC (New England Fishery Management Council). 2000. Stock assessment and fishery evaluation report (SAFE). NEFMC. Available: *www.nefmc.org* (October 2003).

NMFS (National Marine Fisheries Service). 1996. Magnuson-Stevens Fishery Conservation and Management Act. NOAA Technical Memorandum NMFS-F/SPO-23.

NMFS (National Marine Fisheries Service). 1999. Ecosystem-based fishery management. Ecosystem Advisory Panel to NMFS. NOAA Technical Memorandum NMFS-F/SPO-33.

NMFS (National Marine Fisheries Service). 2003. Annual report to Congress on the status of U.S. fisheries—2002. National Oceanic and Atmospheric Administration, NMFS, Silver Spring, Maryland.

NOAA (National Oceanic and Atmospheric Administration). 1990. The potential of marine fishery reserves for reef management in the U.S. southern Atlantic. NOAA Technical Memorandum NMFS-SEFC-261.

NOAA (National Oceanic and Atmospheric Administration). 2000. Atlantic highly migratory species; pelagic longline management, Final Rule. Federal Register 65:148 (1 August 2000). Available: *ww.nmfs.noaa.gov/sfa/hms/timearea.pdf* (October 2003).

NOAA (National Oceanic and Atmospheric Administration). 2003a. Marine protected areas of the United States. NOAA. Available: *www.mpa.gov* (October 2003).

NOAA (National Oceanic and Atmospheric Administration). 2003b. Marine sanctuaries. NOAA. Available: *www.sanctuaries.nos.noaa.gov/oms/oms.html* (October 2003).

NOAA (National Oceanic and Atmospheric Administration). 2003c. Florida Keys National Marine Sanctuary: Tortugas 2000. NOAA. Available: *http://floridakeys.noaa.gov/tortugas/welcome.html* (October 2003).

NRC (National Research Council). 1998. Improving fish stock assessments. NRC, National Academy of Sciences, National Academy Press, Washington, D.C.

NRC (National Research Council). 1999a. Sustaining marine fisheries. NRC, National Academy of Sciences, National Academy Press, Washington, D.C.

NRC (National Research Council). 1999b. Sharing the fish. National Research Council, National Academy of Sciences. National Academy Press, Washington, D.C.

NRC (National Research Council). 2001. Marine protected areas: tools for sustaining ocean ecosystems. NRC, National Academy of Sciences, National Academy Press, Washington, D.C.

Palumbi, S. R. 2002. Marine reserves. A tool for ecosystem management and conservation. Pew Oceans Commission, Arlington, Virginia.

Pauly, D. 1996. One hundred million tonnes of fish, and fisheries research. Fisheries Research 25:25–38.

Polunin, N. V. C. 2002. Marine protected areas, fish and fisheries. Pages 293–318 *in* P. Hart and J. Reynolds, editors. Handbook of fish biology and fisheries, volume 2, fisheries. Blackwell Scientific Publications, Oxford, UK.

Restrepo, V. R., G. G. Thompson, P. M. Mace, W. L. Gabriel, L. L. Low, A. D. McCall, R. D. Methot, J. E. Powers, B. L. Taylor, P. R. Wade, and J. F. Witzig. 1998. Technical guidelines on the use of the precautionary approach to implementing National Standard 1 of the Magnuson-Stevens Fishery Conservation and Management Act. NOAA Technical Memorandum NMFS-F/SPO 31.

Roberts, C. M., S. Andelman, G. Branch, R. H. Bustamante, J. C. Castilla, J. Dugan, B. S. Halpern, K., D. Lafferty, H. Leslie, J. Lubchenco, D. H. McArdle, H. P. Possingham, M. Ruckelshaus, and R. R. Warner. 2003. Ecological criteria for evaluating candidate sites for marine reserves. Ecological Applications 13(Supplement):S199–S214.

Roberts, C. M., J. A. Bohnsack, F. Gell, J. P. Hawkins, and R. Goodridge. 2001. Effects of marine reserves on adjacent fisheries. Science 294:1920–1923.

Sanchirico, J. N., and J. E. Wilen. 1999. Bioeconomics of spatial exploitation in a patchy environment. Journal of Environmental Economics and Management 37:129–150.

Sanchirico, J. N., and J. E. Wilen. 2001. A bioeconomic model of marine reserve creation. Journal of Environmental Economics and Management 42:257–276.

Shanks, A. L., B. A. Grantham, and M. H. Carr. 2003. Propagule dispersal distance and the size and spacing of marine reserves. Ecological Applications 13(Supplement):S159–S169.

Sladek-Nowlis, J. S., and C. M. Roberts. 1999. Fisheries benefits and optimal design of marine reserves. U.S. National Marine Fisheries Service Fishery Bulletin 97:604–616.

Walters, C. 2000. Impacts of dispersal, ecological interactions, and fishing effort dynamics on efficacy of marine protected areas: how large should protected areas be? Bulletin of Marine Science 66:745–758.

American Fisheries Society Symposium 42:37–61, 2004

Marine Protected Areas in Fisheries: Design and Performance Issues

TREVOR J. WARD[1]

M004 Institute for Regional Development,
Faculty of Natural and Agricultural Sciences,
University of Western Australia, 35 Stirling Highway,
Crawley, Western Australia 6009, Australia

Abstract.—Various forms of area protection have long been used in the management of marine capture fisheries, and there are now increasing calls for wider use of marine protected areas (MPAs), including no-take areas (reserves), to resolve issues of fisheries sustainability. However, global policy has preceded the development of generic models and experience to underpin the design of effective MPAs for sustainable fisheries. Four main classes of potential benefits can be identified for MPAs in fisheries—biological support for the exploited stock, socioeconomic benefits for the fishers, biodiversity benefits for the fished ecosystems, and increased fishery stability. The extent to which any MPA will achieve these benefits depends on its design, its integration with the broader fishery management system, and the effectiveness of the existing stock management. An MPA may deliver only limited additional support for a stock that is already well managed, but an appropriately designed MPA may be able to offer other benefits to the fishery such as an ecological offset for the effects of fishing on habitats. Delivering these benefits depends on establishing a priori design principles and decision rules in relation to each specified objective that an MPA is established to achieve. Measuring the benefits requires a performance assessment and reporting system, including specific performance indicators, that is directly linked to the objectives of the MPA. The biggest single threat to the success of fisheries MPAs is model failure—the situation where MPAs are established that do not reflect the implicit and explicit sustainability objectives of the fishery. Fisheries MPAs established without comprehensive design and performance assessment systems are likely to be neither successful nor secure and are unlikely to meet medium-term or long-term sustainability objectives.

Introduction

Recent revelations about the state of the world's fisheries (Hutchings 2000; Watson and Pauly 2001; Pauly et al. 2002; Myers and Worm 2003) and the environmental impacts of fishing (e.g., Thrush et al. 1998; Cryer et al. 2002) have intensified the debate about the potential utility of marine protected areas (MPAs) in the management of fisheries. The potential benefits of areas that are closed to fishing for fisheries management purposes (known variously as closures, refuges, reserves, sanctuaries, no-take areas, parks, etc.) have been described in numerous reviews over the past decade and increasingly in modeling and analysis of empirical data from closed areas. These include the biological benefits (Murray et al. 1999; Halpern 2003), the economic costs and benefits (Arnason 2001; Sanchirico and Wilen 2001; Sumaila 2001), and the broader role of MPAs in the sustainability of fisheries and the integrated management of marine systems (Botsford et al. 1997; Lubchenco et al. 2003).

This recent upsurge in the debate about the value of MPAs to fisheries seems ironic given the long history of the use of closed areas in fisheries management in many countries and cultures. Area protection has a long history of use in fisheries management, across industrial, community-based tropical and temperate fisheries. For example, closed areas are recorded from the earliest known fishing systems and are a feature of many traditional fishing systems in the Pacific Islands (Johannes 1978), and many contemporary

[1] E-mail: tward@ird.uwa.edu.au

fisheries include closed areas as a tool to assist with the achievement of specific fisheries management objectives (e.g., Beumer et al. 1997).

The expanding scientific literature on MPAs in fisheries is focusing, both explicitly and implicitly, on no-take areas rather than on the full range of types of closed areas that are available and have been used in fisheries management. This literature is also focused on the role of MPAs for stock management purposes. No-take areas represent a strong and effective management intervention, by excluding all forms of fishing and collecting of biological material. But no-take areas lie near one end of what is essentially a continuum of forms of marine area protection and management. This protection continuum is well recognized in both the theory and practice of management of marine systems, and protected areas representing all parts of the continuum are able to contribute to both the conservation of marine biodiversity and to the management of fish stocks (Agardy et al. 2003). However, the extent of this contribution to fisheries objectives, and its cost-effectiveness, is likely to depend heavily on the protected area design process and, to a large extent, on how the benefits are perceived, defined, and evaluated (Roberts et al. 2003a). A focus on the potential benefits of no-take areas for stock management alone risks ignoring other types of area protection that could also assist in resolving a broader range of fishery management issues simultaneously with stock issues.

Area closure is a tool that is common to both conservation and fishing management systems, but their management goals and specific objectives typically are different. An area protected under conservation management may require complete protection to maintain a rare habitat or population of a species, while an area protected under fisheries management may require protection of a habitat and a species for spawning purposes or protection of juveniles of a commercially exploited species. The conservation protection may seek to ensure that the habitats and populations are maintained in a natural condition while the fishery protection would seek to ensure that the spawning biomass and population fecundity was maintained. Such differences in objectives for closed areas may lead paradoxically to the situation where fishing using some low-impact gear types (as well as a range of other activities) may be acceptable within some forms of fishery closed areas, but the same activities may be completely unacceptable in a conservation area designed to protect the same species.

In this paper, I use the term "marine protected area" (MPA) to mean any effective form of protection for an area of seabed, overlying waters, and the biodiversity contained therein and managed using enduring and effective controls and rules. I consider the term "no-take area" to be a specific form of MPA where all extraction of any living or nonliving resources is prohibited. I use "closed area" and "reserve" as forms of MPAs where some types of fishing (and other activities), but perhaps not all, are prohibited. A "fishery MPA" is an MPA designed and implemented by fishery managers or communities to achieve fishery management objectives (where these objectives will also include other objectives, such as for habitats and nontarget species).

The recent World Congress on Aquatic Protected Areas (Cairns, Australia, 2002) heard considerable debate about the breadth, depth, and potential role of MPAs in fisheries from a mixture of policy analysts, managers, scientists, and community groups. Perspectives from government agencies reiterated their legislatively empowered mandate and responsibilities in such matters as fisheries management, conservation of species, and management of a world heritage area; and fisheries managers, nongovernmental organizations, and scientists debated the various models, experiences, and concepts of marine protected areas more broadly. The main themes emerging throughout the congress (see Appendix) emphasized the diversity of roles that MPAs are expected to play depending on the situation and the range of approaches and performance issues that surround the design and implementation of MPAs.

Broadly speaking, the Cairns Congress debate inferred that protected area systems in fisheries management were potentially able to resolve some part, and possibly a major part, of the problem of improving the management of the oceans and their natural resources. But, as expected, the purpose and acceptance of MPAs as a useful tool appeared to depend largely on the sectored background, institutional domain, and the personal perspective of the commentator. A somewhat uncomfortable coexistence of conservation and resource purposes for MPAs was evident throughout the congress. Nonetheless, as observed about the history of fishing in the Pacific Islands, "environmentally destructive practices coexisted, as in most societies, with efforts to conserve natural resources. But the existence of the former does not diminish the significance of the latter" (Johannes 1978).

In response to empirical evidence of problems with fisheries and in oceans management systems, the global marine policy arena is gradually changing. The Plan of Implementation from the recent Johannesburg World Summit on Sustainable Development (WSSD; WSSD 2002) envisages integrated oceans management involving MPAs that are implemented to achieve multiple objectives, including improving fisheries management, in the global oceans. The need for a more comprehensive approach to the management of fisheries than has been provided by the classical single-species assessment approach is also now widely understood and is consistent with modern approaches to the management of a range of natural resources and marine issues (Ward 2000; Campbell et al. 2001).

Modern fisheries management, in well-managed fisheries, focuses on managing marine systems in the light of considerable uncertainty that is well beyond the issues of harvesting a fish population (Botsford et al. 1997; Phillips et al. 2003). The interactions of fishing and target species with the dynamics of the ecosystems at large and small scales, the socioeconomic objectives of local and national communities, and the impact of fishing on nontarget species and habitats are all now considered to be important for effective and sustainable fisheries management (Botsford et al. 1997; FAO 2002; Pauly et al. 2002; Ward 2003a). As a result, management of fisheries is likely to involve two new challenges: managing aspects of ecosystems that interact with fishing and taking a much more integrated approach to the process of fishery management. An integrated approach to management, recognizing the dynamics and interactions with ecosystems and where other users and stakeholders may participate in decisions about fisheries, is now recognized as not only appropriate and responsible public policy, but such a "holistic, ecosystem approach to fisheries management" including MPAs appears to be the best chance for improving the sustainability of fisheries (Botsford et al. 1997; Pauly et al. 2002).

Faced with such a complex and evolving management framework, fishery managers intending to design and implement MPAs need to consider carefully what approach to take, how to ensure that any potential benefits can be efficiently secured, and how any costs can be either fully justified or offset against benefits. In such circumstances, a key requirement will be to assess the efficiency and effectiveness of a network of MPAs expected to contribute benefits to the fishery. This paper, taking the perspective of fishery managers intending to design and implement a network of MPAs, reviews a series of important policy and planning issues associated with fishery MPAs. I argue here that, given the global policy imperatives, contemporary fishery MPAs will usually be required to meet multiple sets of objectives, including contributing to regional biodiversity conservation. Therefore, this review considers the policy and planning issues underlying the broad range of potential benefits that a fishery may be able to secure from MPAs and specific design and performance assessment issues that are inherent in a fishery management system that involves the use of MPAs designed to achieve multiple sets of objectives within an integrated oceans management context.

The Potential Fishery Benefits

The potential benefits of no-take MPAs for both fisheries and conservation of biodiversity have been widely discussed in the mainstream science literature (e.g., Roberts et al. 2001; Halpern and Warner 2002). The use of no-take MPAs set in the context of ecosystem-based management and the precautionary approach, are oft-cited elements of a new fisheries management paradigm that is expected to develop in response to concerns about the effectiveness of classical fisheries management (Pauly et al. 2002; Ward et al. 2002). However, suggested improvements in fisheries management involving the use of MPAs have not been immediately embraced by fisheries scientists, managers, fishers, or the public at large, and, indeed, are actively opposed by some (Roberts and Polunin 1993; Ballantine 1995; Gubbay 1995; Sant 1996; Bohnsack 1997; Williams 1998; Suman et al. 1999; Ward et al. 2002). A key weakness in convincing fishers of the merit of MPAs, and particularly no-take areas, as an effective tool to assist with fisheries management is the lack of successful precedents for design and implementation of closed areas to achieve multiple objectives that include benefits for fisheries. Many demonstrated benefits relate to tropical reef fisheries that have suffered extreme overfishing and habitat degradation (Roberts et al. 2001). Nonetheless, a well-documented series of benefits have been described for the Merritt Island National Wildlife Refuge in Florida, where the exclusion of fishing is considered to have had a major benefit for target species in the MPA and for the fishery for the same species outside the MPA (Johnson et al. 1999; Roberts et al. 2001)

More broadly, many recent analyses of the literature and empirical studies have concluded that MPAs can, in certain situations, provide benefits for species that are fished in terms of increased abundance in closed areas and increased catch in adjacent areas (Edgar and Barrett 1999; Murawski et al. 2000; Cote et al. 2001; Roberts et al. 2001; Ward et al. 2001; Fisher and Frank 2002; Halpern and Warner 2002; Mapstone et al., in press). It is clear from these studies that the benefits that MPAs may be able to bring to a fishery are related to many factors, including the identity of the species being fished, the effectiveness of the fishery management system in place, the matters that are considered to be benefits, and the tools and procedures used to assess those benefits.

In order to help synthesize and assess the possible mechanisms for delivery of potential benefits in any specific fishery, a simple conceptual model of the processes that might operate to deliver such benefits was developed by Ward et al. 2001 (Figure 1). The model enabled the potential benefits often discussed in the literature to be resolved into more specific groups of potential benefits and pathways that might be applicable to a single fishery in the circumstances where a no-take MPA was to be introduced. On the basis of this simple conceptual model, Ward et al. (2001) identified 58 possible indicators, each linked to an associated potential direct or indirect benefit for a fishery. Of these, seven types of benefits (in four broad classes) can be considered to be indicative of the most direct benefits that may accrue to a fishery from the implementation of multiple objective, no-take MPAs (Table 1). They are considered here to be fishery benefits in the broadest sense, given the increasing policy expectations for fisheries management to encompass impacts of fishing on ecosystems and to take a broad view of stakeholder interests and their expectations of an integrated oceans management system that includes wild capture fisheries.

But despite the potential, the benefits above could not be expected to be secured by all fisheries, and not all the benefits could be expected to be achieved in any single fishery. Also, not all fisheries will stand to gain benefits equally across the full geographic scope of the fishery or in all circumstances. Mapstone et al. (in press) describe the benefits from the closed areas (marine national park zones) that accrue to the reef line fishery in the Great Barrier Reef Marine Park, Australia, and document the variability in these benefits across the region ranging from almost zero to several-fold (depending on the location and the specific benefit).

Nonetheless, some types of fisheries may be able to capture many of these benefits. The fisheries that

may be able to gain the most and largest benefits are those that:

(1) are overfished;

(2) are fully exploited;

(3) have substantial fishing effects on ecosystems, habitats, or bycatch species;

(4) exploit species that are associated with specific areas of seabed or that have obligate habitat requirements;

(5) exploit species that can have high levels of recruitment; or

(6) exploit species that have a well-studied life history (Ward et al. 2001).

Ecological Offsets

The ecological and ecosystem impacts of fisheries (Jennings and Kaiser 1998; Hall 1999; Cryer et al. 2002) are increasingly being considered in detail as part of the sustainability assessment of modern fisheries (Botsford et al. 1997; Pauly et al. 2002). In the global eco-labeling program of the Marine Stewardship Council (Phillips et al. 2003), the performance of a fishery is critically examined to determine if it meets a global sustainability standard that includes the ecosystem effects of fishing. And similarly, under Australia's national legislation, fisheries are being assessed, among other bases, on their ecological impacts on ecosystems. These assessments are consistent with the intent of the WSSD and with modern interpretations of the need for the management of fisheries to be based more on ecosystems (Ward et al. 2002).

In contributing to the goals of ecosystem-based management and to stock benefits, MPAs may be able to provide ecological offsets to a fishery. The concept of "ecological offset" (Ward and Hegerl 2003) is that a protected area may be designed to provide a form of compensation as an offset for the ecological impacts of a fishery that cannot be avoided or resolved in any other way (such as through gear modifications, etc.). By protecting areas that provide refuges for fishing-affected communities and species, specifically designed MPAs may be able to assist a fishery to ensure that the unavoidable ecological impacts inherent in their fishing activities can be effectively minimized. Where there is insufficient detailed knowledge about the impacts of fishing, a precautionary approach would

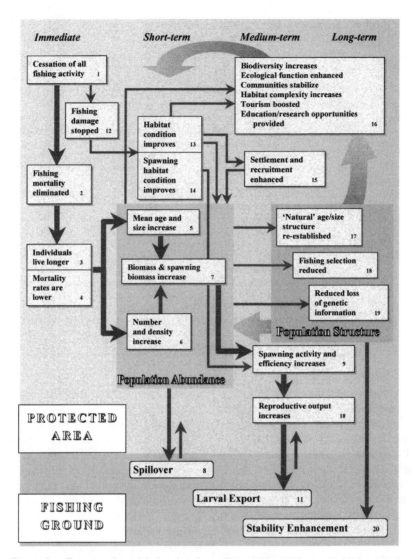

Figure 1.—Conceptual model showing the pathways by which the establishment of a no-take marine protected area (MPA) could lead to environmental enhancement of the area and potentially to enhancement of a fishery outside the MPA through the processes of spillover, larval export, and enhancement of fishery stability. The large, upper box represents a no-take area in an MPA, and the lower box represents the fished areas outside the MPA. Each box within the MPA box represents an event, state, or effect within the hypothesized cause–effect pathways (numbers are referenced in the text of Ward et al. [2001] and more detail provided). The size of arrows indicates roughly the likely importance of that pathway to the potential for fisheries enhancement. Very roughly, the time frame within which these events, states, and processes might be expected to occur, following area protection, increases from immediate on the left to long term on the right. Boxes 5–7 are grouped together to indicate that they are the processes involved in increases in population abundance, the most obvious manifestation of the process of MPA improvement. Boxes 17–19 are grouped because they are the processes responsible for the long-term changes to MPA populations, which along with the short-term abundance changes are responsible for the improvements in population stability and resilience. The very large arrows in the background indicate poorly defined pathways. Improvements to population structure have been hypothesized to provide feedback to improve population abundance, but the mechanisms have not been clearly identified.

Table 1.—Potential benefits for fisheries from no-take marine protected areas (MPAs; Ward et al. 2001).

Benefit type	Examples of specific benefits
Biological outcomes: fishery benefits inside the MPA	Increased size and age of focal species of fish; increased abundance (density) of focal species of fish; increased size of spawning stock; increased reproductive output at age for focal species of fish
Biological outcomes: fishery benefits outside the MPA	Net movement of adult focal species of fish from inside to outside of MPA; increased abundance (density) of focal species (across total fishery); increased individual size of focal species (across total fishery); increased yield of focal species, standardized for fishing effort (across whole fishery); yields in other fisheries in region or district maintained
Biological outcomes: non-fishery benefits inside the MPA	Establishment or maintenance of areas of undisturbed habitat; enhanced habitat complexity; more natural species diversity; enhanced community complexity (e.g., trophic complexity); improved populations of fishing-affected species
Biological outcomes: non-fishery benefits outside the MPA	Maintenance or enhancement of habitat complexity, species diversity, and/or community complexity; maintenance or enhancement of populations of fishing-affected species
Management outcomes	Simplified enforcement; contributes to integrated ecosystem-based management of marine ecosystems; provides sites suitable for benchmarking and minimum disturbance reference conditions; reduced data collection requirements; contributes to increased fishery stability
Economic outcomes	Enhanced and diversified local and regional economic opportunities; enhanced opportunities for employment in local industries
Social outcomes	Contributes to maintenance and enhancement of the social and cultural well-being of local communities

be to identify and fully protect representative samples of habitats that are normally targeted by the fishery (Day et al. 2001). This approach to minimizing the ecological impacts of fishing invokes conservation on a bioregional scale and recognizes that fishing has impacts, some of which may be unavoidable, in fishing grounds. Nonetheless, regional conservation objectives may be achieved through an appropriate set of MPAs that give refuge to populations of species and habitats that might otherwise be degraded in fishing grounds.

Protected areas will never be able to completely offset the ecological impacts of a fishery, and there may be some fishing impacts (such as impacts on rare species or habitats or on highly migratory species) where MPAs can achieve only a limited offset. However, when taken in conjunction with other mitigation activities (such as gear modifications), MPAs may provide a cost-effective contribution to help offset the overall ecosystem impacts of a fishery. The success of an MPA in providing such an offset is completely dependent on the quality of the design processes and how the MPA interacts with the fisheries management system.

In Australia, the northern prawn fishery (NPF), a major demersal trawl fishery that targets nine species of tropical penaeid prawns, has recognized that no-take MPAs are an important management tool that

can benefit the fishery. The MPAs can provide greater protection to critical nursery habitat than can currently be provided by Australian fisheries legislation as well as provide refugia for many of the benthic and bycatch species affected by NPF trawling. The NPF now has a significant research effort underway (in conjunction with Environment Australia) to identify benthic species assemblages, model the performance of existing spatial closures, and identify different MPA configurations that can fully achieve biodiversity conservation objectives, while at the same time maximizing the value of the commercial fishery (Carter et al. 2003). These MPAs are expected to provide a substantive ecological offset for the fishery when they are implemented, assuming performance assessments show they are meeting their objectives.

Multiple-Objective MPAs

To be effective in providing for sustainability, modern fisheries management systems should go well beyond multispecies management and deal with multiple objectives within the context of broader oceans management. Only some of these multiple objectives will be about the target species. In a sustainable fishery, management objectives may range across the familiar stock objectives (e.g., maintaining spawning biomass at 50%

of virgin spawning biomass) into the less-familiar ecosystem objectives (e.g., maintaining 50% of habitat types untrawled) and possibly beyond into the even more complex areas of socioeconomic objectives (e.g., maintaining employment in the fishery at or above a benchmark level).

In the broad context of meeting the expectations of sustainability and ecosystem-based management, determining the objectives and establishing reference levels for fishery MPAs probably will normally include aspects of ecosystem performance. However, fisheries managers generally have little experience in managing ecosystems. Also, objectives for ecosystems are more difficult to specify in terms that are equivalent to the formulation of objectives for fish stock management. Nonetheless, designing effective MPAs relies heavily on having objectives that are well specified, meaningful, and achievable (see performance assessment system below). To establish multiple objectives, including stock and ecosystem objectives that are agreed upon and integrated across the various interests of government, industry, and the community, the MPA design will require a systematic approach (Margules and Pressey 2000) and will need to involve aspects of collaborative and participatory management. Such an approach requires a range of technical expertise including fisheries management and ecosystems management, an intensive engagement with stakeholders, and a supportive policy and operational framework within which the MPA design and implementation activities should operate. To secure effective and enduring MPAs, this design process should be set within a broader marine ecosystems management context and, particularly, an integrated oceans management framework. Without this, and together with systematic processes of design and evaluation, a network of MPAs risks substantive failure (Walters 1997; Ward 2000).

A Supportive Management Context

For multiple-objective fishery MPAs to be effective in the long term, they need to be developed within a policy and planning framework that provides support for the planning and management of effective MPAs and discourages perverse incentives that might act to subvert an MPA. While this may be arranged within a definitive legislative framework, this will often not be possible because of the lack of integration of oceans management in most countries. The most effective and efficient management system for fishery MPAs is likely to be some form of collaborative or participative management system which encourages manage-

ment to be responsive to dynamics in the resource and in the system being managed. Several such management systems are widely applied in non-marine natural resource management (Blumenthal and Jannink 2000). While the technical design for a network of MPAs may be comprehensive, if the MPAs are not set within the appropriate broader management context for integrated oceans management, they may fail, irrespective of the quality of the biophysical or other criteria used to establish the network.

Some countries have recognized the need for more integrated and holistic approaches to management of marine systems, including MPAs (such as Canada, the USA, and Australia), and have initiated various forms of oceans management policy designed to encourage the more efficient and effective management of the ocean resources. While these are broadly based on the concept of integrated management systems and ecosystem-based management that involve MPAs, none are fully effective or operational yet (Alder and Ward 2001).

Although the need for a holistic approach to the management of natural resources (Holling 2000) and fisheries (Botsford et al. 1997) is well recognized, and it is clear that this should involve the broad participation of stakeholders, MPA management is still in a developmental stage (Day et al. 2003). Management models for fishery MPA networks that have full participation of stakeholders have yet to be developed in any detail other than for traditional and subsistence fisheries (Johannes 1978; Adams 1998). Five main types of collaborative management models have been widely used in natural resource management, each of which have strong features (Blumenthal and Jannink 2000) that would appear to be important in a management framework for fishery MPAs, particularly adaptive management (Johnson 1999). But no individual management model appears to be sufficiently well developed to provide a proven model for integrated marine systems management that includes explicit guidance for designing fishery MPAs.

While there is no overall oceans management framework in any region, a number of specific local initiatives in adaptive management and ecosystem-based management have focused on MPAs as a key element of marine management, including fisheries. This is perhaps best demonstrated in the Great Barrier Reef World Heritage Area, where multiple-use zones have been established to provide for a range of activities that do not compromise the natural values of the World Heritage Area (Day et al. 2001; Lawrence et al. 2002). The concept of ecosystem-based management promises to provide a framework for collaborative management while maintaining a focus on the quality of natural systems and identifying a role for MPAs (Ward et al. 2002). Nonetheless, as for several of the possible marine management approaches, ecosystem-based management (sometimes also known as the ecosystem approach) is still a concept in search of a practice and provides only an elementary guide to effective practice for fisheries MPAs.

Design Issues

It is well acknowledged that the design theory and management practice of MPAs is still in its infancy (Botsford et al. 2003). But it is equally certain that the success of fishery MPAs will be critically dependent on their design (Jones et al. 1993). A poorly considered design may mean that few of the benefits will be delivered, and perhaps worse, this may produce negative results for fishery sustainability. There are many issues to be considered in establishing a network of multiple-objective MPAs in a fishery, and since these are likely to be based on complex interactions and dynamics in natural systems, operating at different scales and with different stakeholders and interests, a systematic approach to the design of MPAs is most likely to be efficient and effective in the long term. This is more likely to avoid ad hoc planning decisions leading to ad hoc outcomes. One of the key reasons for this is that the basic assumptions and models used in making choices at each step of a systematic MPA selection process need to be made explicit and agreed upon among the stakeholders to help clarify and agree on the areas of critical uncertainty.

In developing a set of MPAs for a fishery, there are several key design and planning issues that need to be addressed very carefully: setting the objectives, choosing the sites that will meet those objectives most optimally including determining the location and size of the MPAs, determining the MPA rules and ensuring compliance, dealing with fishing effort that may have to be displaced from the MPAs, and establishing agreed decision rules to apply in the MPA identification and selection process. Each one of these involves collaboration with stakeholders and requires expertise, data, and models from a range of disciplines in addition to fisheries science and management.

Setting the Objectives
The first, and potentially most critical, challenge to be faced when designing an MPA network is to define the goals that the MPAs are expected to achieve.

While this is obvious enough, conceptualizing outcomes into objectives at a level that can be expressed in outcome-oriented terms is difficult when the goals are expressed at a high level (strategic objectives), are complex, or are of only a short-term nature. Nonetheless, the setting of clear objectives is the most critical step to get right and is well recognized as a fundamental initial step in establishing MPAs within an ecosystem-based management system that can be adaptive and continuously improved (Ward 2000; Day et al. 2001).

When MPA objectives are simple, as may be the case for a fishery where a specific habitat type is recognized as a critical nursery area for the target species, then, perhaps, operational implementation of an effective MPA is easiest. This is the case, for example, for fisheries habitat reserves, where the objective is to simply prevent physical alienation of the habitat by foreshore reclamation works or by mining or other destructive activity (Beumer et al. 1997).

When objectives for a fishery MPA are more complex, only expressed at a strategic level (such as "protection of biodiversity" or "mitigate the impacts of fishing"), or expressed at a mixture of levels, the design problem is much more complex. When MPA objectives are specified in different ways and expressed at different levels of ecological organization and at different scales of spatial resolution, the risk is that the "easy" objectives will be addressed first and "difficult" objectives deferred for later implementation. For fisheries in this situation, balancing design priorities and the subsequent management tasks and priorities among specific objectives may be a substantial challenge.

Balancing the priorities accorded to the objectives that might be considered to be traditional fishery objectives (such as stock objectives) as opposed to ecosystem objectives and associated tasks that are outside the normal fisheries operating arena brings into focus the values inherently underpinning the management system, and it is here that stakeholder participation can be most critical so that conflicts can be resolved early in the MPA design process. Failing to resolve such fundamental values conflicts can be fatal to resource management systems (Walters 1997). However, it is also fatal to avoid such complexities simply to ensure technical or administrative expediency; "doing the thing right" (precise, local measurement) rather than "doing the right thing" (Walters 1997) is the basis for "pseudo power" (Ward and Jacoby 1995) which may lead to delusionary outcomes, increased risks of disenfranchising stakeholders, and, ultimately, damage to fish stocks and the ecosystem.

Site Criteria

Matters usually considered when designing MPAs can be classified into four main groups:

(1) biophysical values, requirements, conditions, and constraints;

(2) social and cultural values and constraints;

(3) existing uses and obligations; and

(4) practical management constraints, including the interaction with external pressures.

The biophysical design issues are complex and are most often discussed in the technical literature, but practical experience in MPAs shows that there are many other, often related, types of issues involving socioeconomic matters, institutional issues, traditions and customs of local communities, and stakeholder engagement, which will also have an impact of the siting of MPAs and their ultimate success (Walters 1997; Jones 1999; Day et al. 2001). This is particularly true in coastal MPAs, and while the greatest experience of this has been with MPAs for biodiversity conservation, as fisheries planners seek to embrace MPAs with multiple sets of objectives, their design procedures will also need to involve broader consultation with stakeholders and other users (Roberts et al. 2003b).

Biophysical issues, for both conservation and fisheries objectives in MPAs, often revolve around questions of existing conditions, oceanographic patterns; mobility of the main species of interest in their adult, juvenile, and larval stages; and minimum MPA size (Roberts et al. 2003a). The spacing and location of MPAs to ensure ongoing linkages between habitat patches is usually considered crucial to maintaining meta-populations, natural range, age-size structure in populations, and genetic diversity (Acosta 2002; Gerber et al. 2002; Lockwood et al. 2002). However, it is prohibitively costly to attempt to measure or even model such parameters for more than a few marine species, and MPA designs will normally need to be based on predicted (or assumed) distribution characteristics of the main species from high-level models. Typically, there is only very limited empirical biophysical data available to MPA planners, including information on exploited species. It is rare to have a comprehensive baseline of biophysical information, but even data-rich situations (such as Ward et al. 1999b; Sala et al. 2002)

make many assumptions about ecosystem conditions, life history patterns, dispersal patterns, cause–effect relationships, oceanographic dynamics, migratory species, and the distribution of unknown biodiversity in planning for the siting of MPAs. And further, when biophysical data are available for a region, it normally will be of different quality, focus on different spatial scales and taxonomic resolution, and have different sampling characteristics, thereby complicating the internal consistency of data across the region.

In order to account for the vast uncertainty in knowledge about the species found in marine habitats and ecosystems that may be expected to be protected by an MPA, planners almost invariably must resort to models and surrogates in order to ensure that biological factors are incorporated as far as is practical in the siting of MPAs. The surrogates may be higher-order taxonomic resolution of marine species or assemblages or may by biophysical habitats or ocean conditions or even substrate types (Ward et al. 1999b; Sala et al. 2002; Day et al. 2003). Where surrogates do not provide complete or adequate representation of marine populations (the usual situation), MPA planners attempt to encompass a representative range of different types of habitats, substrates, topography, and ocean conditions to attempt to capture biodiversity that is otherwise unknown and has not been specifically modeled or used in the MPA design process.

To achieve high levels of protection for habitats and species within MPAs, because of such high levels of uncertainty, fisheries managers may need to adopt the approach of a graded series of managed uses within a network of MPAs. This is to try to ensure that there are high levels of protection provided by no-take core areas but that these areas are not threatened by surrounding uses that are highly destructive. Using this concept of graded protection zones, fishery MPA planners may be able to use the concept of nesting levels of protection to ensure that MPAs have a broad base of acceptance to the users and stakeholders while still providing for effective fishery benefits and conservation outcomes that may be delivered through a combination of zones.

Less-protective zones can be used (e.g., as a fishing area for low-impact gear types) to provide a buffer that reduces the risk of fishing activities degrading the biodiversity in the core no-take areas of an MPA network. Achieving this outcome in a network of MPAs requires information on the cause–effect relationships between various forms and intensity of fishing and the biodiversity of the various zones of protection as well as a detailed understanding of how the fishery management system interacts with zoned-use MPAs. Such matters are likely to be poorly understood, and so assumptions used in design criteria will always need to be highly precautionary.

In addition to providing a form of buffering for the higher-protection zones of an MPA network, MPAs where some forms of fishing are permitted will contain substantial biodiversity in their own right. Such MPAs may be considered to be important for conservation of larger, more mobile species that cannot be properly sampled and protected fully within smaller no-take core areas in an MPA network. In this way, various forms of regional conservation objectives can be met using a network of zoned uses for MPAs arranged in a spatial manner to provide support for fisheries objectives.

While it is clear that the success of a system of MPAs depends heavily on the quality and extent of design, including matters such as investment in establishing robust site criteria (Hockey and Branch 1997; Roberts et al. 2003a), there have been few documented examples of MPA designs that have failed to support a fishery. There are many examples of MPAs that do not appear to achieve their main objectives (such as Nickerson-Tietze 2000), but few have been assessed in any detail to determine why they fail to achieve useful outcomes. Nonetheless, many instances of failure appear to be related to a failure of management controls, such as failure of controls on access or on the uncontrolled use of destructive fishing gear (Russ and Alcala 1996).

In South Africa, MPAs created to support the fishery for rock lobsters *Jasus lalandii* are considered to have failed, because the densities of lobsters inside the protected areas are similar to those outside and the proportion of egg production in the MPAs is proportional to their area of the overall coast, indicating that there is no overall enhancement of the spawning lobster population by the MPAs (S. Mayfield et al., abstract from World Congress on Aquatic Protected Areas, 2002). This situation is considered to result from a lack of controls on fishing in the MPAs but also occurred because the areas were arbitrarily situated in locations that did not have good habitat for rock lobsters, which is a design failure.

MPA Size

Size of an MPA is often discussed as a key criterion (Agardy et al. 2003; Roberts et al. 2003a). Sometimes the desired size is expressed as a proportion of available area and used to establish targets, such as 20% of the total area for a specific type of MPA. However, size per se is not a useful criterion for selection of areas to include in an MPA. The size of an MPA must be large enough to ensure that the objectives for the MPA can be achieved, accounting for uncertainty in the design process and the varying utility of any surrogates used to determine boundaries and location of the MPA and any zones. Therefore, a more appropriate expression of a size criterion is (typically) the adequacy of the size of the MPA to achieve the objectives. In this context, adequacy refers to the area, location, and boundary arrangement in the context of the specific complement of habitats and ecosystems and the management framework that will allow the objectives of the MPA to be achieved.

Size can be an important parameter to describe an MPA and is a useful comparative descriptor for policy purposes, but it has little biological meaning in isolation from a competent MPA design and so cannot be used as a criterion for choosing areas to include within an MPA. A possible exception to this is in cases where two areas are equally likely to contribute to specified objectives, and then the larger one would likely be preferred for inclusion in an MPA network to minimize the risk of failing to protect unmeasured populations or habitats. Likewise, proportions based on areal comparisons (such as 20% of a region) have little value as criteria for selection of MPAs, although they may be useful as comparative measures across different jurisdictions and bioregions.

Compliance

Apart from biophysical factors, MPA designs need also to take account of the placement of boundaries and the shape of the closed area to ensure that day-to-day management can be conducted in an efficient and effective manner. Clear and easily distinguished boundaries are still very important, even with the modern availability of global positioning systems, to ensure that users can be fully aware of the boundaries and zones of an MPA (Causey 2003). The spatial configurations should be determined in close conjunction with the dynamics of the fishing fleet and other users so that the rules of entry and use, when they are agreed upon, will have the best chance of being respected by the fishing fleet and the other users.

Ideally, in commercial fisheries that operate within or adjacent to MPAs, each fishing vessel should be required to carry a vessel monitoring system that transmits the position to a monitoring agency and a logbook that records fishing sites and catches. In many cases, this can be less expensive than the alternative, which is to provide for a fishery-independent system of compliance monitoring to ensure that violations of the rules of the MPA are reported. Fishery-independent compliance systems may include aerial surveillance and boating patrols, conducted with sufficient intensity to provide a credible deterrent.

Creation of MPAs is often achieved by reducing access of specific user groups, some of whom may have a traditional and long-held access to these areas for fishing, tourism, recreation, etc. Therefore, for both commercial fisheries and for these other stakeholders, it is always important to ensure that MPA performance and any compliance failures are adequately monitored and reported in a manner that is accessible and understandable by each stakeholder group. The crucial role of this monitoring and reporting is to ensure that stakeholders are fully informed about the effectiveness of the MPA and to build stakeholder support for the on-going maintenance and management of facilities and staff. Where stakeholders are not fully engaged and supportive of MPA design and rules, experience shows that the MPA will not be effective because, in time, the stakeholders will be able to either violate the MPA with no fear of reprisal or work to alter the boundaries or rules in their favor and reduce the value of the MPA (e.g., see Russ and Alcala 1996).

The establishment of a feasible and effective performance assessment and reporting system is also important for confirming that the objectives for which the MPA was established are being met, including benefits for both fisheries and biodiversity conservation. This reporting on performance also provides the basis for justifying improvements that may be needed in the MPA, such as change in the boundaries to ensure the objectives can be fully achieved. For commercial activities, this becomes especially important so that the role of the MPA in any economic benefits can be made explicit.

Dealing with Displaced Effort
In many situations where MPAs are to be declared, there are existing uses that will be removed or displaced to other locations outside the MPA. This may create a particularly difficult problem in fisheries management. The creation of areas closed to fishing will naturally result in the concentration of all the existing effort into the remaining (open) areas. Such fisheries may then need to invoke a range of input controls (such as limits of permitted gear type and zone restrictions) and develop appropriate mechanisms to assess sustainable yields and impose controls on the fleet on a spatial basis to prevent fishing in the MPAs and over-fishing outside and close to the MPAs.

Where fisheries MPAs are needed within existing fishing grounds (which may not always be the case; see below), fishers may argue that the declaration of an MPA over ground they have customarily fished is a removal of an existing (or traditional) right of usage. This may then provide the basis for compensation in financial terms or in access to other fishing grounds, species, or fisheries that they cannot currently access. This raises the important issue of compensating for the displacement of fishing effort into the non-MPA areas and how this might interact with the MPA itself, the objectives of the MPA, biodiversity impacts in surrounding areas, or other affected fisheries. Clearly, where a fishery management system is not fully effective, creation and enforcement of an MPA may then lead to detrimental biodiversity and fishery outcomes in the remainder of the fishing grounds caused by inappropriate levels or types of fishing.

While fisheries MPAs may be declared within an overall fishing ground, it does not necessarily follow that this will displace effort. There may be habitats and subregions within fishing grounds that are not fished for a range of reasons, and if these can function effectively to meet the ecosystem offset objectives, these areas may be able to make a substantial contribution to the MPA network without invoking a major displacement of fishing effort.

Also, establishment of an MPA that does reduce access to existing fishing grounds does not necessarily mean that catch from the fishery overall will decline in proportion to the area of the MPA closed to fishing. The magnitude of the impact of the MPA on the fishery will depend on the nature of the fishery management system that is in place, the life history of the exploited species, and, perhaps most importantly, the effectiveness of the MPA in providing direct support for the fishery through spillover and larval export.

In fisheries that are overexploited, the shift of additional effort into non-reserved areas may increase the effort beyond the resilience of the stock and possibly cause the fishery to crash. However, empirical data from overexploited fisheries indicate that the opposite is usually true—highly overexploited fisheries benefit from protection of key remnants of the stock from fishing, enabling them to then provide spillover or larval export that may ultimately provide the basis for (at least a partial) recovery of the fishery. This response has been observed in both tropical and temperate fisheries of various scales. Indeed closing a fishery to all (or most) fishing is the usual response by fishery managers to signals of stock collapse in a fishery (Roberts et al. 2001; Fisher and Frank 2002).

In a well-managed fishery, the effects on a target stock of the increase in effort that may follow the creation of an MPA may be well within the capacity of the stock to withstand, at least for a period of time. Therefore, in a well-managed fishery, any such displaced effort would normally be comfortably managed to ensure that impacts on the stock did not lead to overfishing or other detrimental impacts, because the fishery management system would have suitable procedures in place to detect and respond to such signals. And a similar argument would apply to non-stock impacts of the fishery—if the management system in place is effective, then any detrimental impacts of displaced effort would be readily detected and appropriate responses generated.

For both overexploited and well-managed fisheries, the impact of the displaced effort will depend largely on three factors:

(1) effectiveness of the fishery management system that is in place and its ability to manage the displaced effort to ensure that overfishing does not occur;

(2) life history of the exploited species (benthic phases, pelagic phases, spawning behavior, migratory, etc.); and

(3) effectiveness of the MPA design in providing spillover or larval export into the fishery.

In addition to considering the impact of displaced effort on fished stocks, the potential for ecosystem impacts also needs to be assessed in an MPA design. The declaration of an MPA should not result in displacement of effort that promotes the use of different gear types or more intensive fishing in places previously only lightly fished to improve catch unless this is part of the planned management. A shift in the gear types or pattern of fishing may have increased consequences for biodiversity in the non-reserved areas. Such consequences may include increases in bycatch, additional fishing into deeper unprotected waters, or use of more destructive gear types. Models developed to assess the effects of MPA declaration on fisheries suggest that the way in which displaced effort is managed is crucial to capturing the benefits of the MPA (Apostolaki et al. 2002; Haddon et al. 2003). For modeled overexploited stocks and fully exploited stocks of both sedentary and mobile species (including spiny lobsters *Jasus edwardsii* and Mediterranean hake *Merluccius merluccius*, also known as European hake), there are

yield and productivity benefits that can be secured by introduction of MPAs, but this is only if displaced effort and fishery selectivity are correctly adjusted and managed in the non-reserved areas.

In some circumstances, introduction of an MPA (or a network of MPAs) may be the main tool that is available for recovering a fishery to an acceptable level of productivity (Roberts et al. 2001) and so results in an increase in effort in a fishery. However, in those fisheries that are fully exploited or where there is a marginal overexploitation, the creation of MPAs may precipitate the removal of some fishing capacity from the fishery in order to deal with displaced effort. While this is not necessarily a consequence of all MPA declarations, it is usually a risk perceived by fishers and, along with reduced yield, is a matter that generates significant initial opposition to fishery MPA proposals (Ward 2003b).

Rules for Choosing MPAs

In a systematic approach to choosing MPAs, it is necessary to establish the decision rules that will guide all decisions that are to be made in the design process. These rules may comprise guidance for how to make decisions as well as provide guidance as to the specific types of outcomes that will be expected from the decision-making process. These decision rules should not be confused with criteria, principles, or various other forms of guidance suggested in the literature. For example, the criteria developed by Roberts et al. (2003b) are a mixture of decision rules, objectives, and a selection procedure. Various forms of decision rules, principles, and criteria have been developed to guide the MPA identification and selection process based on research and theoretical grounds (Salm and Price 1995; Nilsson 1998; Botsford et al. 2003; Roberts et al. 2003b), but few have been applied to actual MPA implementation (Day et al. 2001; Airame et al. 2003). Some of these principles and criteria also embed specific objectives, particularly in relation to biodiversity conservation. In the selection processes that use a criterion sieve approach, where sites must meet a set of biological criteria before meeting socioeconomic criteria (Day et al. 2001; Roberts et al. 2003b), the objectives relating to biodiversity are given explicit priority over resource use objectives. This is an appropriate approach for MPA networks designed to provide for biodiversity conservation as the first priority; but for resource uses, the most appropriate form of decision rules will relate to the objectives that are established for the overall MPA system. Otherwise, MPAs will only ever relate to resource uses (such as fisheries) by coincidence rather than by design (Airame et al. 2003).

In the identification and selection of MPAs in the Great Barrier Reef World Heritage Area, the process of choosing an appropriate set of no-take areas has been guided by sets of biophysical and sociocultural operating principles. These provide guidance on what decisions had to be made throughout the process and provided indications about acceptable outcomes (Day et al. 2001). These operating principles established the primacy of biodiversity conservation objectives, as befitting the management of a World Heritage Area. The biophysical operating principles for no-take areas for the Great Barrier Reef World Heritage Area include:

(1) have no-take areas whose minimum size is 20 km along their smallest dimensions (except for coastal bioregions for which 10 km is the minimum dimension);

(2) have larger (versus smaller) no-take areas;

(3) Have sufficient replication;

(4) include only whole reefs within no-take areas;

(5) for reef bioregions, have at least 5 reefs and $x\%$ of reef area included;

(6) for non-reef bioregions, have at least $x\%$ of non-reef area included;

(7) include $x\%$ or x number of each community type and physical environment type in the overall network (e.g., diversity of depths, reef sizes, submerged reefs);

(8) maximize the use of environmental information to determine the best configuration of no-take areas;

(9) include biophysically special or unique places (e.g., significant nursery sites);

(10) consider sea and adjacent land uses in determining no-take areas; and

(11) capture regional diversity across the contintental shelf and latitudinally (Day et al. 2001).

These principles cannot be unconditionally applied in all instances; rather, they are recommendations to be implemented as far as is practicable in the planning process. The principles are not in any order of priority. For principles 5, 6, and 7, percentages are defined by a technical consultative process for each principle.

Integrate with Other Fishery Management Tools

For MPAs to be properly implemented to provide support for fisheries, the fishery management controls may need to be reviewed and assessed, not just the arrangements in the vicinity of the MPA. It may even be necessary to completely revise the management basis of the fishery, such as adding spatial management to a quota management system or shifting to some other form of fishing controls such as gear constraints. Equally, it may also be feasible to lift certain constraints, subject to appropriate assessments and monitoring. For example, in traditional Pacific Island fisheries, access and area closures are the major tools for managing fish stocks, and size limits are rare (Johannes 1978). A review of the management basis of the fishery and specific controls is necessary to ensure that the MPAs are effective and efficient in their delivery of fisheries benefits. But the introduction of MPAs may also be linked to other forms of restructuring initiatives in the fishery, such as economic reforms and rationalization, that may be needed for effort reduction in response to stock management issues alone.

Implementing MPAs in a fishery managed under a quota system may require adjustments to the overall size of the quota or implementation of matching input controls, and the transitional arrangements may be complex. Determining how to apply any quota reduction across the fishery may be controversial and may require external policy intervention (as would possibly be required for quota reductions for any other reason). Also, a fine-scale adaptive system may be used to examine and determine what impact the MPAs had on fishing yields and stocks, recognizing that important design objectives for the MPAs include providing support to stocks (such as in ensuring spawning biomass was maintained above a limit level across the fishery). If the MPAs were not achieving the required stock objective, the design could be adapted to improve performance, and, possibly, reserved power to reduce quotas temporarily might be applied by the management agency to avoid overfishing. Conversely, if the MPAs were shown to be effective in, for example, maintaining spawning stocks in refuges distributed strategically throughout the fishery area, other fishery controls (such as a minimum size rules or gear rules) may be relaxed in an incremental manner. The full combination of area refuges, zoned MPA uses, minimum fish sizes, gear restrictions, seasonal restrictions, and other fishery controls may need to be reviewed in the MPA design process in order to ensure that the fishery achieves each of its specific MPA objectives.

Following an extensive consultation process with fishers and stakeholders in the Reef Line Fishery, a reef fishery in Australia's Great Barrier Reef, the fishery objectives were determined to include preserving near-virgin biomass of coral trout *Plectropomus leopardus* on the reefs closed to fishing, ensuring satisfactory levels of populations available for harvest, maintaining economically viable commercial catch rates and recreationally rewarding recreational catches of coral trout, and minimizing variation in harvests from year to year (Mapstone et al., in press). In this fishery, the existing protection and spatial distribution of the no-take MPAs of the Great Barrier Reef Marine Park (which were established primarily for biodiversity conservation purposes; Lawrence et al. 2002) contribute to the maintenance of spawning biomass in the region. This was considered to be supported by the minimum size rule, set above the age at first spawning, which was assumed to help ensure that reproductive potential in the fishery was maintained (Mapstone et al., in press).

Systematic Approach to Design

Almost any form of MPA no matter where placed, provided it has a strong and effective management regime, will make at least some contribution to the conservation of local marine biodiversity. However, MPAs established through a process of systematic design will be most effective in achieving specific objectives such as conservation goals (protection of the habitat of important species or representative samples of the typical habitats and species assemblages of a region) and in providing protection for a resource species (Margules and Pressey 2000). Nonetheless, the history of MPA establishment is that most MPAs have been established with only limited systematic analysis or with data that relates, typically, to only a few key species or habitat types, and it has been difficult to extract meaningful general principles because of weak design (Jones et al. 1993; Botsford et al. 2003). But even weakly designed MPAs appear to be able to protect species within their boundaries and improve the condition of biodiversity within a few years of reserve establishment (Halpern and Warner 2002; Halpern 2003).

Despite their universal role in protecting biodiversity, no-take MPAs may have a greater value for those species that are fished than for those that are not fished (Cote et al. 2001), although this will depend on the nature of pressures on ocean ecosystems outside the MPAs and the management system within which a fishery operates. Therefore, each fishery needs to take a systematic approach to design of MPAs, considering each specific MPA objective and ensuring that there is a suitable optimization process to deal with competing objectives within the context of the fishery management system. Being systematic about the design means, for example, taking account of interactions between objectives and explicitly incorporating all the constraints into the MPA design process.

Systematic reserve selection procedures are typically used as the basis for detailed planning procedures and also for stakeholder negotiation rather than to make explicit final choices of areas (i.e., they are a form of decision support for planners and managers). Final decisions about the exact boundaries and choices of areas are made in conjunction with stakeholders based on options or candidate areas. In some circumstances (Pressey 1998), it has been of specific advantage to have an approach and algorithm that can be iterated in a few hours or less so that planners, managers, and stakeholders can quickly evaluate the effects of any changes they might be considering to a specific arrangement of areas in a candidate set of protected areas.

Three important concepts have emerged from the systematic procedures for the selection of protected areas—irreplaceability, efficiency, and effectiveness.

Irreplaceability

In the context of conservation networks, Pressey et al. (1994) defined the irreplaceability of an area in two ways: (1) the likelihood that an area will be required as part of a conservation network that achieves a specified set of targets, and (2) the extent to which the options for achieving the set of conservation targets are reduced if the area is unavailable for protection. A number of variants of the original concept of irreplaceability can now be found in the literature: weighted, weighted-summed, global, conditional, and more (e.g., Rebelo 1994; Csuti et al. 1997; Pressey 1998). Belbin (1995) also refers to a multivariate concept of irreplaceability—the degree of isolation of an area from its closest neighbor in multivariate space, indicating its distinctiveness across all the biological or environmental variables being analyzed. Nonetheless, the basic idea behind all expressions of irreplaceability is that it provides an estimate of the overall importance of an area to achieving conservation objectives based on estimates of the conservation features that the area contains.

In practice, irreplaceability of an area is estimated by assessing the conservation features it contains, the regional conservation targets for those features, and the distributions of those features in other areas within the region. The conservation features may be presence–absence of species, abundances, or any other relevant biodiversity parameter that can be estimated spatially, including endemicity, rarity, or species richness. Despite different interpretations and bases for estimation, irreplaceability has proved to be a key concept that is well accepted by stakeholders and provides an effective medium for communicating the comparative values of specific terrestrial protected area configurations. It is a key concept that also may be used in MPA design and selection (Airame et al. 2003), because it recognizes that specific trade-offs are required between conservation and resource use targets and permits sets of optional MPA networks that best meet multiple objectives to be explored and assessed by stakeholders and resource users alike.

Conservation Efficiency
Conservation efficiency is the extent to which the required targets for biodiversity are achieved within the minimum area of reserve (e.g., Rodrigues et al. 2000). This is the specific case of a more general concept of conservation efficiency which considers that the biodiversity targets should be contained within a protected area of minimum cost, where the cost may include purchase price (in the case of land); the option cost of displacing existing uses, which may be measured in terms of the value of harvesting production displaced, for example; or the cost of management. Conservation efficiency is typically reflected in protected area designs through the intention to secure the required conservation goals within the smallest area of reserve. However, Sala et al. (2002), in their marine reserve design, eliminate efficiency expressed as reserve size as an issue, preferring to ensure that the conservation goals are fully achieved.

Conservation Effectiveness
The capacity of a reserve system to provide for persistence of species and other elements of biodiversity is termed effectiveness in reserve design lexicon (Bedward et al. 1992; Vane-Wright 1996; Pressey and Taffs 2001). The conservation effectiveness of reserve design approaches and algorithms is interpreted to be a measure of the extent to which the biodiversity will endure within the protected area. Conservation effec-

tiveness may be measured in a number of ways, including vulnerability bias (Pressey and Taffs 2001), but the essential feature of conservation effectiveness is the extent to which reserve solutions generated from systematic reserve designs provide for both the inclusion and maintenance (through time) of the biodiversity of interest.

Conservation effectiveness depends on the interaction of the characteristics of the original complement of biodiversity with the system used to manage the chosen areas and the surrounding areas. Where a set of areas are chosen for inclusion in a reserve network, the effectiveness of this choice of areas will depend as much on the subsequent management arrangements as it will on the initial choice of areas. In this context, if two sets of areas have a similar complement of biodiversity, then the most effective choice will be the set of areas that can be managed to control any processes or situations that might act to degrade that complement of biodiversity, whether natural or human-induced, and including surrounding area uses or other externalities that might degrade the reserved areas. In this sense, fisheries managers should seek to ensure that MPAs are designed to be highly effective by careful analysis of external processes (including fishing) that might affect fishery MPAs.

Algorithms and Area Selection Approaches
The quest for improved conservation effectiveness and the need to consider a broad range of costs mean that reserve design problems are becoming increasingly complex (Rodrigues et al. 2000). This complexity comprises uncertainty (knowledge, data, process, and model); the need for flexibility, efficiency, and effectiveness; and the need for accountability in order to justify decisions that have important consequences for biodiversity and for existing ocean users. This complexity has focused research attention toward decision support procedures and algorithms that can deliver solutions that attempt to achieve optimal solutions across multiple competing objectives, including both biological objectives and cost constraints.

Iterative heuristic algorithms have been extensively used in terrestrial protected area design problems and appear to be well suited to some aspects of marine reserve design problems. However, with the increasing complexity of the problems to be solved, despite the computer efficiency and logical simplicity of simple heuristic algorithms, they are becoming increasingly limited in their capacity to provide for effective MPA solutions (Ball and Possingham 2000). In contrast, optimization approaches are now being adopted more widely, based on algorithms for simulated annealing and integer programming. Annealing is an iterative approach but offers the prospect of a near-optimal solution to reserve design problems because it can generally achieve solutions that are less prone (than are more simple heuristics) to be trapped in local minima. Integer programming guarantees optimal solutions by definition because of the mathematical formulation (Arthur et al. 1997).

The Great Barrier Reef Marine Park Authority has used MarXan—a marine simulated annealing program (Ball and Possingham 2000)—to expand the existing no-take MPAs in the Great Barrier Reef World Heritage Area (Day et al. 2001). MarXan has been similarly applied to examine the efficiency of reserve allocation in South Australia (Stewart et al. 2003) and in selection of no-take areas in the Channel Islands (California; Airame et al. 2003), and MarXan is currently being used to establish priority areas in Australia's South East region. But beyond these, and a number of research projects (such as Sala et al. 2002; Leslie et al. 2003), there have been few applications of the optimization approaches for MPAs. A similar approach using a geographic information system-based, multiple-criteria analysis has addressed the complexity of designing an MPA in Italy. This analysis included a range of uses and values, including fishing, and was found to be a useful tool in highlighting potential no-take areas that were compatible with stakeholder priorities (Villa et al. 2002).

There are no standard approaches or algorithms for design of MPAs, although simulated annealing appears to be the most effective contemporary algorithm because it appears to be best able to cope with the complex design problems of MPAs, particularly the aspect of costs. Fisheries MPAs will always be faced with the challenge of attempting to achieve the optimal outcome for both the fishery and for biodiversity conservation, and MPA design approaches and supporting algorithms will need to be able to include the considerable spatial complexity that is inherent in biological and fishery objectives and the social, cultural, and economic costs of MPA designs.

Performance Assessment System

A key part of the supporting framework for fishery MPAs is the design for an integrated monitoring and assessment program that evaluates and reports on how the MPAs have influenced both the fishery and conservation in the bioregion. This is important for informing fishers, other users, and stakeholders and for underpinning their ongoing commitment to the MPAs. A monitoring and assessment program for fishery MPAs conceptualized in the design phase and planned to be implemented as part of an integrated approach to assessment of the condition of marine ecosystems is likely to be the most effective and enduring approach. Such assessment programs may be best implemented through partnerships, interagency agreements, and cooperation with private sector users, but always providing for public accountability and reporting. In this way, the responsibility for monitoring and assessment of MPAs in the context of sustainable use of resources may be shared across a range of ocean users and responsibilities and could form an important part of the process of integrated support for fishery MPAs.

Designing the Performance Assessment System

The performance of MPAs is normally evaluated to provide the basis for adaptive management and more broadly to provide for improved planning, accountability, and resource allocation for management purposes. Specifically, a fishery MPA performance assessment system should be designed (Day et al. 2003) to:

(1) determine the level of achievement of the specified objectives;

(2) enable a more systematic linkage between objectives and management activities;

(3) provide direct information about what works and what does not work for review and improvement of MPA strategies and activities;

(4) learn more about how the MPA actually works in supporting the fishery, including the ecological dynamics and interactions;

(5) provide for an informed review of MPA policies, strategies, and activities;

(6) provide for specific outputs of direct concern to stakeholders;

(7) promote openness and accountability in expenditure of management resources, maintenance of ecological values, and delivery of outcomes;

(8) demonstrate appropriate returns on investments; and

(9) provide the basis for systematic improvement

of the management system itself, including resource allocations for management.

An outcomes-based management framework will establish objectives for the resource and the system managing it, and this is an effective basis for assessing performance. The philosophy of such an approach is derived from business systems and global approaches to management using principles of continuous improvement (such as Standards Australia/Standards New Zealand 1996). In protected areas management, the outcomes–assessment approach is advocated for nature conservation (Hockings et al. 2000) and for multiple-objective MPAs (Day et al. 2003). Outcomes-based management is also the management model consistently advocated within existing best practice marine environmental and fisheries management systems (Ward 2000; FAO 2002; Phillips et al. 2003). All of these systems require objectives, reference points, and monitoring data for routine review of performance.

In designing and implementing MPAs, establishing an efficient and effective performance assessment system should always be included as an integral part of the design process. Leaving the performance assessment system till after the MPA design, as is often the case (Day et al. 2003), is fraught with serious risks. When a performance assessment system is added after the design process, the achievement of the MPA objectives may be difficult to measure, the resources needed for assessing performance may not be available, and the opportunity for sharing of the costs of MPA performance assessment with other users and stakeholders may be reduced.

Fisheries MPAs will generally be designed to provide support to a fishery that already exists, and it is critical to ensure that the MPA performance assessment system is well integrated within the overall fishery management as a whole. This will involve ensuring that the MPA objectives are consistent with, and preferably part of, the fishery objectives; the activities in MPA performance assessment are integrated with other activities in the fishery management; and that the regular review and assessment of MPA performance is conducted as part of the review and assessment of stock performance.

An effective performance assessment system for fisheries MPAs, following that developed for other natural resource management systems (Campbell et al. 2001), for marine systems (Ward 2000), and for MPA evaluation (Hockings et al. 2000; Day et al. 2003), would normally involve:

(1) derivation of an appropriate conceptual model of the management system for the fishery and the MPAs;

(2) establishment of outcome-oriented objectives that are measurable;

(3) identification of performance indicators and an effective, achievable system of measurements for each one that reflects the need for management information (this should focus on space–time scales that will enable appropriate levels of change to be detected);

(4) establishment of performance benchmarks or reference levels for each indicator;

(5) implementation of monitoring data collection programs;

(6) periodical assessment of the monitoring data in the context of the benchmarks and reference levels;

(7) reporting of findings to managers and stakeholders, with recommendations for subsequent action;

(8) making appropriate adjustments to the fishery, the monitoring program, or other management measures as recommended; and

(9) revisiting and reviewing to confirm or adjust the conceptual model of the system and the performance indicators and all previous steps in this list.

Indicators of Performance

Indicators of fisheries MPA performance should be determined as part of the systematic process discussed above. The indicators should be constructed to ensure that routinely collected monitoring data provides an effective and efficient basis to assess the achievement of both the specific MPA objectives and the fishery objectives. Where the achievement of an objective is difficult to measure directly, as will often be the case, surrogates should be chosen as part of the monitoring program. In this circumstance, the conceptual linkages among objective, indicator, surrogate, and reference points and benchmarks need to be clearly expressed in an underlying conceptual model of the performance assessment system.

In any MPA performance assessment system, it usually will be important to include a set of indicators that are specifically linked to the objectives (as discussed above) but also to include a set of objectives that may be designed to be the safety net in the event of model failure (see below) or management system failure. The safety net indicators would normally be directed at confirming that highly valued attributes of the fishery and ecosystems are maintained within their normal boundaries and are not being degraded by management actions or externalities that have not been predicted (Ward 2000; Day et al. 2003).

Effective marine performance indicators can only be derived within the context of the management system framework (Ward 2000). In particular, for fishery MPAs, it is important to ensure that any ecosystem surrogates chosen for monitoring have an explicit conceptual linkage to the indicator and to the objective, and that the monitoring programs that are to be implemented have the capacity to resolve any important type and magnitude of change related to the objectives. In order to resolve changes in the MPA performance caused by fishery management interventions, or changes in the fishery caused by the MPAs, it may be necessary to be able to resolve local-scale (MPA level) changes from those caused by regional and global factors. At this level, it is important to ensure that the fishery MPA program has the support and commitment of other ocean users, including government agencies with an interest in long-term broad scale changes in ocean ecosystems, to enable sharing of responsibilities for activities and the costs of monitoring.

Linking the indicators to the objectives has a number of important benefits for the management system, as outlined above. Clearly specifying the benchmarks and reference points forces the management system to clearly establish the values that underpin the fishery management system and identify how these will be recognized in terms of fishery opportunities and constraints. This is crucial for the success of the management process (Walters 1997), guides the management of the fishery, and provides the context for evaluating the performance of the fishery in terms of each specified objective. For this to be effective, a monitoring program for each indicator that is achievable within likely resource constraints must be designed and implemented.

Implementation Issues

Performance evaluations cost money and effort, and there is a natural reluctance to increase fishery management overheads for purposes that may not always be seen to provide secure returns to a fishery. However, best practice business systems always include some forms of performance evaluation, and in natural resource management and protected area management, monitoring of specific indicators is essential for effective management (Hockings et al. 2000; Ward 2000; Campbell et al. 2001; Day et al. 2003).

In fishery MPAs, where some of the objectives are likely to be related to management of ecosystems, substrates, and species that normally have not been the domain of fisheries managers and scientists, there will be new challenges. Detecting change in species, ecosystems, and habitats of the continental shelf, for example, as a result of the implementation of a network of MPAs, will require data on indicators at a range of scales and potentially involve a substantial investment of resources (Bax and Williams 2001). Such resources may be well beyond the scope of a single fishery, depending on its size and stage of development, but if a range of stakeholders and other ocean users are involved in supporting the design of fishery MPAs, there may be the opportunity to design shared data collection programs. In addition, performance assessment may be developed using a rostered approach, where, superimposed over a minimum base of monitoring activity, specific indicators may be given detailed attention according to a roster of research investment to match priorities established in the performance design process. Such detailed attention may, for example, investigate fine-scale dynamics in space and time of each specific indicator, in turn, and develop models to underpin the detection of change, linkage to fishery responses, and the refinement of ongoing monitoring to ensure that the relevant changes were able to be detected at a given level of monitoring effort.

A further opportunity to optimize the costs of monitoring involves the use of innovative designs for the design, capture, and analysis of indicator data. Multivariate statistical tools enable data collection programs to be designed in a more hierarchical manner so that higher levels of data analysis can make highly efficient use of data from finer-scale data. The multivariate synthesis of monitoring data are effective in a range of marine and terrestrial performance assessment applications (Ward et al. 1998, 1999a; Alder et al. 2001; Campbell et al. 2001; Pitcher and Preikshot 2001). These approaches employ a hierarchical multivariate approach to data aggregation to summarize data and evaluate trends for reporting purposes. The data can be analyzed on a range of indicator types and at a range of scales, sensitivity to a range of data uncertainties can be readily explored and displayed, and the relative positioning of the trajectory of the resource relative to benchmarks or reference points can be displayed. These innovative uses of monitoring data open the opportunity for improved cost-effectiveness of monitoring data derived from the finest scales of space and time, and this may reduce the inherent difficulty and cost of designing and implementing cost-effective performance assessment, a major impediment to effective performance assessment in MPAs (Day et al. 2003).

Model Failure

Models of various forms underpin all MPA designs, and they may include process understanding (oceanographic and ecological), conceptual models, and policy contexts. As with all marine systems, uncertainty pervades such models at a number of levels, including:

- model form (the type of model being used as the basis for planning is not appropriate);

- model structure (the underlying conceptual model is incomplete or incorrectly structured);

- unclear communication (lack of a common understanding of concepts and issues);

- description errors (errors in making measurements of a variable);

- variability (heterogeneity in an environmental variable);

- data gaps (measurements not made); and

- uncertainty about a variable's true value (wide distribution of data points).

Each of these forms of uncertainty will affect the design and implementation of fisheries MPAs, as it does for fisheries management and protected areas in general. However, dealing effectively with these uncertainties in designing fisheries MPAs will mean dealing with a much broader science–policy gap (Bradshaw and Borschers 2000) than is recognized at present. Not only are there new objectives (environmental issues) with their own dynamics and uncertainties, but these will interact with the traditional management objectives (stock issues) creating a more complex environment within which fisheries MPAs must be designed.

To meet the requirements of more integrated oceans management, such as expressed in the WSSD

Plan of Implementation, and to be able to secure the broader range of benefits potentially available, fishery MPAs will have to operate in an extended management framework to deal with these new risks and challenges. In particular, designing fisheries MPAs so that they recognize and respond to issues beyond stock management will require a conceptual model of the fishery that includes objectives from the range of potential benefits described in Table 1 and by Ward et al. (2001). These potential benefits may be secured from a fishery in specific circumstances. Where objectives are narrow (such as constrained to stock objectives), the MPAs may not be able to meet the broader expectations of other users and fit within the broader framework of multiple-objective MPAs and integrated oceans management. In this sense, where objectives are constrained to a narrow set of fisheries objectives but, nonetheless, there are broader expectations for biodiversity and conservation outcomes held (at times implicitly) by other users and stakeholders within the broader policy environment, narrowly designed fisheries MPAs can be deemed to be suffering model failure. And specifically, since MPAs can assist with the task of delivering ecological offsets, if these ecological offsets are not explicitly included within the objectives for fisheries MPAs and the design process, in most contemporary circumstances, this will also be considered to amount to model failure. Therefore, model failure in fishery MPAs is defined as the situation where fishery MPAs do not explicitly contribute to offsetting the impact of the fishery on regional biodiversity.

The empirical literature on the impact of MPAs in fisheries is dominated by the experience from situations that have been highly degraded and where the main objective has been to try to recover some semblance of the original fishery and ecosystem productivity. In some of these situations, there has been gross pollution of habitats, habitats have been substantially removed or altered, or fisheries have been grossly overfished. In these circumstances, most forms of remedial activity, including MPAs, will assist a fishery in recovering. For example, any control that was able to either remove or greatly reduce fishing effort for a substantial period probably would assist in a recovery of the fishery to some level of productivity, depending on the target species, timeframes, and issues of regime shift (Hutchings 2000). The experiences and lessons of such MPAs are entirely appropriate as a basis in planning for recovery of other disaster areas, but there are major risks in attempting to extrapolate the lessons learned in these situations to fisheries that are well managed and have ecosystems generally in reasonable condition. Using the experiences of highly degraded systems as the basis for conceptual models and objective setting for MPAs in well-managed fisheries is potentially a highly biased approach, and this could also lead to model failure.

The extent of area protection, the type of protection used, and the nature of habitats protected may need to be different in MPAs implemented in well-managed fisheries. In particular, the linkage between area protection and specific stock management objectives is likely to be expressed in a different way. For example, in a highly degraded fishery, almost any form of area protection is likely to result in fishery improvement, and so areas may be chosen haphazardly to reduce fishing effort. But in a well-managed fishery, specific areas known to have key roles in the fishery management system (spawning grounds, feeding grounds, migratory aggregations, rare species habitats) may be most important for protection and may integrate more effectively with existing management approaches (such as maintaining spawning biomass). However, in all fishery situations, integrating the MPA selection process within the fishery management system will allow for the most optimum design for MPAs, provided the fishery incorporates the ecosystem objectives and strategies in an appropriate way. So, irrespective of the condition of the fishery, model failure can be avoided by ensuring that the scope of the management system is appropriate (e.g., incorporates ecosystem issues and stakeholder values) and by adopting a systematic and comprehensive approach to the design and implementation of MPAs for the fishery.

The most effective approaches to minimizing the risk of model failure are to:

- ensure that all aspects of the problem bounding and conceptual model building process are kept as broad as possible;

- involve a range of stakeholders within the MPA design and implementation process;

- design specific trials and pilot studies to ensure that there is adaptive management and continuous improvement and that uncertainty can be estimated for inclusion in revisions;

- ensure that regional conservation objectives are incorporated into the design of fishery MPAs so that both fishery and conservation objectives can be achieved;

- ensure that targets are well specified for all objectives established for the MPAs, in a manner that permits performance to be established; and

- always build on experience and lessons from other fisheries and MPA activities, suitably adapted and incorporated into the local context.

Conclusions

If sustainability is defined in the broadest sense (Holling 2000; WSSD 2002), then environmental issues and socioeconomic matters must be considered in fisheries management and dealt with a manner similar to that of stock management. This will provide the basis for achieving sustainability of a fishery. As an important part of sustainability, multiple-objective MPAs developed to give effect to an ecosystem-based approach to fishery management are likely to provide a number of benefits for fisheries. The design of an effective network of fishery MPAs needs to be comprehensive in its scope, including regional biodiversity objectives, to provide for ecological offsets and specific fishery and stock objectives and to provide for harvest and related benefits such as fishery stability and local employment. Assessment of the performance of the MPAs should confirm that the specific objectives for the MPA network and for each MPA are being achieved in order to provide ongoing support for MPA and fishery management activities and for adaptive revision of the MPAs.

The crucial aspects of a design and implementation process for MPAs in fisheries appear to be:

(1) the fisheries management system within which the MPA is embedded is comprehensive in its scope and capacity (to avoid model failure);

(2) the principles of ecosystem-based management are used as the design basis for the fishery and the embedded MPA network (involving stakeholders in developing decision rules and setting reference points);

(3) an outcomes-based approach using agreed reference points applies to all management objectives expressed in the fishery (including the MPAs);

(4) a systematic approach to design and implementation of MPAs (drawing on the experiences in reserve planning and conservation ecology);

(5) MPA design approaches that achieve multiple objectives, specifically, site selection optimization approaches and algorithms that deal explicitly with the uncertainty (particularly model failure at all levels in the management system), use realistic assumptions about impacts of the fishery on biodiversity, and incorporate the interactions of area closures with harvest strategies and stock dynamics;

(6) science-based processes of continuous improvement in the fishery;

(7) collaboration with other ocean users and stakeholders to develop joint monitoring programs through strategic partnerships; and

(8) application of the hard-won experience from MPAs that have been used for fisheries or for conservation purposes elsewhere.

But perhaps above all, for any fishery, investment in design and performance assessment is critical to ensure that fishery MPAs are successful and will persist. Securing poorly planned areas for MPA purposes is likely to result in securing weak outcomes for biodiversity and for fisheries and will likely represent the worst possible outcome of the present upsurge of policy and scientific interest in fisheries MPAs, providing a false sense of security for fishery managers and, perhaps, masking negative trends in fisheries and in ecosystems.

Acknowledgments

My colleagues Dennis Heinemann, Nathan Evans, Eddie Hegerl, Tony Smith, Bruce Phillips, Chet Chaffee, and Richard Kenchington have assisted me in crystallizing many of these concepts in recent years, and I am grateful for their constructive contributions and willingness to share these ideas. This paper was presented as an invited keynote address to the Aquatic Protected Areas Symposium at the 2003 American Fisheries Society Annual Meeting in Quebec City. I also acknowledge the comments of two anonymous symposium reviewers who helped to sharpen the focus of my ideas for this review.

References

Acosta, C. A. 2002. Spatially explicit dispersal dynamics and equilibrium population sizes in marine harvest refuges. ICES Journal of Marine Science 59:458–468.

Adams, T. J. H. 1998. The interface between traditional and modern methods of fishery management in the Pacific Islands. Ocean and Coastal Management 40:127–142.

Agardy, T., P. Bridgewater, M. Crosby, J. Day, P. Dayton, R. Kenchington, D Laffoley, P. McBonney, P. Murray, J. Parks, and L. Peau. 2003. Dangerous targets? Unresolved issues and ideological clashes around marine protected areas. Aquatic Conservation: Marine and Freshwater Ecosystems 13:1–15.

Airame, S., J. E. Dugan, K. D. Lafferty, H. Leslie, D. A. McArdle, and R. R. Warner. 2003. Applying ecological criteria to marine reserve design: a case study from the California Channel Islands. Ecological Applications 13(Supplement):S170–S184.

Alder J., U. R. Sumaila, D. Zeller, and T. J. Pitcher 2001. Evaluating marine protected area management: a new modelling approach. Pages 1–10 in U. R. Sumaila, editor. Economics of marine protected areas. UBC Fisheries Centre, Research Report 9(8), Vancouver.

Alder, J., and T. J. Ward. 2001. Australia's oceans policy—sink or swim? Journal of Environment and Development 10:266–289.

Apostolaki, P., E. J. Milner-Gulland., M. K. Mcallister, and G. P. Kirkwood. 2002. Modelling the effects of establishing a marine reserve for mobile fish species. Canadian Journal of Fisheries and Aquatic Sciences 59:405–415.

Arnason R. 2001. Marine reserves: is there an economic justification? Pages 19–31 in U. R. Sumaila, editor. Economics of marine protected areas. UBC Fisheries Centre, Research Report 9(8), Vancouver.

Arthur J. L., M. Hachey, K. Sahr, M. Huso, and A. R. Kiester 1997. Finding all optimal solutions to the reserve site selection problem: formulation and computational analysis. Environmental and Ecological Statistics 4:153–165.

Ball I., and H. P. Possingham. 2000. MarXan, version 1.2. Marine reserve design using spatially explicit annealing: a manual prepared for the Great Barrier Reef Marine Park Authority. Prepared for Great Barrier Reef Marine Park Authority, Australia.

Ballantine, W. J. 1995. The New Zealand experience with "no-take" marine reserves. NOAA Technical Memorandum NMFS-SEFSC-376:C15–C31.

Bax, N. J., and A. Williams. 2001. Seabed habitat on the south-eastern Australian continental shelf: context, vulnerability and monitoring. Marine and Freshwater Research 52:491–512.

Bedward M., R. L. Pressey, and D. A Keith 1992. A new approach for selecting fully representative reserve networks: addressing efficiency, reserve design and land suitability with an iterative analysis. Biological Conservation 62:115–125.

Belbin, L. 1995. A multivariate approach to the selection of biological reserves. Biodiversity and Conservation 4:951–963.

Beumer J., L. Carseldine, and B. Zeller. 1997. Declared fish habitat areas in Queensland. Department of Primary Industries, Brisbane, Australia.

Beumer, J. P., A. Grant, and D. C. Smith, editors. 2003. Aquatic protected areas—what works best and how do we know? Proceedings of the World Congress on Aquatic Protected Areas. University of Queelsland Printery, St Lucia, Queensland, Australia.

Blumenthal, D., and J. L. Jannink. 2000. A classification of collaborative management methods. Conservation Ecology 4:13 [online]. Available: *www.consecol.org/vol4/iss2/art13* (November 2003).

Bohnsack, J. A. 1997. Consensus development and the use of marine reserves in the Florida Keys, USA. *In* Proceedings of the 8th International Coral Reef Symposium, volume 2. Smithsonian Tropical Research Institute, Balboa, Panama.

Botsford, L. W., J. C. Castilla, and C. H. Peterson. 1997. The management of fisheries and marine ecosystems. Science 277:509–515.

Botsford L. W., F. Micheli, and A. Hastings. 2003. Principles for the design of marine reserves. Ecological Applications 13(Supplement):S25–S31.

Bradshaw, G. A., and J. G. Borchers. 2000. Uncertainty as information: narrowing the science-policy gap. Conservation Ecology 4(1):7 [online]. Available: *www.consecol.org/vol4/iss1/art7* (November 2003).

Campbell, B., J. A. Sayer, P. Frost, S. Vermeulen, M. Ruiz Pérez, A. Cunningham, and R. Prabhu. 2001. Assessing the performance of natural resource systems. Conservation Ecology 5:22 [online]. Available: *www.consecol.org/vol5/iss2/art22* (November 2003).

Carter D., E. J. Hegerl, and N. Loneragan. 2003. The role of marine protected areas in the management of the Australian northern prawn fishery. Pages 545–547 *in* Beumer et al. (2003).

Causey, B. 2003. Success factors in the implementation and management of aquatic protected areas. Pages 275–284 *in* Beumer et al. (2003).

Cote, I. M., I. Mosqueira, and J. D. Reynolds. 2001. Effects of marine reserve characteristics on the protection of fish populations: a meta-analysis. Journal of Fish Biology 59(Supplement A):178–189.

Cryer M., B. Hartill, and S. O'Shea 2002. Modification of marine benthos by trawling: toward a generalization for the deep ocean? Ecological Applications 12:1824–1839.

Csuti, B., S. Polasky, P. H. Williams, R. L. Pressey, J. D. Camm, M. Kershaw, A. R. Kiester, B. Downs, R. Hamilton, M. Huso, and K. Sahr. 1997. A comparison of reserve selection algorithms using data on terrestrial vertebrates in Oregon. Biological Conservation 80:83–97.

Day J., L. Fernandes, A. Lewis, G. De'ath, S. Slegers, B. Barnett, B. Kerrigan, D. Breen, J. Innes, J. Oliver, T. Ward, and D. Lowe. 2001. The Representative Areas Program—protecting the biodiversity of the Great Barrier Reef World Heritage Area. Pages 687–696 *in* M. Kasim Moosa, S. Soemodihardjo, A. Nontji, A. Soegiarto, K. Romimohtarto, Sukarno, and Suharsono, editors. Proceedings of the 9th International Coral Reef Symposium. Indonesia Ministry of Environment, Indonesian Institute of Sciences, International Society for Reef Studies, Jakarta, Indonesia.

Day, J., M. Hockings, and G. Jones. 2003. Measuring effectiveness in marine protected areas—principles and practice. Pages 401–414 *in* Beumer et al. (2003).

Edgar, G. J., and N. S. Barrett. 1999. Effects of the declaration of marine reserves on Tasmanian reef fishes, invertebrates and plants. Journal of Experimental Marine Biology and Ecology 242:107–144.

FAO (Food and Agriculture Organization of the United Nations). 2002. FAO guidelines on the ecosystem approach to fisheries. FAO, Final Draft, Rome.

Fisher, J., and K. Frank. 2002. Changes in finfish community structure associated with an offshore fishery closed area on the Scotian Shelf. Marine Ecology Progress Series 240:249–265.

Gerber L. R., P. M. Kareiva, and J. Bascompte. 2002. The influence of life history attributes and fishing pressure on the efficacy of marine reserves. Biological Conservation 106:11–18.

Gubbay, S. 1995. Fisheries and marine protected areas—a UK perspective. Pages 183–188 *in* N. L. Shackell and J. H. M. Willison, editors. Marine protected areas and sustainable fisheries. Acadia University, Wolfville, Nova Scotia.

Haddon M., C. Buxton, C. Gardner, and N. Barrett. 2003. Modelling the effect of introducing MPAs in a commercial fishery: a rock lobster example. Pages 428–436 *in* Beumer et al. (2003).

Hall, S. J. 1999. The effects of fishing on marine ecosystems and communities. Blackwell Science, Oxford, UK.

Halpern, B., and R. Warner. 2002. Marine reserves have rapid and lasting effects. Ecology Letters 5:361–366.

Halpern, B. S. 2003. The impact of marine reserves: do reserves work and does reserve size matter? Ecological Applications 13(Supplement):S117–S137.

Hockey, P. A. R., and G. M. Branch. 1997. Criteria, objectives and methodology for evaluating marine protected areas in South Africa. South African Journal of Marine Science 18:369–383.

Hockings M., S. Stolton, and N. Dudley. 2000. Evaluating effectiveness: a framework for assessing the management of protected areas. The World Conservation Union (IUCN), World Commission on Protected Areas Best Practice Guideline 6, Gland, Switzerland.

Holling, C. S. 2000. Theories for sustainable futures. Conservation Ecology 4:7 [online]. Available: *www.consecol.org/vol4/iss2/art7* (November 2003).

Hutchings, J. 2000. Collapse and recovery of marine fishes. Nature (London) 406:882–885.

Jennings, S., and M. J. Kaiser. 1998. The effects of fishing on marine ecosystems. Advances in Marine Biology 34:201–352.

Johannes, R. E. 1978. Traditional marine conservation methods in Oceania and their demise. Annual Reviews in Ecology and Systematics 9:349–364.

Johnson, D. R., N. A. Funicelli, and J. A. Bohnsack. 1999. Effectiveness of an existing estuarine no-take sanctuary within the Kennedy Space Center, Florida. North American Journal of Fisheries Management 19:436–453.

Johnson, B. L. 1999. The role of adaptive management as an operational approach for resource management agencies. Conservation Ecology 3:8 [online]. Available: *www.consecol.org/vol3/iss2/art8* (November 2003).

Jones G. P., R. C. Cole, and C. N. Battershill 1993. Marine reserves: do they work? Pages 29–45 *in* C. N. Battershill, D. R. Schiel, G. P. Jones, R. G. Creese, and A. B. MacDiarmid, editors. Proceedings of the 2nd Temperate Reef Symposium. National Institute of Water and Atmosphere, Auckland, New Zealand.

Jones, P. J. S. 1999. Marine nature reserves in Britain: past lessons, current status and future issues. Marine Policy 23:375–396.

Lawrence, D., R. Kenchington, and S. Woodley. 2002. The Great Barrier Reef: finding the right balance. Melbourne University Press, Melbourne, Australia.

Leslie H., M. Ruckelshaus, I. R. Ball, S. Andelman, and H. P. Possingham. 2003. Using siting algorithms in the design of marine reserve networks. Ecological Applications 13(Supplement):S185–S198.

Lockwood, D. R., A. Hastings, and L. W. Botsford. 2002. The effects of dispersal patterns on marine reserves: does the tail wag the dog? Theoretical Population Biology 61:297–309.

Lubchenco J., S. R. Palumbi, S. D. Gaines, and S. Andelman. 2003. Plugging a hole in the ocean: the emerging science of marine reserves. Ecological Applications 13(Supplement): S3–S7.

Mapstone, B. D., C. R. Davies, L. R. Little, A. E. Punt, A. D. M Smith, F. Pantus, D. C. Lou, A. J. Williams, A. Jones, A. M. Ayling, G. R. Russ, and A. D. MacDonald. In press. The effects of line fishing on the Great Barrier Reef and evaluation of alternative potential management strategies. Report to the Great Barrier Reef Marine Park Au-

thority and Fisheries Research and Development Corporation, CRC Reef Research Centre, Townsville, Australia.

Margules, C. R., and R. L. Pressey. 2000. Systematic conservation planning. Nature (London) 405:243–253.

Murawski S. A., R. Brown, H. L. Lai, P. J. Rago, and L. Hendrickson. 2000. Large-scale closed areas as a fishery-management tool in temperate marine systems: the Georges Bank experience. Bulletin of Marine Science 66:775–798.

Murray, S. N., R. F. Ambrose, J. A. Bohnsack, L. W. Botsford, M. A. Carr, G. E. Davis, P. K. Dayton, D. Gotshall, D. R. Gunderson, M. A. Hixon, J. Lubchenco, M. Mangel, A. MacCall, D. A. McArdle, J. C. Ogden, J. Roughgarden, R. M. Starr, M. J. Tegner, and M. M. Yoklavich. 1999. No-take reserve networks: sustaining fishery populations and marine ecosystems. Fisheries 24(11):11–25.

Myers, R. A., and B. Worm. 2003. Rapid worldwide depletion of predatory fish communities. Nature (London) 423:280–283.

Nickerson-Tietze, D. 2000. Scientific characterization and monitoring: its application to integrated coastal management in Malaysia. Ecological Applications 10:386–396.

Nilsson P. 1998. Criteria for selection of marine protected areas. Swedish Environmental Protection Agency, Stockholm.

Pauly, D., V. Christensen, S. Guénette, T. Pitcher, R. Sumaila, C. Walters, R. Watson, and D. Zeller. 2002. Towards sustainability in world fisheries. Nature (London) 418:689–695.

Phillips, B., T. Ward, and C. Chaffee, editors. 2003. Eco-labelling in fisheries: what is it all about? Blackwell Scientific Publications, Oxford, UK.

Pitcher, T. J., and D. Preikshot. 2001. RAPFISH: a rapid appraisal technique to evaluate the sustainability status of fisheries. Fisheries Research 49:255–270.

Pressey R. L. 1998. Algorithms, politics and timber: an example of the role of science in a public, political negotiation process over new conservation areas in production forests. Pages 73–87 *in* R. Wills and R. Hobb, editors. Ecology for everyone: communicating ecology to scientists, the public and the politicians. Surrey Beatty and Sons, Sydney, Australia.

Pressey, R. L., I. R. Johnson, and P. D. Wilson. 1994. Shades of irreplaceability: towards a measure of the contribution of sites to a reservation goal. Biodiversity and Conservation 3:242–262.

Pressey, R. L., and K. H. Taffs. 2001. Sampling of land types by protected areas: three measures of effectiveness applied to western New South Wales. Biological Conservation 101:105–117.

Rebelo, A. G. 1994. Iterative selection procedures: centres of endemism and optimal placement of reserves. Pages 231–257 *in* B. J. Huntley, editor. Botanical diversity in southern Africa. National Botanical Institute, Pretoria, South Africa.

Roberts, C. M., S. Andelman, G. Branch, R H. Bustamante, J. C. Castilla, J. Dugan, B. S. Halpern, K. D. Lafferty, H. Leslie, J. Lubchenco, D. McCardle, H. P. Possingham, M. Ruckelshaus, and R. R. Warner. 2003a. Ecological criteria for evaluating candidate sites for marine reserves. Ecological Applications 13(Supplement):S199–S214.

Roberts, C. M., G. Branch, R H. Bustamante, J. C. Castilla, J. Dugan, B. S. Halpern, K. D. Lafferty, H. Leslie, J. Lubchenco, D. McArdle, M. Ruckelshaus, and R. R. Warner. 2003b. Application of ecological criteria in selecting marine reserves and developing reserve networks. Ecological Applications 13(Supplement):S215–S228.

Roberts, C. M., and N. V. C. Polunin. 1993. Marine reserves: simple solutions to managing complex fisheries? Ambio 22:363–368.

Roberts, C. P., J. Bohnsack, F. Gell, J. Hawkins, and R. Goodridge. 2001. Effects of marine reserves on adjacent fisheries. Science 294:1920–1923.

Rodrigues, A. S., J. O. Cerdeira, and K. J. Gaston. 2000. Flexibility, efficiency, and accountability: adapting reserve selection algorithms to more complex conservation problems. Ecography 23:565–574.

Russ, G., and A. Alcala. 1996. Marine reserves: rates and patterns of recovery and decline of large predatory fish. Ecological Applications 6:947–961.

Sala, E., O. Aburto-Oropeza, G. Paredes, I. Parra, J. Barrera, and P. Dayton. 2002. A general model for designing networks of marine reserves. Science 298:1991–1993.

Salm, R., and A. Price. 1995. Selection of marine protected areas. Pages 15–31 *in* S. Gubbay, editor. Marine protected areas: principles and techniques for management. Chapman and Hall, London.

Sanchirico, J. N., and J. E. Wilen. 2001. The impacts of marine reserves on limited entry fisheries. Pages 212–222 *in* U. R. Sumaila, editor. Economics of marine protected areas. UBC Fisheries Centre, Research Report 9(8), Vancouver.

Sant, M. 1996. Environmental sustainability and the public: responses to a proposed marine reserve at Jervis Bay, New South Wales, Australia. Ocean and Coastal Management 32:1–16.

Standards Australia/Standards New Zealand. 1996. Environmental management systems—specification with guidance for use. Standards Australia/Standards New Zealand, AS/NZS ISO 14001, Sydney, Australia.

Stewart, R. R., T. Noyce, and H. P. Possingham. 2003. Opportunity cost of ad hoc marine reserve design decisions: an example from South Australia. Marine Ecology Progress Series 253:25–38.

Sumaila, U. R. 2001. Marine protected area performance in a game theoretic model of the fishery. Pages 229–235 *in* U. R. Sumaila, editor. Economics of marine protected areas. UBC Fisheries Centre, Research Report 9(8), Vancouver.

Suman, D., M. Shivlani, and J. W. Milon. 1999. Perceptions and attitudes regarding marine reserves: a comparison of stakeholder groups in the Florida Keys National Marine Sanctuary. Ocean and Coastal Management 42:1019–1040.

Thrush, S., J. E. Hewitt, V. J. Cummings, P. K. Dayton, M. Cryer, S. J. Turner, G. A. Funnell, R. G. Budd, C. J. Milburn, and M. R. Wilkinson. 1998. Disturbance of the marine benthic habitat by commercial fishing: impacts at the scale of the fishery. Ecological Applications 8:866–879.

Vane-Wright, R. I. 1996. Identifying priorities for the conservation of biodiversity: systematic biological criteria within a socio-political framework. *In* K. J. Gaston, editor. Biodiversity—a biology of numbers and difference. Blackwell Scientific Publications, Oxford, UK.

Villa, F., L. Tunesi, and T. Agardy. 2002. Zoning marine protected areas through spatial multi-criteria analysis: the case of the Asinara Island National Marine Reserve of Italy. Conservation Biology 16:515–526.

Walters, C. 1997. Challenges in adaptive management of riparian and coastal ecosystems. Conservation Ecology 1:1 [online]. Available: *www.consecol.org/vol1/iss2/art1* (November 2003).

Ward, T., and E. Hegerl. 2003. Marine protected areas in ecosystem-based management of fisheries. Department of Environment and Heritage, Canberra, Australia.

Ward, T. J. 2000. Indicators for assessing the sustainability of Australia's marine ecosystems. Marine and Freshwater Research 51:435–446.

Ward, T. J. 2003a. Principle 2—effects of fishing on the ecosystem. Pages 41–56 *in* B. Phillips, T. Ward, and C. Chaffee, editors. Eco-labelling in fisheries: what is it all about? Blackwell Scientific Publications, Oxford, UK.

Ward, T. J. 2003b. Giving up fishing grounds to reserves: the costs and benefits. Pages 19–29 *in* Beumer et al. (2003).

Ward, T. J., D. Heinemann, and N. Evans. 2001. The role of marine reserves as fisheries management tools: a review of concepts, evidence and international experience. Bureau of Rural Sciences, Canberra, Australia.

Ward, T. J., and C. A. Jacoby. 1995. Deciphering spatiotemporal dynamics: what can mesocosm experiments do to improve predictions of environmental impacts? Australasian Journal of Ecotoxicology 1:51–54.

Ward, T. J., R. A. Kenchington, D. P. Faith, and C. R. Margules. 1998. Marine BioRap guidelines: rapid assessment of marine biological diversity. CSIRO Australia, Perth.

Ward, T. J., F. Kingstone, and S. Siwatibau. 1999a. Indicators of success for the South Pacific Biodiversity Conservation Programme, volume 1. Technical Report to the South Pacific Regional Environment Programme, Apia, Samoa.

Ward, T. J., D. Tarte, E. J. Hegerl, and K. Short. 2002. Policy proposals and operational guidance for ecosystem-based management of marine capture fisheries. World Wildlife Fund, Sydney, Australia.

Ward, T. J., M. A. Vanderklift, A. O. Nicholls, and R. A. Kenchington. 1999b. Selecting marine reserves using habitats and species assemblages as surrogates for biological diversity. Ecological Applications 9:691–698.

Watson, R., and D. Pauly. 2001. Systematic distortions in world fisheries catch trends. Nature (London) 414:534–536.

Williams, M. J. 1998. Fisheries and marine protected areas. Parks 8:47–53.

WSSD (World Summit on Sustainable Development). 2002. World Summit on Sustainable Development in Johannesburg: plan of implementation. Available: *www.johannesburgsummit.org/html/documents/summit_docs/2309_planfinal.htm* (November 2003).

Appendix: The Main Discussion Themes in the World Congress on Aquatic Protected Areas (Cairns, Australia, 2002)

Main themes (arranged alphabetically)	Key associated words
Biological benefits	Variable regionally and locally, design dependent
Community expectations	Oceans stewardship, management, environmental protection
Displaced effort	Whole of fishery, restructuring, environmental impact, quota
Empirical data/evidence	Lacking, the need for broader base of examples, quality technical data and analysis, example for MPA benefits, examples against benefits
Incentives	Role of supportive economic instruments (such as eco-labeling, fishery restructuring)
Lasting and effective solutions to fisheries management issues	Need for lasting and effective solutions to fisheries management problems that are affordable and equitable
MPAs are an adjunct, not replacement	Area protection as support for fisheries management, not a replacement management system
Objectives defined	Getting clear and agreed objectives that reflect community expectations
Partnerships	Costs, multiple objectives, whole-of-community expectations, stewardship
Performance assessment	Design, objectives, indicators, measuring, reporting, stakeholders
Planning is crucial	Importance of planning procedures, particularly siting and optimization tools, stakeholder engagement
Scales are important	Scales of nested activities, connectedness of systems, need to be accommodated
Shifting from single species to multiple-objectives management	Complex systems management, performance indicators
Socioeconomic benefits and costs	Broadly based, key drivers
Stakeholders	Importance of community and stakeholder engagement procedures and the empowering role of community support for fisheries management
Terminology agreed upon	Well-defined terminology is crucial to underpin clear and explicit communications
Things change	Need for recognition of the dynamics of ecosystems and industry–economic activities in planning for area-based management

American Fisheries Society Symposium 42:63, 2004

Abstract Only

On the Need for a Global Network of Large Marine Reserves

DANIEL PAULY[1]

*Fisheries Centre, 2204 Main Mall, University of British Columbia,
Vancouver, British Columbia V6T 1Z4, Canada*

Abstract.—The sustainability of fisheries, historically, is largely a matter having no access to the bulk of an exploited population. Fisheries persisted when most of the targeted fishes were in deep, offshore waters or in areas adjacent to lands with low human populations. Thus, earlier fisheries had large no-take marine reserves. Modern fishing technology relies on methods that originated with submarine tracking and other forms of warfare (acoustic fish finders, radar) and preservation technology (artificial ice, blast freezing) that immensely expanded the reach of distant water fleets. Combining this with Cold War technology (geo-positioning systems; detailed, real-time maps of oceanographic features; detailed maps of the sea bottom), fishing vessels now can and will catch, unless restrained, the last fish concentrations in the world ocean. Thus, marine reserves are not a new invention of armchair ecologists, designed to torment hardworking, overregulated fishers. Rather, they would only reestablish the natural structures which have enabled earlier fisheries to maintain themselves for centuries. Getting this message across to a wide public is urgent, given the hardening stance of the fishing industry against marine reserves and the fact that, presently, they cover an infinitesimally small fraction of the world ocean.

[1] E-mail: d.pauly@fisheries.ubc.ca

Part III:
Design and Selection

American Fisheries Society Symposium 42:67–74, 2004

The No-Take Research Natural Area of Dry Tortugas National Park (Florida): Wishful Thinking or Responsible Planning?

ROBERT J. BROCK[1]

Everglades/Dry Tortugas National Parks, South Florida Natural Resources Center,
Marine Resources Branch, Homestead, Florida 33034, USA

BRIEN F. CULHANE

Everglades/Dry Tortugas National Parks, Environmental Planning and Compliance Branch,
Homestead, Florida 33034, USA

Abstract.—Established in 1992, Dry Tortugas National Park (DRTO) is one of the most pristine and remote parks in the National Park System. Located approximately 109 km from Key West, Florida, one of the purposes of establishing DRTO was to "protect and interpret a pristine subtropical marine ecosystem, including an intact coral reef ecosystem" (Public Law 102–525, 102nd Congress, 26 October 1992). Fulfilling this purpose has become increasingly difficult as visitation to DRTO has increased 400% over the last two decades, and boat registrations increased 50% during the 1990s. Clearly, potential threats to DRTO's natural and cultural resources have significantly increased since the last General Management Plan for DRTO was completed in 1983. An interdisciplinary team of scientists assessing the area's lush seagrass beds and corals as well as fishery resources undertook a site characterization. It was concluded that the snapper–grouper–grunt complex was overfished, anchor damage was evident, and water quality had at times degraded beyond acceptable state standards for bathing beaches. Clearly, these conditions were unacceptable according to DRTO's Congressional enabling legislation. Guided by National Park Service (NPS) policies pertaining to natural resource management (NPS-77) and presidential Executive Orders 13089 (coral reefs) and 13158 (marine protected areas), DRTO developed a list of draft management alternatives that would better protect the natural and cultural resources of the park, increase educational and scientific research opportunities, and improve the visitor experience. Here we discuss the rationale for developing, siting, and adopting the largest no-take Research Natural Area in the National Park System.

Introduction

Dry Tortugas National Park (DRTO) is one of the most pristine and remote parks in the National Park System. Located approximately 109 km west of Key West, Florida, the park encompasses seven small islands and 160 km² of the Gulf of Mexico. Established as Fort Jefferson National Monument in 1935, the area was designated a national park by Congress in 1992. The park's enabling legislation directs the National Park Service (NPS) to protect fish and wildlife populations and to preserve the pristine subtropical marine ecosystem including an intact coral reef community. In 1990, much of the surrounding waters gained protection with the establishment of the Florida Keys National Marine Sanctuary (FKNMS), managed jointly by the National Oceanic and Atmospheric Administration (NOAA) and the state of Florida.

The Dry Tortugas are renowned for historic Fort Jefferson and spectacular natural resources, including clear waters, abundant coral reefs, and a diverse array of marine and bird life. These resources play a vital

---

[1] Corresponding author: robert.brock@noaa.gov; present address: NOAA Fisheries, Office of Science and Technology, Marine Ecosystems Division (F/ST7), Silver Spring, Maryland 20910-3282, USA.

role in the dynamics of the larger Florida Keys coral reef ecosystem. Larvae and fish spawned here may be dispersed by currents throughout the Keys and along the southeastern coast, helping to sustain fisheries in Florida and beyond (Lee and Williams 1999).

Although the park is isolated, its visitation quadrupled from 23,000 in 1994 to more than 95,000 in 2000 (DRTO Division of Resource and Visitor Protection, personal communication). The rapid increase in popularity resulted in crowding, noise, strained facilities, and a decline in the quality of the visitor experience. Coral reefs, fisheries, and water quality also began to show the effects of concentrated use (Ault et al. 1998). Although increased visitation was due in large part to the advent of commercial ferry service from Key West, the number of private boaters visiting the park increased as well. The number of private recreational boats registered in Monroe County increased by 50% between 1988 and 1999 (Monroe County Tax Collector, personal communication). To ensure that resources and quality visitor experiences are protected, park management initiated a general management planning process in 1998. A General Management Plan for Fort Jefferson National Monument was completed in 1983, but by 1998, this plan was deemed inadequate to address current conditions at the park. At the same time, the FKNMS was initiating a plan to establish the Tortugas Ecological Reserve, a no-take reserve adjacent to the park (FKNMS 2000). Although the NPS and NOAA have different missions and statutory requirements, they share common goals for Tortugas ecosystem health. By coordinating science, planning, and public involvement, and through collaboration with state agencies, park and sanctuary managers sought to minimize public confusion and maximize participation in both planning processes.

To support planning and decision-making, the park and sanctuary commissioned a Site Characterization for the Dry Tortugas Region that synthesized current knowledge of physical oceanography, benthic (bottom-dwelling) communities, and fisheries. Populations of lutjanids (snappers), serranids (groupers), and pomadasyids (grunts) were found to be significantly overfished, threatening the structure and dynamics of the reef fish community (Schmidt et al. 1999). Anchor damage was evident, and water quality at times had degraded beyond acceptable state standards for bathing beaches. Increase in the size and number of vessels on the water and improvements in navigation and fishing gear likely contributed to these trends. Clearly, these conditions were unacceptable given the explicit mandate in the park's enabling legislation.

To gain additional information, in 1999, the Department of the Interior and NOAA asked the National Research Council of the National Academy of Sciences to examine the utility of marine reserves and protected areas for conserving fisheries, habitats, and biological diversity. The council's report, *Marine Protected Areas: Tools for Sustaining Ocean Ecosystems*, endorsed the increased use of no-take reserves, in concert with conventional management approaches, as tools for managing ocean resources.

Guided by NPS management policies, and presidential Executive Orders 13089 (coral reef protection) and 13158 (marine protected areas) (U.S. Office of the Federal Register 1998, 2000), an interdisciplinary planning team developed draft management alternatives for the park. Executive Order 13089 directs federal agencies whose actions may affect coral reef ecosystems to utilize their programs and authorities to protect and enhance conditions of such ecosystems. The National Action Plan to Conserve Coral Reefs, issued by the U.S. Coral Reef Task Force (2000) in March 2000, includes a goal of designating 20% of all U.S. coral reefs as no-take ecological reserves by 2010.

The goal of the alternatives was to better protect the park's natural and cultural resources, increase educational and scientific research opportunities, and improve the visitor experience, consistent with park purposes. The alternatives included the proposed establishment of a Research Natural Area (RNA) zone where extractive activities, including fishing, would be prohibited. The RNA would help achieve park purposes by protecting a representative range of near-pristine terrestrial and marine resources; protecting biological diversity; ensuring the replenishment and protection of fish stocks; and providing outstanding opportunities for nonmanipulative research, visitor education, and other nonconsumptive uses. These attributes are consistent with the benefits derived from marine protected areas (NRC 2001).

Federal land management agencies have been actively developing a national system of RNAs since 1927. By 2001, eight federal agencies administered approximately 460 RNAs ranging in size from less than 4.047 km^2 (1 acre) to more than 404,700 km^2 (100,000 acres) and totaling nearly 20,235,000 km^2 (5 million acres) (NPS 1991). The NPS administers approximately 86 RNAs in 28 national park units ranging in size from 60.7 km^2 (15 acres) to approximately 283,290 km^2 (70,000 acres) (NPS 1991). From the inception of the program, there have been two primary purposes for developing a comprehensive system of RNAs: (1) to preserve a representative array of significant natural ecosystems and their inherent processes as baseline areas, and (2) to obtain, through scientific research and education, information about natural system components, inherent processes, and comparisons with manipulated systems.

Objectives and Design of the Dry Tortugas Research Natural Area

The objectives of RNAs are: (1) to preserve a wide range of representative areas or natural situations that possess unique characteristics; (2) to preserve and maintain genetic diversity; (3) to protect against deleterious environmental disturbance; (4) to provide student and professional education; (5) to serve as baseline areas for measuring long-term ecological changes; and (6) to serve as control areas for comparing results from research conducted elsewhere. The criteria that the DRTO science team used to develop the DRTO RNA

were: (1) must include the full range of DRTO habitats; (2) must include monitoring sites currently in existence; (3) must be of sufficient size; (4) must be large enough to be self-sustaining; (5) must be easily identifiable and enforceable; (6) must consider socio-economic impacts; and (7) must consider the larger ecosystem.

To begin deliberations on what areas would best fulfill the RNA criteria, a geographic information system (GIS) map depicting the known benthic resources (in 1992) of the park was produced (Figure 1). This area continues to be groundtruthed and the map updated today. As one RNA criterion required that all known habitat types be represented in any RNA developed, a GIS map of this type was necessary. Comprehensive monitoring stations measuring (1) water quality; (2) seagrass diversity distribution, abundance, and density; (3) lobster abundance, size, and fecundity; and (4) reef fish community structure and habitat preference were plotted on a GIS map and overlain on the benthic resources map. This ensures that not

Figure 1.—Benthic resources of Dry Tortugas National Park (from Dry Tortugas National Park Management Plan).

only are the monitoring station locations known, but the habitat associated with each station is known as well.

To quantify the amount of habitat types (criterion 1) and monitoring stations (criterion 2) included in possible RNA configurations, Table 1 was developed to show how each alternative fulfilled these criteria as well as where any shortcoming may occur. The sufficient size (criterion 3) and self-sustaining (criterion 4) RNA requirements were the most difficult to fulfill and open to the most interpretation. It was decided that sufficient size meant that any RNA configuration proposed include biota from all trophic levels. In other words, any RNA configuration proposed must "support" the highest trophic levels such as carcharhinids (sharks), dasyatids (rays), and marine mammals. This support meant that these species had to be "regularly" sighted within the proposed RNA area. This requirement would also aid in fulfilling the larger goal of protecting biodiversity. In order to be considered self-sustaining, the proposed RNA configuration must include known spawning aggregation sites. Due to some hydrodynamic information that suggests regional counter-clockwise gyres develop (Lee et al. 1994), it was hypothesized that an unknown percentage of the eggs and larvae would have an acceptable chance of being produced and settling out within the proposed RNA (Lee and Williams 1999).

Table 1.—The percent of each habitat type included within three possible research natural area (RNA) configurations (from Dry Tortugas National Park [DRTO] General Management Plan). ALT = alternative.

Community type	RNA ALT "C"	RNA ALT "D"	RNA ALT "E"
Seagrass (27%)	44%	71%	99.6%
Bank barrier/ fringing reef (<1%)	15%	0%	100%
Patch reefs (<1%)	70%	34%	99.5%
Algal communities (<1%)	48%	<1%	100%
Staghorn reefs (1.8%)	77%	12.4%	91%
Octocoral hard-bottom (16%)	33%	36%	99.9%
Sedimentary habitats (54%)	49%	30%	99.6%
Land (<1%)	49%	4%	76%
Total area of DRTO included in proposal	46%	41%	99.5%

One of the most important aspects of developing no-take areas involves enforceability (criterion 5). First, the boundaries of the area must be easily recognizable by the various stakeholder groups. Figure 2 shows that the proposed boundary lines of the no-take area were drawn along straight longitude–latitude lines. It was hypothesized that with the global positioning systems available today, boaters would be able to easily confirm whether they are in the restricted area or not. The distance from the NPS ranger headquarters at Ft. Jefferson is shorter to the outermost boundary of the proposed RNA than any of the alternatives. This will likely enhance speed and ease of enforceability.

While the park legislation clearly mandated protecting natural and cultural resources, it also cited the importance of the visitor experience as well as responsible educational and recreational activities. With that in mind, an analysis of recreational and commercial activities in and adjacent to the park was undertaken. Activities such as locations where stakeholders fish (Figure 3) and scuba dive were plotted on a GIS map to ascertain how much of these activities would be lost, with possible losses in economic activities as well, if the proposed no-take area was implemented. For example, 54% of park waters (54 square nautical mi) will remain open for recreational fishing under the proposed RNA alternative. This area includes five of the park's seven islands and the most popular fishing area, the waters surrounding Garden Key (Fort Jefferson). This area is not in the proposed no-take zone so it was hypothesized that any negative socioeconomic consequences (criterion 6) would be minimal, if any at all.

While an analysis to determine if an RNA would fulfill the NPS goals and objectives for protecting natural resources was occurring, the adjacent waters of the FKNMS were being assessed to determine criteria for developing a no-take Tortugas Ecological Reserve. It seems logical that much would be lost if shallow water habitats such as patch reefs and seagrass meadows were protected within the park but adjacent deeper water reefs with large pelagic species and their known spawning aggregation sites within the FKNMS were left unprotected. It was determined that it would be most effective and best fulfill RNA goals and objectives (criterion 7) to protect connecting shallow and deepwater ecosystems wherever possible. These areas can have very different ecological functions, where shallow park waters protect juvenile life stages within their extensive seagrass beds while the deeper waters in the adjacent FKNMS support more adult stages of pelagic and highly migratory species. By protecting both the shallow wa-

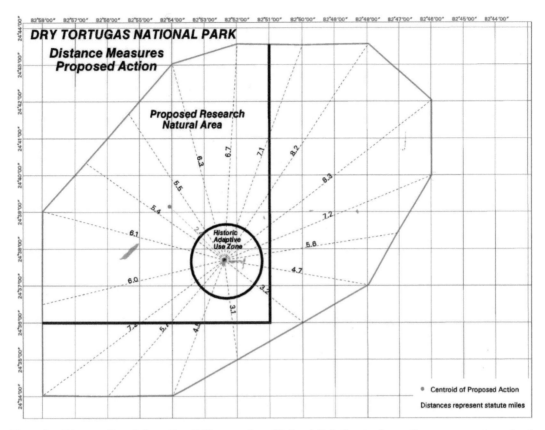

Figure 2.—Distance (in mi) from Fort Jefferson, where National Park Service law enforcement agents are head-quartered, to the park boundaries (from Dry Tortugas National Park Management Plan).

ters of the park and the deeper waters of the FKNMS, it is hypothesized that the habitat and depth requirements for the entire life history of many marine organisms are included. The selected no-take zones of the FKNMS and DRTO are depicted in Figure 4.

Consistent with the park's legislated purposes and NPS policies, Alternative "C" was determined to best achieve the objectives for resource protection with the establishment of a large research natural area zone while continuing to accommodate a wide range of appropriate visitor uses. Although Alternative "E" would have maximized the area of resource protection with the RNA zone extended to the majority of the park, the diversity of visitor experiences would have been significantly restricted as a consequence. Available scientific evidence also indicates that the size and configuration of the RNA zone in Alternative "C" would reliably achieve the primary objectives for increased biological diversity and protection of a vital range of habitats (NPS 2001a, 2001b).

RNA Implementation

The Record of Decision for the Dry Tortugas General Management Plan was signed in July 2001 (NPS 2001b). Implementation of the RNA is currently on hold pending resolution of two issues: (1) a dispute over ownership of the submerged lands within park boundaries, and (2) completion of a rule-making process to implement the management plan, including changing the park's fishing regulations. For several years, there has been discussion between the Florida Department of Environmental Protection and the federal government on the issue of ownership of the submerged lands within the park. Each party claims they are the rightful owner, and discussions and negotiations continue to this day. The question of ownership has far-reaching implications for managing the park in accordance with laws and polices established by Congress and the federal government. There are also specific implications related to establishment of the

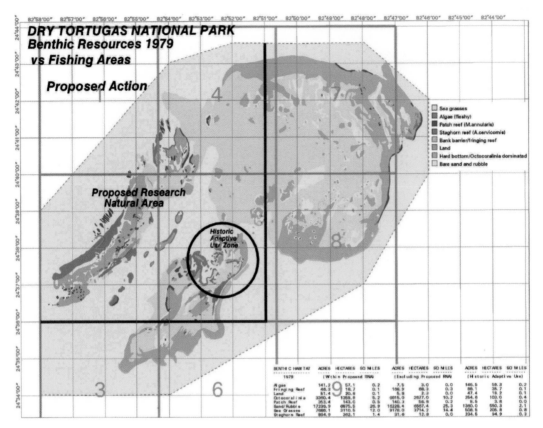

BENTHIC HABITAT 1979	ACRES	HECTARES	SQ MILES	ACRES	HECTARES	SQ MILES	ACRES	HECTARES	SQ MILES
	(Within Proposed RNA)			(Excluding Proposed RNA)			(Historic Adaptive Use)		
Algae	141.2	57.1	0.2	7.5	3.0	0.0	146.5	59.3	0.2
Fringing Reef	46.3	18.7	0.1	166.9	68.3	0.3	88.1	35.7	0.1
Land	51.1	20.8	0.1	5.8	2.3	0.0	47.4	19.2	0.1
Octocoralinia	3350.4	1369.8	5.2	6615.0	2677.0	10.2	264.6	103.0	0.4
Patch Reef	353.4	143.0	0.5	140.3	56.8	0.2	9.5	3.8	0.0
Sand/Rubble	17235.9	6975.5	26.9	10228.4	6507.4	25.3	1360.0	550.3	2.1
Sea Grasses	7685.1	3110.5	12.0	9176.0	3714.2	14.4	508.5	205.8	0.8
Staghorn Reef	894.9	362.1	1.4	31.6	12.8	0.0	234.6	94.9	0.3

Figure 3.—Assessment of preferred fishing areas for anglers (from Dry Tortugas National Park General Management Plan).

RNA. Until the ownership question and related management implications can be resolved, the pursuit of rulemaking cannot occur, as the NPS must have the legal authority and jurisdiction to manage the natural, cultural, and recreational resources within park waters. Without agreement of the parties to both the ownership or management rights and responsibilities, rules to describe future management conditions cannot be fully drafted or finalized.

Discussions between the Department of the Interior and the state of Florida continue to hopefully reach, if not an agreement, at least an understanding of how park resources would be managed by the NPS in the future. It is currently unclear whether such an agreement or understanding will be reached or what a reasonable timeframe for resolution might be.

Summary and Conclusions

Statutory requirements clearly emphasized that the natural and cultural resources of DRTO be protected

and that the coral reef ecosystem remain "pristine and intact." Taking into account population estimates of southern Florida, trends in boat registrations and visitors to the park, and the best available scientific information showing significant declines in reef fish species, the park made a proactive responsible decision to better protect resources entrusted to its care. Once the decision was made that an RNA would best fulfill DRTO's goals and objectives, a transparent public process was put into place describing the selection criteria and why boundary lines were drawn where they were. Only a legally and scientifically defensible monitoring program to ascertain the response of the ecosystem to these management actions will determine the efficacy of this undertaking.

Acknowledgments

The discussions and the development of the General Management Plan (including the Research Natural

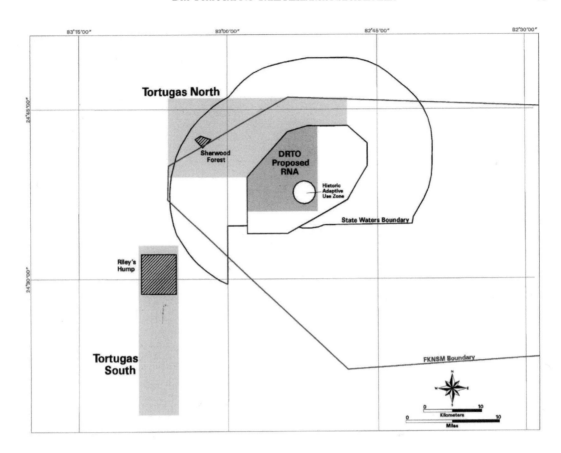

Figure 4.—View of the selected Dry Tortugas National Park (DRTO) no-take and surrounding Tortugas Ecological Reserve of the Florida Keys National Marine Sanctuary (FKNMS; from DRTO and FKNMS general management plans).

Area designation) was a total team effort by several employees of Everglades and Dry Tortugas National parks and the NPS Denver Service Center. Special thanks must be afforded to wildlife biologist Skip Snow and marine biologist Thomas Schmidt. Skip was instrumental in bringing together much of the information necessary to develop the GIS maps, and Tom developed several of the tables depicting the potential impacts of each alternative on the biological resources of that area.

References

Ault, J. S., J. A. Bohnsack, and G. A. Meester. 1998. A retrospective (1979–1996) multispecies assessment of coral reef fish stocks in the Florida Keys. U.S. National Marine Fisheries Service Fishery Bulletin 96(3):395–414.

FKNMS (Florida Keys National Marine Sanctuary). 2000. Final supplemental environmental impact statement/final supplemental management plan. National Oceanic and Atmospheric Administration, National Ocean Service, Office of National Marine Sanctuaries, Washington, D.C.

Lee, T. N., M. E. Clarke, E. Williams, A. F. Szmant, and T. Berger. 1994. Evolution of the Tortugas Gyre and its influence on recruitment in the Florida Keys. Bulletin of Marine Science 54(3):621–646.

Lee, T. N., and E. Williams. 1999. Mean distribution and seasonal variability of coastal currents and temperature in the Florida Keys with implications for larval recruitment. Bulletin of Marine Science 64:35–56.

NPS (National Park Service). 1991. Natural resources management guidelines (NPS-77). U.S. Department of the Interior, National Park Service, Washington, D.C.

NPS (National Park Service). 2001a. Dry Tortugas National Park, Final general management plan amendment/environmental impact statement. Denver Service Center, Denver.

NPS (National Park Service). 2001b. Record of decision, Final general management plan amendment/environmental impact statement, Dry Tortugas National Park, Florida. NPS, Homestead, Florida.

NRC (National Research Council). 2001. Marine protected areas: tools for sustaining ocean ecosystems. National Academy Press, Washington, D.C.

Schmidt, T. W., J. S. Ault, and J. A. Bohnsack. 1999. Site characterization for the Dry Tortugas region: fisheries and essential habitats. NOAA Technical Memorandum NMFS-SEFSC-425.

U.S. Coral Reef Task Force. 2000. The national action plan to conserve coral reefs, March 2, 2000. U.S. Coral Reef Task Force, Washington, D.C.

U.S. Office of the Federal Register. 1998. Executive order 13089. Coral reef protection. Federal Register 63(16 June 1998):32701.

U.S. Office of the Federal Register. 2000. Executive order 13158. Marine protected areas. Federal Register 65(31 August 31 2000):34909.

American Fisheries Society Symposium 42:75–86, 2004
© 2004 by the American Fisheries Society

Sustainability and Yield in Marine Reserve Policy

LOUIS W. BOTSFORD[1] AND DAVID M. KAPLAN

Department of Wildlife, Fish, and Conservation Biology,
University of California, Davis, California 95616, USA

ALAN HASTINGS

Department of Environmental Science and Policy,
University of California, Davis, California 95616, USA

Abstract.—In the process of implementing marine reserves, policy makers typically are occupied with (1) choosing the spatial configuration of areas to protect, and (2) addressing the concerns of fishermen regarding the effects of proposed reserves on fishery yield. The spatial configuration is typically set by choosing the habitat, species, and ecosystems to protect, assuming that the associated species will be sustained in that configuration. The concerns of fishermen are typically addressed by describing various spillover mechanisms and suggesting that yields will increase. There is a growing scientific understanding of the effects of reserves on the sustainability of populations and fishery yield, and the practical implications of those results should be incorporated into policy decisions. While there are exceptions, analytical and simulation results from models with sedentary adults indicate that yield will increase only if a population has been fished hard enough to cause a substantial decline in recruitment. This is consistent with the rough equivalence between yields possible with marine reserves and conventional management. This equivalence is a useful benchmark in the absence of information on larval advection and spatial variability in productivity. With reserves, lower fishery yields will be obtained from species dispersing shorter distances. Both yield and preservation goals depend on species in reserves being sustainable. Sustainability of species in reserves will depend on their dispersal distances and the spatial configuration of reserves. Species will be sustained in marine reserves if the alongshore dimension of the reserve is greater than its mean dispersal distance (assuming little alongshore displacement of the dispersal pattern), but species dispersing all distances will be sustained in networks of reserves, if a specific fraction of the coast is covered. Yield will be greater as the size of individual reserves in that network becomes smaller. Shorter-distance dispersers are always more likely to persist. Sustainability and yield in marine reserves depend on three categories of uncertainty: (1) uncertainty in population response to management is less when employing marine reserves than in conventional management, (2) uncertainty in the slope of the stock–recruitment relationship at low abundance affects both reserves and conventional fishery management, and (3) uncertainty in the pattern of larval dispersal affects management by reserves much more than conventional fishery management. Most of the available results are modeling results, and there is a need for better empirical information on both sustainability and yield. We need to know more than just whether marine reserves sustain populations and increase yield; rather, we need to know which kinds of reserves (i.e., size and spacing) sustained populations and which kinds of species (i.e., dispersal distance) were sustained and showed increased yields.

Introduction

Marine reserves are being proposed and implemented worldwide with two fundamental goals: preservation

of natural ecosystems and fishery management (Murray et al. 1999; Lubchenco et al. 2003). Reserves for fishery management seek to increase the yield to fisheries, while reserves for preservation are put in place to preserve areas with a natural, functioning ecosystem. The design of these reserves, or systems of

---

[1] Corresponding author: lwbotsford@ucdavis.edu

76 BOTSFORD ET AL.

such reserves, involves policy choices regarding how much area to protect and where it should be. The biological basis for these decisions typically has involved identification of the types of ecosystems, habitats, or species of concern to be protected, then a systematic procedure for choosing specific areas to meet preservation or fishery goals (e.g., Leslie et al. 2003; Roberts et al. 2003a, 2003b). The promise in the end is sustainable fisheries or ecosystems. Greater fishery yield is frequently promised on the basis of a description of the spillover mechanism that could bring about that greater yield.

Two key issues that need to be addressed in this planning process are: (1) how the design of reserves will affect sustainability of populations in a system of marine reserves, and (2) how yield will change in affected fisheries. To date, both of these have been addressed in only a qualitative fashion by reserve planners, and projected outcomes have tended to be optimistic. This presents a problem in that if there is to be any truly lasting value from marine reserves, the reserves being designed and implemented now need to be able to live up to expectations.

Here, we describe current progress in assessing the sustainability afforded by various marine reserve designs and their effect on fishery yield. Sustainability involves the question of whether species in marine reserves will continue to persist under specific reserve designs, an issue that is relevant to both preservation and fishery goals. The effect of reserves on fishery yield is often treated as an issue of interest only for reserves designed with the goal of improving fishery yield. However, the effect of reserves on fishery yield is a central issue in the policy decisions for all marine reserves, including those intended purely for conservation. In virtually all marine reserves, fishing is the primary action being limited by the implementation of reserves; hence, any loss in fishery yield is a dominant cost in any cost–benefit analysis. In the first part of this presentation, we summarize evolving modeling results regarding sustainability and yield (see Gerber et al. 2003 for a comprehensive review of the models of marine reserves). In the second part, we present new simulation results that demonstrate how sustainability and yield interact in an example of marine reserves in a typical size-structured, fished population distributed over space.

Our discussion here is limited by three fundamental assumptions that must be kept in mind: (1) sedentary adults, (2) Laplacian larval dispersal, and (3) no movement of fishers. Assuming species with sedentary adults means that we focus on movement in the larval stage. These are the species that will be afforded the most protection in reserves, and uncertainty in larval dispersal patterns makes understanding their possible effects important. We are beginning to understand how coastal currents shape dispersal patterns (e.g., Jones et al. 1999; Swearer et al. 1999; Cowen et al. 2000; Warner et al. 2000; Botsford 2001; Shanks et al. 2003) but do not yet have enough information for practical reserve design. Here, we use the simplest dispersal pattern that contains the spatial scale of dispersal, Laplacian dispersal (exponential decay in settlement in both directions). The movement of fishers and consequent shift in effort with the advent of marine reserves is a key element in both sustainability and yield. For each of these assumptions, we point out where results would differ if they were included, and we provide relevant references. For a description of modeling results regarding the effects of fish movement on the efficacy of marine reserves see Botsford et al. (2003) and Gerber et al. (2003). In addition to the limitations imposed by these assumptions, we also do not address all of the possible differences between marine reserves and conventional management, such as differences in habitat damage, bycatch, and political ease of implementation. For a more comprehensive review of issues beyond sustainability and yield, see Hilborn et al. (in press).

Sustainability

Sustainability of fisheries is an issue that has received increasing attention since the late 1980s, resulting in explicit focus on the persistence of fished populations. Persistence has been described in terms of equilibrium conditions for age-structured populations (Sissenwine and Shepherd 1987). These conditions specify that a population with density-dependent recruitment will have a nonzero equilibrium as long as the number of eggs produced in the lifetime of an individual exceeds the inverse of the slope of the relationship describing the number of recruits produced by a specified number of eggs spawned (i.e., the stock–recruitment relationship with stock depicted in terms of total egg production) (Figure 1). Note that lifetime egg production (LEP) is a quantity known as R_0 in ecology and eggs per recruit or spawning biomass per recruit in fisheries (Goodyear 1993). To compare that condition across species in general terms, fishery biologists express it as the fraction of the natural, unfished LEP, which we will denote FLEP (this quan-

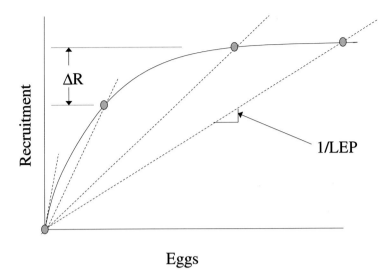

Figure 1.—A schematic plot of the number of recruits produced for each value of total population egg production (solid line). Population equilibria for each level of fishing lie at the intersection of the dashed lines, which have slope 1/LEP (lifetime egg production). When 1/LEP equals the slope of the egg-recruit line at the origin, the population collapses to zero recruitment. Also shown is an example of the reduction in recruitment (ΔR) by the fishery that typically must be present for the implementation of reserves to increase yield.

tity is referred to as spawning potential ratio in fisheries). Considerable effort has gone into determining the value of that quantity required for persistence of marine fish (e.g., Clark 1990; Mace and Sissenwine 1993), with more recent efforts on meta-analysis to determine generic consistencies based on taxonomy (Myers et al. 1999) and upward adjustment of previously low values (Clark 2002; Ralston 2002).

The reason that this persistence condition is not useful in the assessment of population persistence in marine reserves is that a population distributed across a system of marine reserves is not the single, well-mixed population assumed above, but rather a meta-population of such populations distributed over space, connected by dispersing larvae (Botsford et al. 1994). The recruits produced at each location end up distributed along the coast through the process of larval dispersal, and much of the recruitment at each location is produced elsewhere. Because of this additional complexity, a general expression for stability of these marine meta-populations has been difficult to obtain (Armsworth 2002). Researchers have, however, developed useful ways of describing population dynamics in terms of source–sink dynamics (Crowder et al. 2000; Lipcius et al. 2001).

Some results regarding persistence of simple meta-populations have been developed in the context

of marine reserves. Initial results assumed a population with sedentary adults, post-settlement density-dependent recruitment of the Beverton–Holt type (Beverton and Holt 1957) and larvae with a Laplacian dispersal pattern (i.e., exponentially decaying with distance in both directions) in a system of uniformly spaced reserves of width w and spacing s, along a coastline with uniform habitat, with fishing removing all fish between reserves (Botsford et al. 2001). The result was that for a single reserve of a certain width, species with mean dispersal distances less than or roughly the same as that width would persist. However, a system of uniformly spaced reserves of any width that covered a specific fraction of the coastline would allow persistence of all species, regardless of dispersal distance. The latter result indicated that a network of reserves could function in a way that was greater than the sum of the workings of the individual reserves (i.e., it was a "network result"). The specific fraction of coastline that needed to be covered was the value of FLEP required for persistence of the single, well-mixed population, as discussed in the previous paragraph. In this simple case, one can gain some intuitive understanding of why FLEP sets the minimum area in reserves by observing that the role of fishing in reducing LEP in the single, well-mixed population can be thought of as being replaced in the case of the meta-population by reduction in the area in which larvae can settle and grow to maturity, due to fragmentation of the coastline by intense fishing between reserves. Alongshore advection made the reserve areas required for sustainability much larger (Botsford et al. 2001).

Subsequent analyses have refined and extended these results. Lockwood et al. (2001) has shown that the initial results hold for a variety of shapes of dispersal patterns centered on the origin, with the mean dispersal distance being the important characteristic. This is an important result, as dispersal patterns are generally poorly known. Also, these results have been extended to the case in which there is a specific level of fishing between reserves, and there are discontinuities in the quality of benthic habitat that create species boundaries (D. L. Lockwood, A. Hastings, and L. W. Botsford, University of California, unpublished). These results show how fishing at levels less than complete removal lead to population persistence with lower fractions of the coastline in reserves, while losses of larvae across species boundaries lead to requirements for a greater fraction of coastline in reserves.

Fishery Yield

In its simplest form, the question in the minds of policy makers of how implementation of a marine reserve will affect fishery yield is essentially whether the loss of fishable area will be compensated for by changes in the population brought about by implementation of the reserve. Assuming that reserves are not going to affect individual growth and mortality rates outside the reserve, they will not affect yield per recruit, and we need consider only the effect on recruitment. For yield to increase with the implementation of reserves, the increase in recruitment due to the increased egg production from the reserves (ΔR) must be large enough to compensate for the fraction of the area placed in reserves (ΔA), in other words,

$$(1 - \Delta A)(1 + \Delta R) > 1.$$

This requires not just an increase in egg production, but that the larvae produced are able to reach the fished areas to increase recruitment by the required amount. Thus, if recruitment has not been substantially reduced by the current fishery, there is little scope for recruitment to be increased by the additional egg production supplied by a system of marine reserves (Figure 1). In turn, the amount by which recruitment will have been

reduced by fishing a certain amount will depend to some degree on the slope of the egg–recruitment relationship at the origin (Figure 1).

This observation regarding the potential for reserves to increase yield raises the policy question of whether yield can be increased as much by changing management in a conventional way (i.e., by decreasing fishing effort). That question was addressed by analysis of a simple model with no adult movement, larval settlement equally distributed across the population, and post-dispersal density-dependent recruitment only (Hastings and Botsford 1999). The answer, that the maximum yield problem for conventional management was mathematically the same as maximizing yield using reserves, indicates that there is a rough equivalence between reducing effort in conventional management and implementing marine reserves. This conclusion is also indicated by the invariant noted by Mangel (1998). This rough equivalence is consistent with results from a number of simulation studies of more complex models. The typical result is that marine reserves produce greater yield only for fishing mortality rates greater than a certain minimum value (e.g., Holland and Brazee 1996), or conversely, marine reserves are a means of guaranteeing sustainability even if the fishing mortality increases to very high values (e.g., Quinn et al. 1993).

The rough equivalence between conventional management and management by reserves provides policy makers with an easily computed benchmark estimate of the yield possible with marine reserves. Whether reserves or conventional management are actually superior depends on further detail, usually in an obvious way. For example, if compensatory density-dependence occurs prior to dispersal (e.g., density-dependent fecundity or indirect effects on fecundity such as density-dependent growth), reserves will have less advantage because reserves will increase density. On the other hand, if pre-dispersal density dependence is depensatory, such as in broadcast spawning, reserves will have greater advantage. Another example, fishermen shifting effort from inside reserves to outside, rather than simply leaving as assumed implicitly or explicitly in virtually all models of marine reserves (Gerber et al. 2003), is treated in several recent publications (Smith and Wilen 2003; Halpern et al., in press). In that case also, reserves have less advantage.

Situations involving substantial movement and heterogeneity in productivity can lead to exceptions to this rough equivalence. One example is populations with ontogenetic movement, in which specific life history stages can be protected (e.g., reserves can be placed to protect spawning and rearing areas; Apostolaki et al. 2002). Other examples involve differences in larval productivity and substantial alongshore advection. When there are areas with excess larval production, and other areas with populations at less than the benthic carrying capacity, then reserves can provide greater yield than can any scheme with the same fishing effort at all locations. Morgan and Botsford (2001) showed that increasing protection of a single source population coupled to three sink populations could increase yield by an amount greater than that possible through conventional fishery management of all four populations. However, that advantage was not possible unless the source was known. Gaines et al. (2003) showed another example in which populations in the middle of a bounded area along a coastline with reversing currents could produce greater yield with reserves than with conventional fishing throughout. In the cases of these exceptions to the rough equivalence between conventional management and management by reserves, yield could increase even when the fishery has not diminished recruitment.

Consideration of Both Sustainability and Yield

We know of only one general analysis that included aspects of both sustainability and yield. A comparison of the spatial configuration likely to be best for conservation with that best for yield indicated that for conservation one could use a small number of large reserves of a size that would allow persistence of the longest-distance disperser one desired to protect (Hastings and Botsford 2003). This was contrasted with the best configuration for yield, which was that configuration that sustained the population but also supplied the greatest export of larvae from reserves, a system of many reserves as small as practicable. For sustainability, this system needed to cover a certain fraction of the coastline, the minimum FLEP required for sustainability of the fished species, as noted above (Botsford et al. 2001). This analysis assumed the cost of a reserve for conservation was proportional to the shoreline placed in reserve and did not include the cost of fishery yield lost due to displaced effort. While that analysis provides valuable insight into the benefits supplied by different spatial considerations, it is now clear that cost of foregone yield is a significant issue in real policy decisions.

To demonstrate the spatial distributions of recruitment, catch, and biomass, and how they provide for sustainability and yield, we present here the results of simulations of a system of marine reserves along a coastline at different levels of fishing. The model consists of 100 size-structured populations with individuals growing according to von Bertalanffy growth with a truncated Gaussian distribution of values of L_∞ as in Smith et al. (1998) and Morgan et al. (2000; see Table 1 for parameter values). Density-dependent recruitment is represented as the Beverton–Holt type, with the slope (a_{BH} in Table 1) set so that collapse occurs when LEP is 35% of the natural, unfished value. Reserves are equally spaced at four locations, and each covers either 2, 5, or 10 populations, so that 8%, 20%, or 40%, respectively, of the coastline is covered. We focus here on the consequences of larval dispersal distance, so adults are considered to be sedentary. Larval dispersal distance varies from 0 to 25 spatial units (each of the 100 populations is considered to occupy 1 spatial unit, s.u.). The model "wraps" dispersal at the boundaries to the other boundaries, so that there are no effects of species boundaries due to a change in suitability of benthic habitat or specific circulation features (Gaylord and Gaines 2000; Lockwood, Hastings, and Botsford, unpublished).

The changes in recruitment, yield, biomass, and LEP with fishing mortality rate for a single population without reserves are shown in Figure 2. Biomass and recruitment decline monotonically, while catch increases, then decreases. All are zero at the point

Table 1.—Parameter values for population model. s.u. = spatial unit.

Parameter	Symbol	Value
Growth		
Maximum size	L_∞	118 mm
Standard deviation of maximum size	σ_L	10 mm
Reproduction		
Size of first reproduction	l_m	60 mm
Fecundity versus weight		
Coefficient	a	5.47×10^{-6}
Exponent	b	3.45
Mortality		
Natural mortality	M	0.08/year
Fishing mortality	F	(0.05 – 0.2)/year
Size limit	L_F	60 mm
Reserve size		0–10 s.u.
Number of reserves		4
Dispersal		
Dispersal distance		1–20 s.u.
Recruitment		
Beverton-Holt slope	a_{BH}	0.00662
Beverton-Holt capacity	C_{BH}	12,000,000

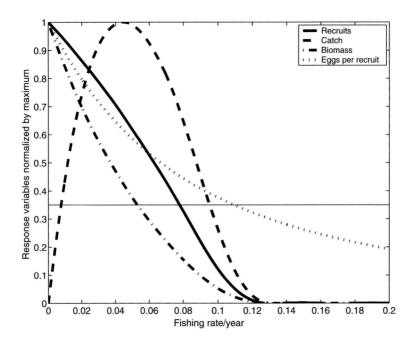

Figure 2.—Performance of a single population of the model, used here without reserves, versus fishing mortality rate. Parameter values are set so that as fishing increases, recruitment, biomass, and catch go to zero near where FLEP drops below 0.35. This occurs at a fishing mortality rate of 0.11/year. All variables are given as a fraction of their maximum values, which are recruitment = 2.35×10^9 individuals, biomass = 7.25×10^{12}, and catch = 1.33×10^{11}.

where FLEP is 0.35, which occurs at approximately $F = 0.11$/year.

The response of catch to increasing fishing mortality rate changes as reserve area is increased (as suggested by the yield results above), but that response depends critically on dispersal distance (as suggested by persistence results; Figure 3). For this model, peak catch generally declines as more reserves are added, but the nature of that dependence changes with dispersal distance. The plot of catch with no reserves is, of course, the same as Figure 2, independent of dispersal distance. Having just 8% of the coastline in reserves (reserves of width 2 s.u.) is enough to provide persistence, and some catch, for low-distance dispersers. However, any response for species dispersing 10 s.u. requires at least 20% in reserves (reserves of width 5 s.u.), which provides persistence and higher catch out to a fishing rate of 0.2/year. Persistence of individuals dispersing long distances requires 40% in reserves. Note, however, that as the fraction in reserves increases, catch at lower dispersal distances always remains low.

These characteristics follow from the spatial distribution of catch as dispersal distance varies (Figure 4). Catch is the highest of any location or dispersal distance at low dispersal distance just outside the reserves. However, for these short-distance dispersers it quickly drops to very low values as distance from the reserve increases. For dispersal distances greater than 10 s.u., on the other hand, catch between reserves varies little with space and dispersal distance. At a fishing mortality rate of 0.1/year, populations of long-distance dispersers would be sustained at high catch with 20% in reserves, as indicated by Figure 3, but for a fishing mortality rate of 0.2/year, catch would be much lower at high dispersal rates.

The spatial distribution of catch is largely determined by the spatial distribution of recruitment (Figure 5). Recruitment is highest for species dispersing short distances but extends only a short distance outside reserves. Recruitment inside reserves is less for long-distance dispersers, but it is evenly distributed over space between reserves. Note from the right hand side of Figure 5 that recruitment levels inside reserves decline faster

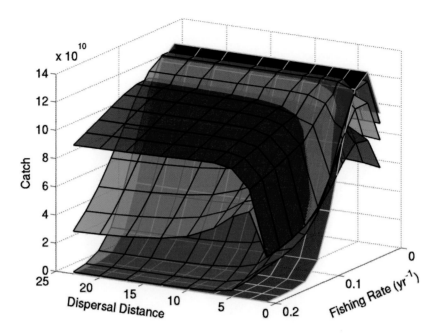

Figure 3.—Yield for the meta-population of 100 size-structured populations linked by Laplacian dispersal with various mean dispersal distances as fishing mortality rate increases. The four surfaces indicate yield for no reserves (gray) and four equally spaced reserves of size 2 spatial units (s.u.) (8% in reserves; red), 5 s.u. (20% in reserves; green), and 10 s.u. (40% in reserves; blue).

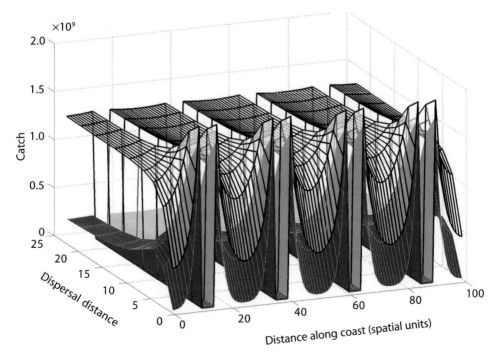

Figure 4.—The spatial distribution of catch for the model in Figure 3, at different mean dispersal distance, and two fishing mortality rates, $F = 0.1$ (top mesh surface) and $F = 0.2$ (lower solid surface).

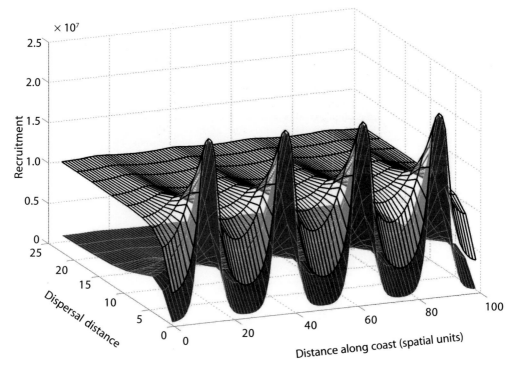

Figure 5.—The spatial distribution of recruitment for the model in Figure 3, at different mean dispersal distance, and two fishing mortality rates, $F = 0.1$ (top mesh surface) and $F = 0.2$ (lower solid surface).

at a fishing rate of 0.2/year than at 0.1/year. This follows from reduced connectivity, as fishing reduces populations to lower values.

The spatial distribution of biomass at each dispersal distance depends on the distributions of recruitment and fishing (Figure 6). Note that for this spatial configuration of reserves, the distribution of biomass over space is uniform for dispersal rates greater than 10 s.u. For lower dispersal rates, biomass declines rapidly, with distance from reserves reflecting the effect of spatial variability in recruitment. Differences between spatial distributions of biomass with fishing appear to follow the differences in recruitment.

Discussion

The primary message to decision makers from these results is that the spatial configuration of reserves and the spatial scale of dispersal of different species make a difference in the advisability and efficacy of marine reserves. Choosing reserve size and location solely because they contain the species, habitats, and ecosystems we want to protect will not guarantee their protection. Rather, we must set the spatial configura-

tion to sustain the species we want to protect. The design of the spatial configuration of the reserves will depend on the spatial scale of dispersal and the minimum value of FLEP required for that species.

A second message to policy makers involved in implementing marine reserves regards the potential loss in fishery yield, a dominant issue in formulating marine reserve policy. There need to be better attempts to assess the change in yield; pointing out that there will be greater egg production in reserves and that there may be spillover is not an adequate answer for decision making. The effects of reserves on yield of fished species will depend on how much FLEP has been reduced by fishing. This is consistent with the rough equivalence between the yield possible through reserves and conventional management. In the absence of detailed information on larval transport and benthic productivity, this equivalence is a valuable benchmark indicator of the effects of reserves on yield. It provides a link between the reserve option and conventional management options. If reducing effort in the fishery is an option, its effect on yield is, to a first approximation, the same as implementing reserves.

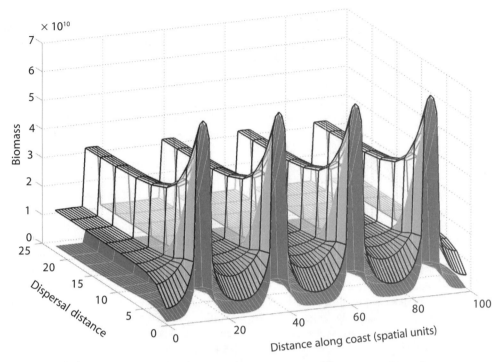

Figure 6.—The spatial distribution of biomass for the model in Figure 3, at different mean dispersal distance, and two fishing mortality rates, $F = 0.1$ (top mesh surface) and $F = 0.2$ (lower solid surface).

A third point for decision makers is the answer to the question of whether reserves are less susceptible to uncertainty than conventional fishery management. Reserves have been shown to be less affected by certain kinds of uncertainty (Lauck et al. 1998), and they can allow populations to persist and produce high yields even as uncertain fishing mortality rate increases to values that would cause collapse if there were no reserves (Figure 3; Quinn et al. 1993; Botsford et al. 2003). However, the above dependencies of sustainability and yield on the virtually unknown dispersal patterns and the uncertain value of FLEP suggest there is considerable uncertainty in projecting the effects of reserves. The best current, practical answer to the question is that managing a fishery by reserves instead of through conventional management decreases dependence on some kinds of uncertainty but increases dependence on others and does not change our dependence on a third kind of uncertainty. Susceptibility to uncertainty in dispersal distances is a new source of uncertainty incurred in spatial management. Uncertainty in the value of FLEP required for persistence is a dominant source of uncertainty in conventional fishery

management, and it has the same effect in management by reserves. The other dominant source of uncertainty in conventional fishery management, uncertainty in the effects of fishing regulations on the population, is alleviated in management by reserves.

A fourth message to policy makers is to be aware of the difference between conditions for sustainability (population persistence) and conditions for yield. This is due to the fact that persistence, by definition, requires only a small part of the population to be extant at a single location, while catch depends on the abundance of the population over all space outside reserves. The consequence of this difference is that results regarding the effects of reserves on persistence cannot be used to infer the effects of reserves on catch and vice versa. An example of the differences is the fact that persistence and yield depend on dispersal distance in different ways; persistence is greater at shorter dispersal distances, but greater catch is possible at longer dispersal distances (Figure 3).

The above discussion of various dependencies on uncertainty raises the question of whether we know enough about dispersal and the required FLEP to begin to make use of each of them in the design of reserves. As noted in the Introduction, there is an aware-ness of the importance of knowing the minimum value of FLEP required for persistence, which stems from its role in conventional fishery management. While much remains to be learned regarding the way in which this parameter might vary with species, habitat (e.g., whether it is higher in an upwelling zone), and climate, research is ongoing (Myers et al. 1999, 2002; Ralston 2002).

While knowledge of the physical and biological factors that determine dispersal patterns is accumulat-ing rapidly, our specific, usable knowledge of how species vary is probably limited to knowing the spa-tial scale of dispersal. Even with that simplification, it may be most prudent to limit our classification to short-distance and long-distance dispersers. In that regard, for the model used here, there seems to be a differ-ence in behavior between species dispersing greater than 10 s.u. and those dispersing less than 10 s.u. The effects of reserve design on short-distance dispersers could be evaluated by considering single reserves, while the effects on long-distance dispersers could be evaluated using the network of reserves and the frac-tion of the coastline covered. Much attention has been drawn recently to the fact that dispersal distances are much shorter for many species than we usually have thought them to be (Warner et al. 2000). The implica-tion of that observation for reserves is commonly taken to be that reserves will work more often than not and that our uncertainty in dispersal distance is not a con-cern. The results here are a reminder that while re-serves will "work," they will sustain those species with short dispersal distances, which provides less yield to fisheries than species dispersing longer distances.

Since the results here were obtained primarily from modeling, a natural question is how they com-pare with empirical observations. These models as-sume an increase in biomass and mean age inside marine reserves, an effect that is consistent with em-pirical observations (e.g., Halpern and Warner 2002; Halpern 2003). Recent further probing into the avail-able empirical data has shown that, as a category, ex-ploited species increased in abundance in reserves, while unexploited species did not (Micheli et al., in press), consistent with the expectations outlined here. There are fewer systematic observations of effects outside reserves, such as increases in yield (e.g., Rob-erts et al. 2001), and even those rarely include the obvious dependent variables identified here. Reserves are often cited as having increased yield, but no infor-mation is presented on the level of fishing prior to reserve implementation. Since the most dramatic in-creases are typically cited, this often leads to confu-sion among the public, and among policy makers, by giving them the impression that all reserves will in-crease fishery yield.

Further confusion regarding these somewhat com-plex issues occurs because of a lack of differentiation between scientific advice and advocacy of conserva-tion. Most biologists have a conservation bias, and many advocate conservation of marine resources. However, marine scientists have a responsibility to identify ad-vice given to policy makers as to whether it is science or advocacy. While there is probably wide agreement with this point, maintaining a level playing field when comparing reserves to conventional management is not always easy. For example, comparisons that assume a complex source–sink structure can show that reserves outperform conventional management, but to take ad-vantage of that capability we would need to know the specific structure (i.e., the dispersal patterns, which are currently, typically unknown) (Morgan and Botsford 2001). We need to differentiate between the ultimate potential of marine reserves and their current potential given existing knowledge.

In summary, marine reserves present a valuable new tool in the conservation of marine ecosystems and the management of fisheries. We have a good start on developing an understanding of their effects and the techniques for analyzing their efficacy and relating it to conventional management. We need to begin to integrate the understanding and techniques into marine policy making if we are going to make achievable projections regarding the promise of ma-rine reserves.

Acknowledgments

This work was supported by the Partnership of Edu-cation in Marine Resource and Ecosystem Manage-ment (PEMREM), California Sea Grant R/F 179, and the working group "Design of Tools for the Practical Design of Marine Reserves" at the National Center for Ecological Analysis and Synthesis that is supported by U.S. National Science Foundation (Grant DEB-94–21535), University of California, California Re-sources Agency, and California Environmental Pro-tection Agency. We thank Churchill Grimes and Ken-neth Frank for their reviews.

References

Apostolaki, P., E. J. Milner-Gulland, M. K. McAllister, and G. P. Kirkwood. 2002. Modelling the effects of establishing a marine reserve for mobile fish species. Canadian Journal of Fisheries and Aquatic Sciences 59:405–415.

Armsworth, P. 2002. Recruitment limitation, population regulation, and larval connectivity in reef fish metapopulations. Ecology 83:1092–1104.

Beverton, R. J. H., and S. J. Holt. 1957. On the dynamics of exploited fish populations. Fisheries Investigations (Series II) 19:1–533.

Botsford, L. W. 2001. Physical influences on recruitment to California Current invertebrate populations on multiple scales. ICES Journal of Marine Science 58:1081–1091.

Botsford, L. W., A. Hastings, and S. D. Gaines. 2001. Dependence of sustainability on the configuration of marine reserves and larval dispersal distances. Ecology Letters 4:144–150.

Botsford, L. W., F. Micheli, and A. Hastings. 2003. Principles for the design of marine reserves. Ecological Applications 13:S25–S31.

Botsford, L. W., C. L. Moloney, A. Hastings, J. L. Largier, T. M. Powell, K. Higgins, and J. F. Quinn. 1994. The influence of spatially and temporally varying oceanographic conditions on meroplanktonic metapopulations. Deep-Sea Research Part II Topical Studies in Oceanography 41:107–145.

Clark, W. G. 1990. Groundfish exploitation rates based on life history parameters. Canadian Journal of Fisheries and Aquatic Sciences 48:734–750.

Clark, W. G. 2002. $F_{35\%}$ revisited ten years later. North American Journal of Fisheries Management 22:251–257.

Cowen, R. K., K. M. M. Lwiza, S. Sponaugle, C. B. Paris, and D. B. Olson. 2000. Connectivity of marine populations: open or closed? Science 287:857–859.

Crowder, L. B., S. J. Lyman, W. F. Figueira, and J. Priddy. 2000. Source-sink population dynamics and the problem of siting marine reserves. Bulletin of Marine Science 66: 799–820.

Gaines, S. D., B. Gaylord, and J. L. Largier. 2003. Avoiding current oversights in marine reserve design. Ecological Applications 13:S32–S46.

Gaylord, B., and S. D. Gaines. 2000. Temperature or transport? Range limits in marine species mediated solely by flow. American Naturalist 155:769–789.

Gerber, L. R., L. W. Botsford, A. Hastings, H. P. Possingham, S. D. Gaines, S. R. Palumbi, and S. J. Andelman. 2003. Population models for marine reserve design: a retrospective and prospective synthesis. Ecological Applications 13:S47–S64.

Goodyear, P. 1993. Spawning stock biomass per recruit in fisheries management: foundation and current use. Canadian Special Publication of Fisheries and Aquatic Sciences 120:67–81.

Halpern, B. 2003. The impact of marine reserves: do they work and does reserve size matter? Ecological Applications 13:S117–137.

Halpern, B. S., S. D. Gaines, and R. R. Warner. In press. Export of production and the displacement of effort from marine reserves: effects on fisheries and monitoring programs. Ecological Applications.

Halpern, B. S., and R. R. Warner. 2002. Marine reserves have rapid and lasting effects. Ecology Letters 5:361–366.

Hastings, A., and L. W. Botsford. 1999. Equivalence in yield from marine reserves and traditional fisheries management. Science 284:1537–1538.

Hastings, A., and L. W. Botsford. 2003. Are marine reserves for fisheries and biodiversity compatible? Ecological Applications 13:S65–S70.

Hilborn, R., K. Stokes, J.-J. Maguire, T. Smith, L. W. Botsford, M. Mangel, J. Orensanz, A. Parma, J. Rice, J. Bell, K. L. Cochrane, S. Garcia, S. J. Hall, G. P. Kirkwood, K. Sainsbury, G. Stefansson, and C. Walters. In press. When can marine reserve areas improve fisheries management? Oceans and Coastal Management.

Holland, D. S., and R. J. Brazee. 1996. Marine reserves for fishery management. Marine Resource Economics 11:157–171.

Jones, G. G. P., M. J. Milicich, M. J. Emslie, and C. Lunow. 1999. Self-recruitment in a coral reef fish population. Nature (London) 402:802–804.

Lauck, T., C. W. Clark, M. Mangel, and G. R. Munro. 1998. Implementing the precautionary principle in fisheries management through marine reserves. Ecological Applications 8:S72–S78.

Leslie, H., M. Ruckelshaus, I. R. Ball, S. Andelman, and H. P. Posssingham. 2003. Using siting algorithms in the design of marine reserves. Ecological Applications 13:S185–S198.

Lipcius, R. N., W. T. Stockhausen, and D. B. Eggleston. 2001. Marine reserves for Caribbean spiny lobster: empirical evaluation and theoretical metapopulation recruitment dynamics. Marine and Freshwater Research 52:1589–1598.

Lockwood, D. L., A. Hastings, and L. W. Botsford. 2001. The effects of dispersal patterns on marine reserves: does the tail wag the dog? Theoretical Population Biology 61:297–309.

Lubchenco, J., S. R. Palumbi, S. D. Gaines, and S. Andelman. 2003. Plugging a hole in the ocean: the emerging science of marine reserves. Ecological Applications 13:S3–S7.

Mace, P. M., and M. P. Sissenwine. 1993. How much spawning per recruit is enough? Canadian Special Publication of Fisheries and Aquatic Sciences 120:101–118.

Mangel, M. 1998. No-take areas for sustainability of harvested species and a conservation invariant for marine reserves. Ecology Letters 1:87–90.

Micheli, F., B. S. Halpern, L. W. Botsford, and R. R. Warner. In press. Trajectories and correlates of community change in no-take marine reserves. Ecological Applications.

Morgan, L. E., and L. W. Botsford. 2001. Managing with reserves: modeling uncertainty in larval dispersal for a sea urchin fishery. Alaska Sea Grant College Program Report AK-SG-01-02:667–684.

Morgan, L. E., L. W. Botsford, S. R. Wing, and B. D. Smith. 2000. Spatial variability in growth and mortality of the red sea urchin, *Strongylocentrotus franciscanus* in northern California. Canadian Journal of Fisheries and Aquatic Sciences 57:980–992.

Murray, S. N., R. F. Ambrose, J. A. Bohnsack, L. W. Botsford, M. H. Carr, G. E. Davis, P. K. Dayton, D. Gotshall, D. R. Gunderson, M. A. Hixon, J. Lubchenco, M. Mangel, A. MacCall, D. A. McArdle, J. C. Ogden, J. Roughgarden, R. M. Starr, M. J. Tegner, and M. M. Yoklavich. 1999. No-take reserve networks: sustaining fishery populations and marine ecosystems. Fisheries 24(11):11–25.

Myers, R. A., N. J. Barrowman, R. Hilborn, and D. G. Kehler. 2002. Inferring Bayesian priors with limited direct data: applications to risk analysis. North American Journal of Fisheries Management 22:351–364.

Myers, R. A., K. G. Bowen, and N. J. Barrowman. 1999. Maximum reproductive rate of fish at low population sizes. Canadian Journal of Fisheries and Aquatic Sciences 56:2404–2419.

Quinn, J. F., S. R. Wing, and L. W. Botsford. 1993. Harvest refugia in marine invertebrate fisheries: models and applications to the redsea urchin, *Strongylocentrotus franciscanus.* American Zoologist 33:537–550.

Ralston, S. 2002. West Coast groundfish policy. North American Journal of Fisheries Management 22:249–250.

Roberts, C. M., S. Andelman, G. Branch, R. H. Bustamante, J. C. Castilla, J. Dugan, B. S. Halpern, K. D. Lafferty, H. Leslie, J. Lubchenco, D. McArdle, H. P. Possingham, M. Ruckelshaus, and R. R. Warner. 2003a. Ecological criteria for evaluating candidate sites for marine reserves. Ecological Applications 13:S199–S214.

Roberts, C. M., J. A. Bohnsack, F. Gell, J. P Hawkins, and R. Goodridge. 2001. Effects of marine reserves on adjacent fisheries. Science 294:1920–1923.

Roberts, C. M., G. Branch, R. H. Bustamante, J. C. Castilla, J. Dugan, B. S. Halpern, K. D. Lafferty, H. Leslie, J. Lubchenco, D. McArdle, M. Ruckelshaus, and R. R. Warner. 2003b. Application of ecological criteria in selecting marine reserves and developing reserve networks. Ecological Applications 13:S215–S228.

Shanks, A. L., B. Grantham, and M. H. Carr. 2003. Propagule dispersal distance and the size and spacing of marine reserves. Ecological Applications 13:S159–169.

Sissenwine, M. P., and J. G. Shepherd. 1987. An alternative perspective on recruitment overfishing and biological reference points. Canadian Journal of Fisheries and Aquatic Sciences 44:913–918.

Smith, B. D., L. W. Botsford, and S. R. Wing. 1998. Estimation of growth and mortality parameters from size frequency distributions lacking age patterns: the red sea urchin (*Strongylocentrotus franciscanus*) as an example. Canadian Journal of Fisheries and Aquatic Sciences 55:1236–1247.

Smith, M. D., and J. E. Wilen. 2003. Economic impacts of marine reserves: the importance of spatial behavior. Journal of Environmental Economics and Management 46:183–206.

Swearer, S. E., J. E. Caselle, D. W. Lea, and R. R. Warner. 1999. Larval retention and recruitment in an island population of a coral-reef fish. Nature 402:799–802.

Warner, R. R., S. E. Swearer, and J. E. Caselle. 2000. Larval accumulation and retention: implications for the design of marine reserves and essential habitat. Bulletin of Marine Science 66:821–830.

American Fisheries Society Symposium 42:87, 2004

Abstract Only

Laying the Scientific Foundation to Evaluate Ecological Recovery in California's Cowcod Conservation Areas

MARY YOKLAVICH[1]

NOAA Fisheries, Southwest Fisheries Science Center, Santa Cruz Laboratory,
110 Shaffer Road, Santa Cruz, California 95060, USA

MILTON LOVE

University of California, Santa Barbara, Marine Science Institute,
Santa Barbara, California 93106, USA

Abstract.—Along much of the Pacific Coast, populations of many groundfish species (including cowcod *Sebastes levis*) are at historically low levels. In an unprecedented effort to protect these species from incidental harvest and to assist with stock rebuilding, the Pacific Fishery Management Council established two Cowcod Conservation Areas (CCAs) in the Southern California Bight in 2001, encompassing 14,1750 km² (4,300 mi²) and including key groundfish habitat. Targeted fishing for groundfishes is prohibited year-round in depths greater than 37 m. Evaluating the effectiveness of the CCAs depends on timely, accurate assessment of the response of target species to increased protection. With multi-institute support, we have initiated a monitoring program to collect baseline data on abundance, size, and distribution of the benthic fishes inside and around the CCAs and on the status and use of protected habitats. This nonextractive survey approach is based on video transect methodologies and direct observations of groundfishes, macroinvertebrates, their habitats, and incidence of fishing gear using an occupied research submersible and incorporates information from seafloor habitat maps and past and recent groundfish catch and effort records. Survey design and results will serve as the foundation for a long-term monitoring program for the CCAs as well as a model for monitoring future deepwater marine protected areas off California.

[1] E-mail: mary.yoklavich@noaa.gov

American Fisheries Society Symposium 42:89–103, 2004
© 2004 by the American Fisheries Society

Marine Managed Areas Designated by NOAA Fisheries: A Characterization Study and Preliminary Assessment

LISA WOONINCK[1]

National Oceanic and Atmospheric Administration Fisheries, Southwest Fisheries Science Center,
Santa Cruz Laboratory, 110 Shaffer Road, Santa Cruz, California 95060, USA

CARLI BERTRAND[2]

University of Rhode Island, Graduate School of Marine Affairs,
Washburn Hall, Kingston, Rhode Island 02881, USA

Abstract.—National Oceanic and Atmospheric Administration (NOAA) Fisheries (National Marine Fisheries Service) is the primary agency responsible for management of fisheries and protection of endangered species within coastal and ocean waters of the United States. To this end, NOAA Fisheries, often in conjunction with fishery management councils, has designated various marine protected areas or marine managed areas (MMAs). We present here the results of a characterization study of 67 NOAA Fisheries MMAs that are currently part of the national MMA database. For a subset of 32 sites (48%), we evaluated their effectiveness and determined whether the design and management of the MMA included goals, targets, timelines, and monitoring practices. Large MMAs (>1,000 km²) with year-round protections and restrictions are managed by NOAA Fisheries. The MMAs are frequently comanaged with other regional MMAs as part of programmatic systems (88%), such as fishery management plans or recovery plans for endangered and threatened species. Far fewer MMAs (38%) function as biologically linked and connected networks. Nearly half of the MMAs promulgate fishing regulations under the Magnuson-Stevens Act combined with laws for the protection of an endangered or threatened species, and more than half of the sites have been established since 1996. All sites in the subset have goals, but only 63% have specific targets and timelines associated with the goals. Monitoring, most frequently in the form of stock assessments, is routinely performed at 87% of the sites within the subset. Lastly, 50% of the sites were either effective or part of an effective program, as evaluated against an MMA's ability to achieve the goals associated with its designation.

Introduction

An ecosystem-based approach to conserving living marine resources, including fishery resources, is a distinct shift from the traditional focus on single species or stocks toward a focus on biological communities and the interactions that occur among species in their physical environment (NMFS 1999; Lubchenco et al.

2003). There is a general consensus among managers, scientists, and policy makers that healthy and viable fisheries depend on healthy and resilient ecosystems. In the past decade, marine protected areas (MPAs) have been valued as an effective method for maintaining and increasing ecosystem function and protection (NRC 1999, 2001). They have been designated worldwide to promote and conserve biodiversity, promote sustainable use of particular resources, protect and enhance key species, preserve traditional uses and cultural heritage, and provide educational and recreational opportunities (NRC 2001; Agardy et al. 2003). Thus, MPAs have been established not only to enhance conservation but to im-

---

[1] E-mail: lisa.wooninck@noaa.gov
[2] Present address: NOAA Fisheries, Sustainable Fisheries International, 1315 East-West Highway, Silver Spring, Maryland 20910, USA.

prove the use and management of marine resources as well.

For purposes of this paper, the terms marine protected area and marine managed area (MMA) are interchangeable. They are both used to signify the wide range of spatial protections and objectives legally afforded marine habitats and resources. No-take MPAs, also referred to as marine reserves, are a type of MMA that prohibits all forms of extractive activity. Scientific studies of marine reserves have provided evidence of enhancement of biodiversity, biomass build-up, increased size–age composition inside reserve boundaries (reviewed in Halpern 2003), habitat improvement (Bradshaw et al. 1999), and spillover of populations outside reserve boundaries (McClanahan and Mangi 2001; Roberts et al. 2001). A small fraction of MMAs occurs in the form of marine reserves, in which less than 1% of the world's oceans are protected (NRC 1999). The majority of MMAs worldwide are implemented as multiple-use areas with management zones of mixed harvest strategies, ranging from no-harvest to restricted harvest (Agardy et al. 2003).

National Oceanic and Atmospheric Administration (NOAA) Fisheries (National Marine Fisheries Service, NMFS), in collaboration with the eight fishery management councils (councils), are mandated through the Fishery Management and Conservation Act of 1976 to manage fisheries and the complex marine systems that support them in the exclusive economic zone (EEZ) of the United States. The act was amended 20 years later by the Sustainable Fisheries Act (SFA) and is known as the Magnuson-Stevens Act (MSA). All fishery management plans (FMPs), their amendments, and regulations are required to comply with 10 national standards as mandated in the MSA. A few of these standards have a direct impact on the scope of marine spatial management applied by NOAA Fisheries. The FMPs are required under National Standard 1 to "prevent overfishing while achieving, on a continuing basis, the optimum yield from each fishery for the United States fishing industry." In the MSA, "optimum" is defined as the amount of fish that "will provide the greatest overall benefit to the Nation . . . taking into account the protection of marine ecosystems." Furthermore, an FMP is required, as mandated by National Standard 8, to be "consistent with the conservation requirements of this Act, take into account the importance of fishery resources to fishing communities in order to (a) provide for the sustained participation of such communities and (b) to the extent practicable, minimize adverse economic impacts on such communities." The courts have rec-

ognized that National Standard 1 is the most important because the fundamental purpose of the MSA is to preserve the continued viability of fishery resources (NAPA 2002).

The combined effect of National Standards 1 and 8 is that NOAA Fisheries often manages fisheries and other living marine resources by setting optimum yields that maximize economic and social benefits while simultaneously attempting to protect ecosystems. One method for striking a balance between these competing interests is the adoption of an MMA. For example, the establishment of an MMA with bottom trawl restrictions could assist in rebuilding groundfish stocks that are overfished by increasing the protection of spawning aggregations and fragile nursery habitats within the MMA. Simultaneously, managers may reduce the exploitation rate of depleted stocks through implementing harvest and effort controls, both inside and outside the MMA. Besides affecting exploitation rates and protecting critical life stages and their habitat, the adoption of an MMA could provide other important fishery benefits, such as reducing bycatch of nontarget species, ensuring against scientific and management uncertainty, conserving the social and genetic structure of natural populations, and creating reference areas for improved research opportunities and stock assessments (Coleman et al. 1996; Yoklavich 1998; NRC 2001)

In addition to implementing the MSA for sustainable fisheries, NOAA Fisheries is also responsible for implementing the Marine Mammal Protection Act (MMPA) of 1972 for cetaceans and pinnipeds under its jurisdiction. Furthermore, NOAA Fisheries is authorized through the Endangered Species Act (ESA) of 1973 to protect and manage a number of threatened and endangered marine turtles and anadromous fish stocks. Harm to marine mammals, also referred to as "take of marine mammals," and bycatch of threatened and endangered species in their critical habitat may be reduced by the establishment of MMAs that restrict certain types of fishing gear and activity within their boundaries.

Lastly, in May of 2000, President Clinton signed Executive Order (EO) 13158 on MPAs, which was endorsed by the Bush administration in 2001. The EO is a significant milestone for marine conservation that directs the Department of Commerce, Department of Interior, and other federal agencies to strengthen the conservation, protection, and sustainable use of U.S. ocean and coastal resources through the effective use of MPAs.

As illustrated, there are a host of legislative mandates and a presidential directive that authorize and encourage the creation of MMAs. The regional councils and NOAA Fisheries implement MMAs having multiple-use strategies for three main purposes: to rebuild and maintain sustainable fisheries, to promote the recovery of protected species, and to protect and maintain the health of coastal marine habitats for the conservation and sustainable use of living marine resources. When nesting MMA design within fisheries and protected resources management regimes, NOAA Fisheries must consider and balance the various legal mandates in support of these goals. Marine managed area practitioners are frequently called upon by stakeholder groups, which may be for or against the increased use of marine managed areas, to adhere to the following guidelines when designing and implementing new MMAs: (1) gain the advice and support of all stakeholder groups by adopting an open and transparent process during the evaluation and design phase of the MMA (Bohnsack 1997); (2) establish clear goals and targets for the management problem that is supposed to be addressed by the creation of the MMA (Murray et al. 1999; C. Syms and M. H. Carr, University of California, Santa Cruz, unpublished data prepared for a workshop sponsored by the Commission for Environmental Cooperation, 2001); (3) design and implement a monitoring plan to determine effects and performance of the MMA (Rudd 2002); (4) use science and scientific methods to inform decision-making during all phases of MMA design and management; and (5) apply adaptive management to modify and improve MMA function and effectiveness within the mix of management options (NRC 2001).

To understand and improve MMA function for our nation's fisheries and protected resources, it is imperative that we first characterize the nature and benefits of the current collection of MMAs adopted by NOAA Fisheries through a systematic review. The review should contain categories of information that readily allow for synthesis and comparative analyses across the diverse types of MMAs (Hockey and Branch 1997). In this paper, we characterize the collection of MMAs managed by NOAA Fisheries that are part of a national database. For a subset of the collection, we also determined whether the MMA's design included goals and monitoring practices. Specifically, we determined whether targets and timelines had been set for an MMA's goals because an MMA's performance is more accurately gauged against a defined "target" level of effectiveness and trajectory (Syms and Carr, unpublished). For example, the objective for which an MMA is estab-lished may be to increase spawning biomass of a commercially important stock that has a target level of effectiveness set at 10% increase in spawning biomass within 5 years. Finally, we evaluated the effectiveness of an MMA within the subset at achieving the goals associated with its design.

Methods

The MMA characterization and evaluation study was developed and conducted in the format of a survey that queried the collection of NOAA Fisheries MMAs in four broad categories of information: (1) basic attributes, including establishment date, regional distribution, legal mandates, and size; (2) level and duration of protection and type of gear restriction; (3) network type; and (4) goals, targets and timelines, monitoring, and effectiveness for a subset of the collection of MMAs.

The survey categories were developed with experts from program offices of NOAA Fisheries that manage or implement MMAs, including Sustainable Fisheries, Habitat and Conservation, Protected Resources, Science and Technology, and NOAA Enforcement. The survey was initially implemented in a test mode on a few select sites to determine how appropriate, practical, and user friendly the survey was in reviewing site attributes.

The answers to the survey were obtained through information sources such as the Code of Federal Regulations, Federal Register postings of proposed and final rules for designation of sites, fishery management plans, take reduction plans, recovery plans, stock assessment reports, journal publications, and other papers documenting the history, goals, and management of NOAA Fisheries MMAs. When necessary, we consulted regional science and management staff of NOAA Fisheries to evaluate the accuracy of the individual site review because of their familiarity with the history, context, and management of particular sites.

The terminology surrounding MMAs can be confusing. Restricted area, sanctuary, conservation zone, closed area, protection area, and conservation district are just a few terms used by NOAA Fisheries and the regional councils to describe areas with a range of protective measures and fishery restrictions afforded to a single species, complex of species, or habitat. Although NOAA Fisheries and the councils use many types of spatial management to accomplish a variety of goals, ranging from gear conflict reduction to harvest strategy management, we narrowed the focus of this study to those spatial measures that are currently part of the national database of MMAs. The database of MMAs is being developed by the National MPA Center as required by EO 13158. It will be a useful database for a broad range of user groups interested in spatial management of marine resources. A better understanding of the distribution, characteristics, and effectiveness of existing U.S. marine managed areas is expected as a result of analyses of the completed database.

The EO defines an MPA as "an area of the marine environment that has been reserved by Federal, State, territorial, tribal laws or regulations to provide lasting protection for part or all of the natural and cultural resources therein." The EO further defines "marine environment" as "those areas of coastal and ocean waters, the Great Lakes and their connecting waters, and submerged lands there under, over which the United States exercises jurisdiction, consistent with international law." The EO, however, does not define other important terms within the MPA definition, such as "area," "reserved," "regulation," "lasting protection," and "cultural resources." Given the wide variety of spatial management measures that could potentially fit these terms, NOAA and the Department of Interior have developed working definitions (Appendix A) to consistently interpret the definition of an MMA to be included in the national database.

A NOAA Fisheries national database team, using these working definitions, identified a collection of spatial management measures for inclusion to the national database of MMAs. Currently, all of these sites have been reviewed for their basic attributes, protection, and network categories. A subset of these sites, which was randomly selected, was additionally reviewed for information pertaining to goals, targets and timelines, monitoring, and effectiveness. It is our intent to review the complete database of MMAs using the full set of survey categories at a future date.

Information on the boundaries and spatial coordinates of the MMAs were obtained using the Code of Federal Regulations. Spatial maps of those data pertaining to regional location, and protections and restrictions of MMAs were generated using ArcView (Environmental Systems Research Institute, Inc. 2003) geographic information system software.

Results and Discussion

The national database of MMAs currently contains 67 NOAA Fisheries sites (Figure 1; Appendix B). All

these sites have been reviewed in terms of their basic attributes, type of protection, and network features. A subset of 32 sites (48%) was reviewed using all survey categories, including the category focused on effectiveness. The dynamic nature of fishery management ensures that the current collection of MMAs will change as new sites are included and existing sites are excluded. Site attributes will also change with revisions to goals and boundaries of current sites. The results presented in this paper are accurate to the best of our knowledge as of December 31, 2002.

Characterization of the Complete National Database of NOAA Fisheries MMAs

(1) Basic Attributes
Establishment date.—More than half of all MMAs managed by NOAA Fisheries (52%) have been established since 1996. A collection of nine sites was implemented in the 1980s, with the first MMA established in 1981 (Figure 2). The first two Habitat Areas of Particular Concern (HAPCs) were established in the Gulf of Mexico in 1984, a decade before adoption of the essential fish habitat (EFH) provisions in the 1996 Sustainable Fisheries Act. The HAPCs are a subset of EFH, and the councils are encouraged to give special consideration to adverse impacts of fishing on HAPCs. The 1990s were characterized by an increase in the number of protected sites designated each year. A large spike in number of sites in 1997 coincided with implementation of the Atlantic Large Whale Take Reduction Plan. These MMAs were created primarily for protection of western stocks of the Northern Atlantic right whale *Eubalaena glacialis*. The Harbor Porpoise Take Reduction Plan accounted for six MMAs established in waters of the Atlantic EEZ in 1999. Finally, the early years of the new millennium marked the establishment of spatial management for protection of juvenile swordfish *Xiphias gladius*, Atlantic sharks, and other highly migratory species caught with pelagic longline gear.

Regional distribution.—Marine managed areas are most concentrated in the region of the Northeast Fishery Management Council (19 sites), comprising 28% of NOAA Fisheries MMA database. The North Pacific (14 sites) and the Gulf of Mexico (10 sites) contain the next highest number of MMAs and comprise 21% and 15% of the total, respectively. The other four regions have fewer MMAs, ranging from 6 sites in the Mid-Atlantic to 3 sites in the Pacific (Figure 1). The councils did not propose all MMAs. For example,

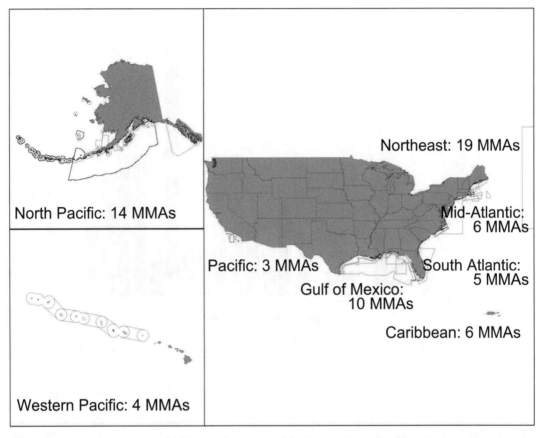

Figure 1.—The 67 marine managed areas (MMAs) managed by National Oceanic and Atmospheric Administration Fisheries, according to regional fishery management council representation.

spatial protective measures for juvenile swordfish and large Atlantic whales were implemented through the Highly Migratory Species and Protected Resources offices of NOAA Fisheries, respectively. In addition, the Atlantic States Marine Fisheries Commission designated two sites off the Mid-Atlantic coast: the Carl N. Shuster, Jr., Horseshoe Crab Reserve and the Flynet Closure.

Legal mandates.—Twenty-five MMAs (38%) have been designated by NOAA Fisheries to protect a threatened or endangered species. While the ESA and MMPA were frequently invoked to justify establishing these MMAs, the MSA was used to promulgate the protective measures. These measures are regulations that affect fishing activity and gear types allowed within the boundaries of the site. The remaining 42 MMAs (62%) were designed to protect fish stocks and their habitats and were justified and promulgated through the MSA alone.

Size.—A majority of sites (69%) are larger than 1,000 km², with a minority of sites (6%) smaller than 10 km². The seven no-take MMAs (marine reserves) are located in the Gulf of Mexico and Caribbean and range in size from less than 10 km² to 518 km², which is somewhat larger than the worldwide median size of 4 km² (Halpern 2003). The aerial coverage of the seven no-take MMAs makes up less than 0.05% of all NOAA Fisheries MMAs.

(2) Protection

Level of protection.—While general maps illustrating the location and dimensions of NOAA Fisheries MMAs may leave the impression that large expanses have spatial bans on fishing (Figure 1), in reality, fishing is only prohibited in small areas of the EEZ. The number of MMAs that prohibit gear types that impact the bottom habitat (e.g., bottom trawls, dredges, or bottom tending gear) is similar to the number that restrict gear types with effects to the surface and

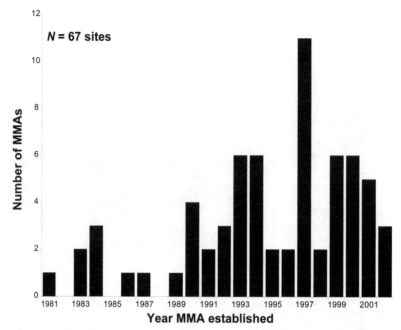

Figure 2.—The number of marine managed areas (MMAs) managed by National Oceanic and Atmospheric Administration Fisheries, according to year of establishment.

midwater habitats (43% and 46%, respectively; Figure 3). Few sites are considered as no-take MMAs (11%), where all forms of recreational and commercial fishing activity are prohibited (Figure 4 shows a close-up view). The first no-take MMA was established in 1991, as a 3-month seasonal closure for protection of spawning aggregations of red hind *Epinephelus guttatus* at Hind Bank, U.S. Virgin Islands. The other six no-take MMAs are also within Caribbean waters and similarly designed to protect coral reef habitat and spawning aggregations of coral reef fishes.

Duration.—A majority of sites (72%) are managed to provide protections year-round. The remaining sites are managed with seasonal protections. Of the seven no-take MMAs, five are managed with seasonal and two with year-round protections.

(3) Network

The EO on MPAs directs federal agencies to work closely with state, local, and nongovernmental partners to create a comprehensive network of MPAs "representing diverse U.S. marine ecosystems, and the Nation's natural and cultural resources." The National Academy of Sciences (NRC 2001) defines an MPA network as "an array of sites chosen for their complementarity and ability to support each other based on connectivity . . . through dispersal of either adults or larvae to ensure persistence and maintenance of genetic diversity for the resident protected species." Using this definition, 38% of NOAA Fisheries MMAs were designed and coestablished to support connectivity through either adult or larval dispersal, often for a single species. In contrast, 88% of the MMAs are comanaged with other regional MMAs through a programmatic system, such as a fishery management plan or take reduction plan for marine mammals. Because such a large proportion of NOAA Fisheries MMAs are already implemented in tandem with other regional MMAs, increased ecosystem benefits and function may be gained by locating any additional MMAs based on their ability, in combination with existing sites, to support long-term stability of resident communities through dispersal corridors of larvae and adults. However, our understanding of the role of connectivity for MMA design or for marine ecosystem function is limited (Jones et al. 1999; Swearer et al. 1999; NRC 2001; Palumbi 2003a, 2003b) and awaits future investigations in the fields of oceanography, population genetics, otolith microchemistry, and larval and adult marine species behavior.

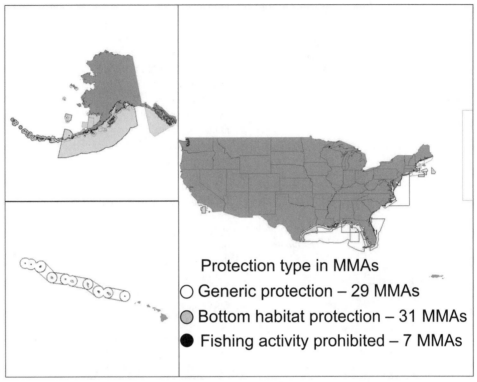

Figure 3.—Level of protection and types of fishing gear restrictions in 67 marine managed areas (MMAs) managed by National Oceanic and Atmospheric Administration Fisheries.

Subset of 32 MMAs Reviewed for Goals, Targets and Timelines, Monitoring Practices, and Effectiveness

Additionally, a subset of 32 randomly selected sites (48%) was evaluated in terms of goals, targets and timelines, monitoring practices, and effectiveness at achieving the objectives that were described when the MMA was established.

Goals

The 32 MMAs within the subset had specific goals at the time of site designation. For many of the sites (75%), changes were made to the goals, which were reflected in the amendments or framework adjustments made later to the management plans that support the MMA. For example, no-take regulations at Hind Bank were extended to year-round in 1999 after 8 years of seasonal protection for red hind only had proven to be successful (Beets and Friedlander 1999). The current year-round restrictions at Hind Bank are to protect spawning aggregations of other fish species and important marine resources, such as corals. Eighty-one percent of the subset of MMAs was established in concert with other sites sharing the same goal. For example, three closures in the Northwest Atlantic were established at the same time to reduce fishing mortality and to rebuild stocks of New England groundfish, such as Atlantic cod *Gadus morhua*, haddock *Melanogrammus aeglefinus*, and yellowtail flounder *Limanda ferruginea*. Only 5 of the 32 MMAs were established with exclusive goals and as solitary sites.

Targets and Timelines

Targets and timelines are essential for evaluation of an MMA's effectiveness (Vanderklift and Ward 2000; Syms and Carr, unpublished) and were associated with the goals of 20 MMAs (63%) within the subset. The targets and timelines were frequently part of rebuilding plans for overfished stocks or efforts to reduce or maintain minimum levels of catch of non-target species. Overfished stocks are now required by the SFA amendment to the MSA to include targets, recovery periods, and trajectories (Powers 1999).

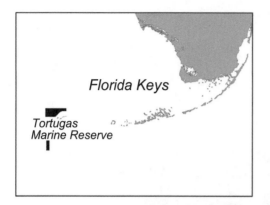

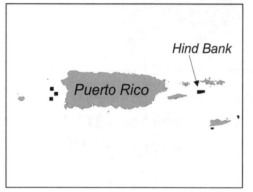

● Fishing activity prohibited - 7 MMAs

573 km² in year-round no-take MMAs

111 km² in seasonal no-take MMAs

685 km² total aerial coverage

Figure 4.—The location and size of the seven no-take marine managed areas (MMAs) managed by National Oceanic and Atmospheric Administration Fisheries.

Monitoring

Monitoring to evaluate management performance is conducted at each of the 32 MMAs, and stock assessments of either target or nontarget species are conducted at 28 sites (87%). As stated under the "Network" and "Goals" sections of this paper, MMAs managed by NOAA Fisheries are frequently designated as part of programmatic systems that include harvest and effort controls outside the MMAs. The use of stock assessments provides valuable information on the performance of those measures taken as a whole. It does not, however, provide an accurate performance measure of the MMA exclusively. To evaluate MMA ef-

fectiveness conclusively, monitoring should include baseline surveys for both community-level and population-level effects inside the MMA, before and after establishment, with a statistically appropriate number of control sites outside the MMA depending on levels of variation within the variable of interest (NRC 2001). Responses in species abundance or habitat variables were measured in 10 sites (32%). The effects and magnitude of responses were determined with baseline or benchmark studies conducted at the time of MMA establishment.

Effectiveness

Effectiveness was measured against the targets and timelines associated with an MMA's goal. Greater than 80% of NOAA Fisheries MMAs are managed with other regional MMAs through programmatic systems that have other control measures directed at reducing fishing mortality for target and nontarget species. Given the difficulty in measuring effectiveness of an MMA as a unique site or management measure, 50% of the MMAs (16 out of 32 sites) were determined to be part of an effective program (i.e., the management goals connected with an MMA were partly or wholly attained). For example, the Northeast Multispecies Closures are considered a major success. The goals of these MMAs, in concert with fishing controls outside the closures, were to rebuild overfished groundfish stocks (Atlantic cod, haddock, and yellowtail flounder) over 5–7 years by reducing fishing mortality. Biomass trends in the key groundfish species have all become positive after the closures were established in 1994. However, effectiveness of any one individual closure is difficult to assess because fishing controls were established outside the MMA simultaneously with the MMAs. Also, increases in groundfish densities were measured through standard stock assessments, not through comparative assessments of population parameters using baseline studies at individual MMAs. An added conservation benefit inside the Multispecies Closures was that populations of sea scallops *Placopecten magellanicus* on Georges Bank (41°N, 68°W) increased fourteenfold over those outside the closures (Murawski et al. 2000).

There have been increases in target population levels and enhanced habitat structure at 6 of the 10 MMAs with baseline information. The ability to detect and evaluate MMA effectiveness was often obscured or rendered infeasible due to various combinations of factors: management targets were not stated or unrealistic, the site had not been protected long enough to detect significant changes, lack of enforcement and compliance, and confounding environmental conditions. For example, the Experimental Oculina Research Reserve (EORR) off the eastern Florida coast was established with the goals to protect corals from destructive fishing practices and protect grouper spawning aggregations of gag *Mycteroperca microlepis* and scamp *M. phenax*. Inadequate enforcement and compliance rendered the EORR ineffective at protecting corals and spawning aggregations, and lack of baseline habitat assessments led to misguided goal setting prior to the MMA's designation. Coral habitat was severely damaged prior to the designation of the EORR, so instead of "coral protection" the goal for establishing the EORR should have been "coral recovery" (Koenig et al. 2000; C. Grimes, NMFS, Santa Cruz Laboratory, personal communication). Recent studies, however, indicate increases in grouper numbers and sizes (Koenig 2001).

There is a growing body of research demonstrating that the use of MMAs can protect habitats, spawning stocks, and endangered species. However, few managers assess the magnitude of these effects and determine whether the effects are tied to the MMA's intended objectives (Rowley 1994). The adoption of realistic goals and targets that include methods to measure effectiveness is crucial to the successful management of MMAs. Increasingly, resource management agencies are required to provide clear statements of the outcome they expect to achieve by creating MMAs and about how they will demonstrate their effectiveness over time (Syms and Carr, unpublished). The Madison Swanson and Steamboat Lumps MMAs in the Gulf of Mexico were created with "sunset clauses" that require demonstrable conservation results for troubled stocks of fish after a fixed period of operation. Evaluation of MMAs with respect to their goals and targets is essential to understanding how the MMA is functioning and if the site should be altered to improve effectiveness or eliminated if proven to be completely ineffective. This information ultimately enlightens regional and national management, enforcement agents, the scientific community, and others through an adaptive management approach.

Ecosystem-Based Management

The primary responsibility for managing and conserving our nation's domestic fisheries, marine mammals, endangered marine and anadromous species, and marine habitats in the context of fishery management belongs to NOAA Fisheries. The application of MMAs may provide a unified and ecosystem-based approach to managing these marine resources compared to the traditional management approaches focused on single species. More importantly, current marine resource management is plagued by uncertainty derived from our incomplete understanding of how marine stocks interact with other species, how environmental variability affects these interactions, and how human actions, such as fishing, influence ecosystem health. Given these uncertainties, the adoption of an MMA within the mix of management tools provides managers with a precautionary approach to conserving resilient ecosystems and economically viable fisheries that depend on them (NRC 1999; NAPA 2002). The NOAA Fisheries MMAs have been embedded within programs and plans that are directed toward single-species or mixed-species management and not for ecosystem benefits. There are multiple examples of MMAs with single-species focus that share similar areas and boundaries with MMAs designed to protect or manage other species (Figures 1, 3). Marine resources within these areas of overlap may benefit from an ecosystem-based management approach that combines the individual goals of separate MMAs into one goal with multiple targets. This approach applies not only to MMAs within NOAA Fisheries but also to MMAs across federal and state programs. An advantage to this approach is the integration and cost sharing of educational programs, scientific research, monitoring plans, and enforcement. Integrating MMAs managed by various programs may also potentially improve compliance by a beleaguered fishing community that has been required to comply with a myriad of fishing regulations and a patchwork of spatial restrictions.

Acknowledgments

We acknowledge the assistance provided by the NOAA Fisheries MMA inventory team: Ralph Lopez, Tanya Dobrzynski, Cheri McCarty, Robert Gorrell, and Robert Brock. Geographic information system support was provided by Mark Amend, John Calgiano, and Julia Brownlee. Mark Gleason assisted with data collection. Mary Yoklavich, Churchill Grimes, and Richard Neal provided insightful editorial comments. We appreciate the kind encouragement and support from Rebecca Lent.

References

Agardy, T., P. Bridgewater, M. P. Crosby, J. Day, P. K. Dayton, R. Kenchington, D. Laffoley, P. McConney, R. S. Murray, J. E. Parks, and L. Peau. 2003. Dangerous targets? Unresolved issues and ideological clashes around marine protected areas. Aquatic Conservation: Marine and Freshwater Ecosystems 13:353–367.

Beets, J., and A. Friedlander. 1999. Evaluation of a conservation strategy: a spawning aggregation closure for red hind, *Epinephelu guttatus*, in the U. S. Virgin Islands. Environmental Biology of Fishes 55:91–98.

Bohnsack, J. A. 1997. Consensus development and the use of marine reserves in the Florida Keys, USA. *In* Proceedings of the 8th International Coral Reef Symposium, volume 2. Smithsonian Tropical Research Institute, Balboa, Panama.

Bradshaw, C., L. O. Veale, A. S. Hill, and A. R. Brand. 1999. The effect of scallop-dredging on Irish Sea benthos: experiments using a closed area. Hydrobiologia 465:129–138.

Coleman, F. C., C. C. Koenig, and L. A. Collins. 1996. Reproductive styles of shallow-water groupers (*Pices: Serranidae*) in the eastern Gulf of Mexico and the consequences of fishing spawning aggregations. Environmental Biology of Fishes 47:129–141.

Environmental Systems Research Institute, Inc. (ESRI). 2003. ArcView, version 8.3, geographic information system software. ESRI, Redlands, California.

Halpern, B. S. 2003. The impact of marine reserves: do reserves work and does reserve size matter? Ecological Applications 13(1):117–137.

Hockey, P. A. R., and G. M. Branch. 1997. Criteria, objectives and methodology for evaluating marine protected areas in South Africa. South African Journal of Marine Science 18:369–383.

Jones, G. P., M. J. Milicich, M. J. Emslie, and C. Lunow. 1999. Self-recruitment in a coral reef fish population. Nature (London) 402:802–804.

Koenig, C. C. 2001. *Oculina* banks: habitat, fish populations, restoration, and enforcement. Report to the South Atlantic Fishery Management Council, Charleston, South Carolina.

Koenig, C. C., F. C. Coleman, C. B. Grimes, G. R. Fitzhugh, K. M. Scanlon, C. T. Gledhill, and M. Grace. 2000. Protection of fish spawning habitat for the conservation of warm-temperate reef-fish fisheries of shelf-edge reefs of Florida. Bulletin of Marine Science 66(3):593–616.

Lubchenco, J., S. Palumbi, S. D. Gaines, and S. Andelman. 2003. Plugging the hole in the ocean: the emerging science of marine reserves. Ecological Applications 13:3–8.

McClanahan, T. R., and S. Mangi. 2001. The effect of a closed area and beach seine exclusion on coral reef fish catches. Fisheries Management and Ecology 8(2):107–122.

Murawski, S. A., R. Brown, H.-L. Lai, P. J. Rago, and L. Hendrickson. 2000. Large-scale closed areas as a fishery management tool in temperate marine systems: the Georges Bank experience. Bulletin of Marine Science 66:775–798.

Murray, S. N., R. F. Ambrose, J. A. Bohnsack, L. W. Botsford, M. H. Carr, G. E. Davis, P. K. Dayton, D. Gotshall, D. R. Gunderson, M. A. Hixon, J. Lubchenco, M. Mangel, A. MacCall, D. A. McArdle, J. C. Ogden, J. Roughgarden, R. M. Starr, M. J. Tegner, and M. M. Yoklavich. 1999. No-take reserve networks: sustaining fishery populations and marine ecosystems. Fisheries 24(11):11–25.

NAPA (National Academy of Public Administration). 2002. Courts congress and constituencies: managing fisheries by default. National Academy Press, Washington, D.C.

NMFS (National Marine Fisheries Service). 1999. Ecosystem-based fishery management. NOAA Technical Memorandum NMFS-F/SPO-33.

NRC (National Research Council). 1999. Sustaining marine fisheries. National Academy Press, Washington, D.C.

NRC (National Research Council). 2001. Marine protected areas: tools for sustaining ocean ecosystems. National Academy Press, Washington, D.C.

Palumbi, S. 2003a. Population genetics, demographic connectivity, and the design of marine reserves. Ecological Applications 13:146–158.

Palumbi, S. 2003b. Marine reserves: a tool for ecosystem management and conservation. Pew Oceans Commission, Washington, D.C.

Powers, J. E. 1999. Requirements for recovering fish stocks. NOAA Technical Memorandum NMFS-F/SPO-40.

Roberts, C. M., J. A. Bohnsack, F. Gell, J. P. Hawkins, and R. Goodridge. 2001. Effects of marine reserves on adjacent fisheries. Science 294:1920–1923.

Rowley, R. J. 1994. Case studies and reviews: marine reserves in fisheries management. Aquatic Conservation: Marine and Freshwater Ecosystems 4:233–254.

Rudd, A. M. 2002. An institutional framework for designing and monitoring ecosystem-based fisheries management policy experiments. Fisheries and Oceans Canada, Policy and Economics Branch, Maritimes Region, Dartmouth, Nova Scotia.

Swearer, S. E., J. Caselle, D. Lea, and R. R. Warner. 1999. Larval retention and recruitment in an island population of coral-reef fish. Nature (London) 402:799–802.

Vanderklift, M. A., and T. J. Ward. 2000. Using biological survey data when selecting marine protected areas: an operational framework and associated risks. Pacific Conservation Biology 6:152–161.

Yoklavich, M., editor. 1998. Marine harvest refugia for west coast rockfish: a workshop. NOAA Technical Memorandum NOAA-TM-NMFS-SWFSC-255:142–148.

Appendix A

Executive Order 13158 defines an MPA as "an *area* of the *marine* environment that has been *reserved* by Federal, State, territorial, tribal laws or regulations to provide *lasting protection* for part or all of the natural and cultural resources therein." Working definitions for italicized terms clarify those area-based management measures designated by NOAA Fisheries that qualify for inclusion to the national database of MMAs (*http://mpa.gov/inventory/criteria.html#table*). We summarize here the working definitions of those terms.

Area

Must have defined geographical boundaries. This working definition excludes generic broad-based resource management without specific locations and species-specific conservation regulations that are not focused on a defined geographic area. For example, this working definition excludes areas with fishing regulations that cover the EEZ of a particular fishery management region.

Marine

Must encompass: (a) an area of the ocean or coastal waters, including intertidal areas, bays and estuaries; or (b) an area of the Great Lakes or their connecting waters. The term "intertidal" is understood to mean the shore zone between mean low water and the mean high water mark. For purposes of the MMA database, the term "estuary" is defined as "part of a river or stream or other body of water having unimpaired connection with the open sea, where salts derived from sea water measure no less than 0.5 ppt during the period of average annual low flow." This working definition excludes, for example, river areas above the fall line of a dam.

Reserved

Must be established by and currently subject to some form of federal, state, territorial, local, or tribal law or regulation. This working criterion excludes, for example, privately created or maintained marine sites. All NOAA Fisheries MMAs fit within this working definition because they are promulgated through federal laws.

Lasting

Must provide protection for a minimum of 3 months continuously within a year at the same location. The site must also be established with the expectation of at least 2 years continuous protection. For example, areas with sunset clauses must provide a minimum of 2 years of continuous protection, and must have a specific mechanism to renew protection at the expiration of the sunset period. This working definition excludes rolling closures in place less than 3 months or areas protected only by emergency fishery regulation under the MSA, which expires after 180 days.

Protection

Must have existing laws or regulations that are designed and applied to afford the site with increased protection for part or all of the natural and cultural resources therein beyond any general protections that may apply outside the site. This working criterion excludes, for example, areas closed to avoid fishing gear conflicts and areas established to limit fisheries by quota management.

Appendix B

A table of the 67 MMAs that are managed by NOAA Fisheries and currently in the national database. The 32 sites with an asterisk were randomly selected to be analyzed for goals, targets and timelines, and effectiveness.

ALWTRP = Atlantic Right Whale Take Reduction Plan; HPTRT = Harbor Porpoise Take Reduction Plan; HAPC = Habitat Area of Particular Concern; and NWHI = North Western Hawaiian Islands.

Table B.1.—The 67 marine managed areas that are managed by NOAA Fisheries.

Marine managed area	Date	Purpose	Duration	Restrictions or requirements
			Northeast region: 19 sites	
*Northeast Multispecies Closed Area I	1994	Stock, spawning habitat protection	Year-round	Gear capable of multispecies catch prohibited.
*Northeast Multispecies Closed Area II	1994	Stock, spawning habitat protection	Year-round	Gear capable of multispecies catch prohibited.
Northeast Multispecies Nantucket Lightship Closed Area	1994	Stock, spawning habitat protection	Year-round	Gear capable of multispecies catch prohibited.
*Great South Channel Northern Right Whale Critical Habitat Lobster Waters	1997	ALWTRP	Year-round	9 months weaklinks required; 3 months no lobster traps/fish pots.
Great South Channel Restricted Gill Net Area	1997	ALWTRP	Year-round	9 months no anchored gill nets; 3 months no gill nets.
*Great South Channel Sliver Area (gill nets)	1997	ALWTRP	Year-round	Anchored gill nets allowed but only with weaklinks.
*Cape Cod Northern Right Whale Critical Habitat (lobster waters/gill nets)	1997	ALWTRP	Year-round	Weaklinks required.
*Northern Inshore State Lobster Waters Area	1997	ALWTRP	Year-round	Must abide by at least 1 option from lobster take reduction list.
*Northern Nearshore Lobster Waters Area	1997	ALWTRP	Year-round	Must have weaklinks for lobster traps/fish pots.
Stellwagen Bank/Jeffreys Ledge Restricted Area (lobster/gill net)	1997	ALWTRP	Year-round	Must have weaklinks for lobster traps/fish pots; anchored gill nets.
*Offshore Lobster Waters	1997	ALWTRP	Year-round	Must have weaklinks for lobster traps/fish pots.
*Western Gulf of Maine Area Closure	1998	Depleted stock protection	Year-round	Gear capable of multispecies catch prohibited.
*Cape Cod South Closure Area	1999	Cod; protection of harbor porpoise Phocoena phocoena	Dec–May	Gill nets with pingers required; 1 month closure to gill nets.
*Massachusetts Bay Closure Area	1999	Cod protection; HPTRP	Dec–May	Gillnets with pingers required; 1 month closure to gill nets.
*Mid-Coast Closure Area	1999	Cod protection; HPTRP	Sep–May	Gill nets with pingers required.
Offshore Closure Area	1999	Cod protection; HPTRP	Nov–May	Gill nets with pingers required; 1 month closure to gill nets.
Northeast Distant Closed Area	2001	Protection of juvenile swordfish and sea turtles Caretta caretta and Dermochelys coriacea	Year-round	No pelagic longline gear.
Cashes Ledge Closure Area	2002	Cod/groundfish protection	Year-round	Gear capable of multispecies catch prohibited.
Special Area Management East	2002	ALWTRP	May–July	Must use low risk weaklinks on lobster traps; anchored gill nets.

Table B.1.—Continued.

Marine managed area	Date	Purpose	Duration	Restrictions or requirements
		Mid-Atlantic region: 6 sites		
*Flynet Closure	1997	Protection of weakfish *Cynoscion regalis*	Year-round	No use of flynets.
*Mid-Atlantic Coastal Waters	1997	ALWTRP	Dec–Mar	Must use weaklinks for anchored gill nets; restrictive night gill nets.
Southern Nearshore Lobster Waters	1997	ALWTRP	Year-round	Must have weaklinks for lobster traps/fish pots.
*New Jersey Waters Closure	1999	Cod protection; HPTRP	Jan–Apr	Must use modified gill nets; closed to gill nets Apr 1–20.
Southern Mid-Atlantic Waters Closure Area	1999	Cod protection; HPTRP	Feb–Apr	Must use modified gill nets; 1 month no large mesh.
Carl N. Shuster, Jr., Horseshoe Crab Reserve	2001	Protection of horseshoe crab *Limulus polyphemus*	Year-round	Prohibits horseshoe crab takes.
		South Atlantic region: 5 sites		
Oculina Bank HAPC	1984	Coral protection	Year-round	No fishing for snapper or grouper; no anchoring.
*Experimental Oculina Research Reserve	1994	Coral and grouper spawning protection	Year-round	No fishing for snapper or grouper; no trawling or anchoring.
*Desoto Canyon Closed Area	2000	Juvenile swordfish protection	Year-round	No pelagic longlining; no use of live bait.
Charleston Bump	2001	Juvenile swordfish protection	Feb–Apr	No pelagic longlining.
Southeastern U.S Restricted Area	2002	ALWTRP	Nov 15–Mar	Restrictive gillnet gear at night; observers may be required.
		Gulf of Mexico region: 10 sites		
Tortugas Shrimp Sanctuary	1981	Protection of juvenile shrimp *Penaeus duorarum*	Year-round	Commercial trawling prohibited.
*Florida Middle Ground HAPC	1984	Fragile coral protection	Year-round	No use of bottom gear.
*West and East Flower Garden Banks HAPC	1984	Fragile coral protection	Year-round	No bottom gear.
Longline/Buoy Gear Area Closure	1990	Reef fish; habitat protection	Year-round	Longline and buoy gear restrictions; bag limits.
*Reef Fish Stressed Area	1990	Reef fish protection	Year-round	No fish traps, roller gear, or use of powerheads.
Alabama Special Management Zone	1994	Reef fish protection on artificial reefs	Year-round	Commercial and recreational fish gear restricted.
*Madison Swanson Spawning Site (in review)	2000	Spawning protection for gag *Mycteroperca microlepis*	Year-round	Only highly migratory species trolling allowed.

Table B.1.— Continued.

Marine managed area	Date	Purpose	Duration	Restrictions or requirements
*Steamboat Lumps Spawning Site (in review)	2000	Gag spawning protection	Year-round	Only highly migratory species trolling allowed.
Tortugas Marine Reserves	2000	Fishery and habitat protection	Year-round	No extractions, no anchoring.
East Florida Coast Closed Area	2001	Juvenile swordfish protection	Year-round	Pelagic longline prohibited.
Caribbean region: 6 sites				
*Hind Bank Marine Conservation District	1991	Habitat protection	Year-round	No extractions; no anchoring.
Mutton Snapper Spawning Aggregations	1993	Spawning protection	Mar–Jun	No-take closure to all species; fishing is prohibited.
Red Hind Spawning Aggregations (Lang Bank)	1993	Spawning protection	Dec–Feb	No-take closure to all species; fishing is prohibited.
Red Hind Spawning Aggregations (Tourmaline Bank)	1993	Spawning protection	Dec–Feb	No-take closure to all species; fishing is prohibited.
Red Hind Spawning Aggregations (Abrir La Sierra Bank)	1996	Spawning protection	Dec–Feb	No-take closure to all species; fishing is prohibited.
Red Hind Spawning Aggregations- (Bajo de Cico)	1996	Spawning protection	Dec–Feb	No-take closure to all species; fishing is prohibited.
Alaska region:14 sites				
Stellar Sea Lion Protection Area Bering Sea/ Aleutian Islands (in review)	1990	Protection of Stellar sea lions *Eumetopias jubatus*	Year-round	Trawling prohibited.
Walrus Island	1990	Protection of walrus *Odobenus rosmarus*	Apr–Sep	No access.
*Stellar Sea Lion Protection Area Aleutian Island Subarea (in review)	1992	Stellar sea lion protection	Year-round	Trawling prohibited.
*Stellar Sea Lion Protection Area Bering Sea Subarea Bogolof District (in review)	1992	Stellar sea lion protection	Year-round	No directed fishing for pollock *Pollachius virens*, cod, Atka mackerel *Pleurogrammus monopterygius*.
Kodiak Red King Crab Closure Area, Type I	1992	Promote rebuilding of red king crab *Paralithodes camtschaticus*	Year-round	No trawl gear, other than pelagic gear.
Kodiak Red King Crab Closure Area, Type II	1993	Promote red king crab rebuilding	Feb 15–Jun 15	No trawl gear, other than pelagic gear.
Pribilof Islands Area Habitat Conservation Zone	1994	Habitat protection for blue king crab *P. platypus*	Year-round	No trawl gear, other than pelagic gear.
*Red King Crab Savings Area	1995	Red king crab habitat protection	Year-round	No trawl gear, other than pelagic gear.
Nearshore Bristol Bay Closure Area	1995	Red king crab habitat protection	Year-round	No trawl gear.

Table B.1.—Continued.

Marine managed area	Date	Purpose	Duration	Restrictions or requirements
*Area 512	1987	Red king crab habitat protection	Year-round	No trawl gear.
Southeast Alaska Outside District	1998	Deepwater coral protection	Year-round	Commercial groundfishing prohibited.
*Area 516	1989	Red king crab habitat protection	Mar–Jun	No trawl gear.
*Sitka Pinnacles	2000	Ecosystem protection	Year-round	No anchoring; no bottom fishing.
*Stellar Sea Lion Protection Area Gulf of Alaska (in review)	2000	Stellar sea lion protection	Year-round	Trawling prohibited.
Pacific region: 3 sites				
Pacific Whiting Columbia River Salmon Conservation Zone	1993	Salmon protection	Year-round	No trawling for Pacific hake *Merluccius productus* (also known as Pacific whiting).
*Pacific Whiting Klamath River Salmon Conservation Zone	1993	Salmon protection	Year-round	No trawling for Pacific hake.
Cowcod Conservation Area	2001	Protection of cowcod *Sebastes levis*; bycatch reduction	Year-round	All groundfish catch prohibited.
Western Pacific region: 4 sites				
Lobster Closed Area	1983	Protection of monk seal *Monachus shauinslandt*	Year-round	No lobster fishing within 10 fathom line in NWHI.
Westpac Bed	1983	Precious coral protection	Year-round	Coral fishing prohibited.
Hancock Seamount	1986	Protection of pelagic armorhead *Pseudopentaceros richardsoni*	Year-round	No fishing on seamount for groundfish or bottomfish.
*Longline Protected Species Zone (3–50 nm)	1991	Monk seal protection	Year-round	Longline prohibited in NWHI.

American Fisheries Society Symposium 42:105–122, 2004

The Importance of Retention Zones in the Dispersal of Larvae

JOHN LARGIER[1]

*Scripps Institution of Oceanography, University of California,
San Diego, California 92093-0209, USA*

Abstract.—Dispersal of young life stages is often the predominant connection between spatially separated populations. Thus, one can expect that dispersal plays a critical role in determining the effectiveness of aquatic protected areas as fishery management tools. This piece focuses on two primary aspects of dispersal that result in dispersal not being a simple product of mean flow and time in the plankton: flow features (retention) and behavior (swimming). Recently, increasing attention is being given to retention zones—regions where currents act to retain propagules (regions with limited exchange with surrounding waters). Both observations and models of meroplankton and of recruitment suggest the importance of retention zones in population dynamics and that these may be good sites for aquatic protected areas. Retention zones may occur in a variety of environments, each exhibiting specific circulation patterns, time scales, and opportunities for import or export. In addition, each type of retention zone offers a different habitat type and different exposure to external influences. The value of these retention zones to fisheries depends on the population of interest and the time scale, size, location, and "retentiveness" of the retention zone. Typically, retention zones are not "closed"—they are not isolated from the main flow. Nonzero exchange between retention zones and surrounding waters is critical for both dispersal of propagules and the nature of the habitat. Swimming can have a large influence on retention processes and how retention zones are used by early life stages. Even very weak behavior can exert control on export or import rates and timing. Age-specific behavior may be critical to populations using these retention zones to enhance local recruitment and spillover effects at the same time.

Introduction

Dispersal of young life stages plays a critical role in metapopulation dynamics, specifically in the connectivity between spatially separated populations with sedentary adults (Palumbi 2001). Here, I explore the role of retention zones in shaping the dispersal of larvae of coastal fish and invertebrate populations with sedentary adults—populations for which spatial management strategies such as aquatic protected areas (APAs) can be very effective. It is argued that these oceanographic structures should be a key factor in the design of APAs. In addition to being a primary influence on the dispersal of planktonic larvae, retention zones may also provide critical habitat for larvae and

juveniles (Warner et al. 2000). Further, larval swimming may be important in the movement of larvae into and out of retention zones, thus controlling the effect of coastal retention zones on dispersal of the population.

Retentive regions are characterized by a slower through-flow of water and water-borne material relative to adjacent regions (e.g., Penven et al. 2000). In other words, the residence time is increased and the alongshore or offshore transport of propagules is reduced. Although at times subtle in form, retention zones play a key role in the probabilistic world of larval dispersal, and thus, the location and nature of retention zones should be a primary factor in determining the location, size, and spacing of aquatic protected areas. Retention has been discussed previously by Bakun (1996) and Sinclair (1988), among others. The underlying concept is that there are regions in which larvae may be retained and that this diminishes larval

[1] E-mail: jlargier@ucsd.edu

mortality (and thus enhances recruitment) as "larval loss" from appropriate areas is the dominant mortality term for the early life stages of many populations ("Hjort's second hypothesis," Sinclair 1988). In other words, "retention" means retaining larvae in regions where they can subsequently recruit to adult habitat. This raises the possibility that, for some populations, recruitment will be dominated by larvae that have been retained for some part of their planktonic period ("Dooley's hypothesis," Bakun 1996) and, thus, that retention zones are the foundation of successful recruitment for these populations. While Sinclair (1988) explores the spatial structure of retention in the context of the population richness of Atlantic herring *Clupea harengus*, this paper is directed at the structure of coastal populations with sedentary adults and, thus, addresses the importance of these coastal sites in shaping the alongshore dispersal outcome (i.e., in determining the dispersal kernel). Further, the focus of this paper is on the retention zones and the importance of swimming behavior in the use of these regions, with only passing comment on the associated metapopulation dynamics and structure.

In this comment, attention is primarily directed at coastal marine protected areas (MPAs)—a major subset of aquatic protected areas. It is now generally accepted that some coastal waters need to be set aside in the interests of conservation or fisheries management. Here, attention is given to the fisheries benefits of MPAs, obtained through protecting the population from extinction or massive decline and through providing spillover of enhanced productivity from MPAs. This discussion addresses flow patterns and the transport aspects of population connectivity. While little attention is given to population dynamics and the distribution of adults or adult habitat, the intention is not to ignore these effects but, rather, to develop our understanding and appreciation of the complexity of dispersal patterns so that these can be better represented in metapopulation models and paradigms (e.g., Gaines et al. 2003) and better integrated into policy and management decisions.

The role of retention zones in metapopulation dynamics depends on the relative length scales of the retention zone, the population domain, and the dispersal of individuals within this domain (Gaines et al. 2003; Largier 2003). The importance of retention zones is most notable in the case of populations with sedentary adults, where dispersal is primarily or entirely through dispersion of planktonic propagules (e.g., reef fish but also benthic invertebrates). This has been a central question in marine ecology over the last few

decades (Botsford et al. 1994; Palumbi 2001). While larval dispersal was initially neglected, and then exaggerated, present discussion recognizes that it may be over short or long distances (Cowen et al. 2000; Largier 2003; Shanks et al. 2003) and attention is turning toward how specific aspects of spawning or larval behavior affect specific dispersal outcomes in a given flow field and habitat distribution. For these sedentary populations, the only way that fishery benefits will be realized outside the MPA is through larval dispersal and recruitment beyond the borders of the no-take MPA (i.e., "spillover").

While it is a truism that every population sustains itself (and therefore must retain sufficient larvae), it is not clear at what scale subsets of the population (portions of the habitat) are self-sustaining. Further, there may be specific locations in which a subset of the population is self-sustaining (i.e., local recruitment is sufficient). These regions are commonly known as "sources" (regions that cannot sustain themselves are known as "sinks"). However, unless a source region also exports enough propagules to affect recruitment in other regions, its role as a source is insignificant. Hence, a true source region must exhibit significant export of larvae in addition to adequate local recruitment (i.e., this subpopulation must have elements of being "open" and elements of being "closed" at the same time; Armsworth 2002), suggesting that the dichotomy of "open" versus "closed" oversimplifies the problem (Cowen et al. 2000). Clearly, true sources would be the best candidates for MPA designation. Further, Armsworth (2002) notes that metapopulation persistence requires local source regions, and Largier (2003) notes that these source regions are necessary to maintain the upstream edge of a population in an advective environment. Clearly, these regions should receive priority in MPA network design. The loss of population viability in a region that acts as an upstream anchor can be expected to result in washout of the population (Gaines et al. 2003) and downstream movement of the range boundary to the next location where a source region remains viable and can anchor the upstream edge.

Patterns of dispersal of eggs and larvae depend on where and when these propagules are released into the plankton, the nature of water movement, the period in the plankton, and the behavior of the propagules. In this comment, attention is given to specific flow features, in addition to mean flow, and the possible retention of propagules within coastal retention zones. Potentially enhanced by the swimming behavior of larvae, these retentive features act to reduce the dispersal distance of larvae that are entrained or spawned into these features. Meanwhile, other larvae may still move long distances. Hence, these features allow for parallel outcomes, with both significant local recruitment and significant distant recruitment occurring within the same population (Largier 2003). In comparing flow systems with and without retention zones, one can appreciate the role of retention zones in reducing offshore larval wastage and, thus, understand how retention can underwrite a "win–win" scenario with enhancements in both local and distant recruitment. Further, it should be noted that while retentive circulation may enhance local recruitment significantly, any concomitant reduction in distant recruitment is likely to be a very small percentage (i.e., enhanced local recruitment does not imply a reduced spillover effect). In addition to the primary importance of effects on settlement patterns, these regions are also important for fisheries in that they often provide critical habitat.

Coastal Retention Zones

Coastal retention is associated with the coastal boundary layer (CBL) with slower moving waters near the coast. Along exposed, convex coasts, the CBL is typically narrow or nonexistent (e.g., Cape Mendocino, California, Largier et al. 1993), whereas a significant CBL can develop along straight or concave coasts (e.g., Vizcaino embayment in northern California, Largier et al. 1993; off the Outer Banks in North Carolina, Lentz et al. 1999; and off Huntington Beach in Southern California). Where bays are found, the main coastal flow may separate from the coast, leading to regions of stalled flow (e.g., Drakes Bay, California, Wing et al. 1998; Coronado embayment, M. Roughan, E. J. Terrill, and J. L. Largier, unpublished), recirculation (e.g., northern Monterey Bay, California, Paduan and Rosenfeld 1996; Graham and Largier 1997), or isolated waters (e.g., San Diego Bay, California, Largier et al. 1997). In this discussion, our interest is not in isolated waters (semienclosed bays like San Diego Bay) but, rather, in open coastal waters that offer "retention" (i.e., alongshore or offshore flow is reduced) so that water moves through the region in a longer time than if it were advected past the retention zone by the mean flow (cf, retention index used by Penven et al. 2000). Further, while the attention here is on retentive flow features associated with the form of the coastline, retention zones may also be found associated with kelp beds (Jackson and Winant 1983),

island wakes (Wolanski and Hamner 1988), shallow banks (Sinclair 1988), and offshore eddies (Nishimoto 2000). Opportunities for coastal retention are offered through the following.

Coastal Boundary Layer

Coastal flows are generally weaker at the shoreline than offshore (e.g., Huntington Beach, California, Figure 1), and the cross-shore component of flow has to go to zero at the coast (L. Carrillo and J. Largier, unpublished). This slowing of alongshore flow may be due either to the proximity of coast or to the shallowness of the water column (Pettigrew and Murray 1986). Where the CBL is narrow (~100 m), its value as a retention zone is limited to organisms with short planktonic larval duration (PLD), but where it is broader (>1 km), it is an important retention zone, specifically for larvae released near shore, as they may remain in this slow-flow environment for many days due to the notably weak cross-shore exchange. This is particularly true in the case of wind-driven downwelling circulation (Largier and Boyd 2001; Austin and Barth 2002).

Embayments

The CBL is typically broad off concave coasts—in embayments between promontories (e.g., in the Vizcaino embayment between Cape Mendocino and Point Arena in northern California, Largier et al. 1993). So, while alongshore flow moves through broad open bays, remaining attached to the coast, and while this flow correlates well with offshore flow, the alongshore currents are weaker nearshore and are more likely to reverse as forcing relaxes. Thus, the alongshore transport of propagules near to the coast is slowed, and these will take a longer time to move through this region than those following the faster route offshore.

Bays/Headland Wakes

Where there is a more marked headland, or an inward step in the coastline, the alongshore flow may separate from the coast, leaving a body of water enclosed in the bay (e.g., Cape Columbine, South Africa, Figure 2, Penven et al. 2000; Point Loma, California, Roughan, Terrill, and Largier, unpublished; and Monterey Bay, Paduan and Rosenfeld 1996). The separation of near-surface flow may be enhanced by the buoyancy of waters within the bay, either due to a local freshwater source or due to temperature differ-

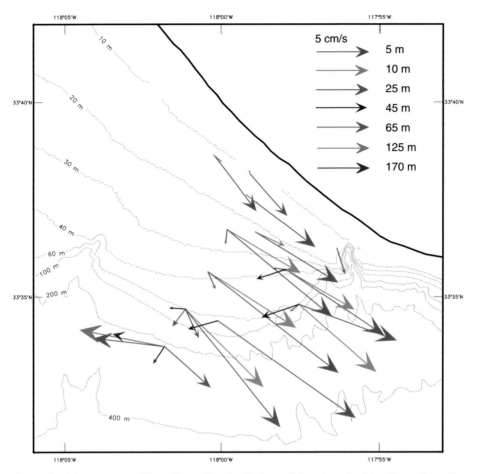

Figure 1.—Mean currents off Huntington Beach, Southern California, exhibiting the tendency for slower flows close to the coast and reversed flow at depth. The same is true for flow variability (further results are given in L. Carrillo and J. L. Largier [unpublished]).

ence associated with residence and surface heating of bay waters. The thermal effect is evident in "upwelling shadows" observed in Monterey Bay, (Graham and Largier 1997); Mejillones Bay, Chile (Escribano et al. 2002); and Lisbon Bay, Portugal (Moita et al. 2003). Retention zones may also form upstream of capes, where stalled surface flow and a thermally stratified water column characterize an "upwelling trap" (e.g., Antofogasta Bay, Chile; Castilla et al. 2002; Largier and others, unpublished). Separation patterns that form downstream of headlands are wakes, comparable with island wakes that tend to be better formed and have been studied in more detail to date (e.g., Wolanski et al. 1984, 1989; Pattiaratchi et al. 1987; Wolanski and Hamner 1992; Swearer et al. 1999). However, the retention effect is similar, with larvae being entrained into or retained within the recirculating wake flows.

While some propagules escape the wake, those that remain within the wake are subject to zero net transport (advection) as they move downstream and back upstream within this flow feature. In a completely retentive wake, residence time goes to infinity and all surviving local spawn settle locally, with zero spillover effect. Persistence of the wake feature depends on persistence of the mean flow past the island or headland. More persistent headland wakes can be expected along the southeastern U.S. coast, associated with the quasi-steady Gulf Stream (e.g., Pietrafesa et al. 1985), or along the eastern coast of South Africa, associated with the quasi-steady Agulhas Current (e.g., Natal Bight, Lutjeharms et al. 2000), than along the western coasts of the United States or South Africa, which are characterized by time-dependent, wind-driven coastal flows (e.g., Monterey Bay, Graham and Largier

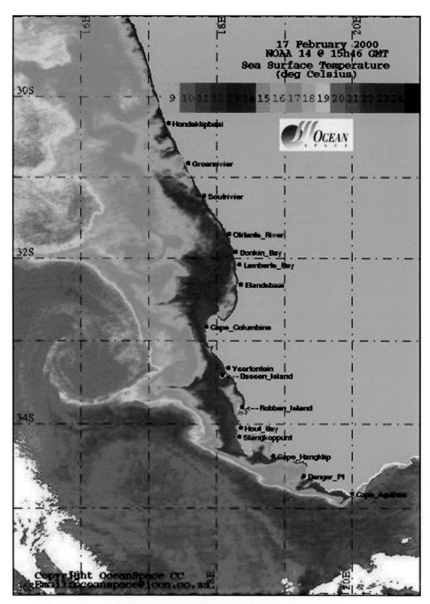

Figure 2.—Satellite-derived sea–surface temperature off the western coast of South Africa (17 February 2000), estimated from AVHRR (Advanced Very High Resolution Radiometer) data. "Upwelling shadow" effects in St Helena Bay (immediately north of Cape Columbine) and Table Bay (see Robben Island) are evident. Colder upwelled water is evident streaming equatorward past these bays, with warmer previously upwelled water resident in the bays. Data courtesy of Scarla Weeks (OceanSpace).

1997). While a neat recirculation pattern has been observed directly in northern Monterey Bay (Paduan and Rosenfeld 1996), in other bays where separation of the shelf flow occurs, the in-bay flow may not be so well organized (e.g., Drakes Bay, Figure 3; and Table Bay, South Africa, Figure 2; F. A. Shillington and A. Johnson, unpublished data). Nevertheless, these bays retain water and retard the export and alongshore advection of larvae such that they exhibit a longer "residence time" (sensu Penven et al. 2000).

Semienclosed Bays
These are bays with narrow mouths so that the alongshore coastal flow bypasses the mouth and does not force flows within the bay. Bay–ocean exchange is

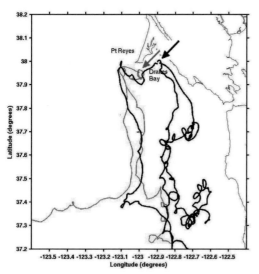

Figure 3.—Tracks of drifters with near-surface (5–10 m) drogues released in Drakes Bay during the upwelling season (May 2001) show a tendency to remain within Drakes Bay, with export either north or south from the bay after 3–6 d. In spite of clear retention, these drifter tracks describe neither a steady nor a coherent recirculation flow structure in the bay. The arrows indicate the locations at which drifters were released (3 black, 2 gray).

limited (e.g., Largier 1996; Largier et al. 1997; Monteiro and Largier 1999) and typically controlled by other processes, such as tidal pumping (e.g., San Diego Bay, Chadwick and Largier 1999). These enclosed bays are a special case, being somewhat isolated from coastal flows, and they are not the subject of our attention here. However, they may also act as retention zones and are mentioned here to show the continuum of bay types, from open coast to enclosed waters.

Retention through Behavior

Where flow is vertical sheared and reverses at depth, swimming larvae can be retained in a given region by vertical migrations between near-surface and near-bottom flows such that these larvae experience near-zero mean transport (e.g., Graham 1972; Peterson et al. 1979; Rothlisberg 1982; Rothlisberg et al. 1983). As larvae do not exhibit perfect diel vertical migration patterns, this retention is likely to be very leaky, with larvae being well dispersed in both directions and only some portion of the larval population remaining in this location; this contrasts with the population

exhibiting no vertical migration in which all larvae are swept downstream together by the mean current at that depth. A similar effect can be achieved in flows that reverse in time (e.g., tidal flows), with swimming larvae migrating into a slow bottom boundary layer during unfavorable flows and up into the water column at other times (e.g., Kimmerer and McKinnon 1987; Laprise and Dodson 1989; DiBacco and Chadwick 2001). Vertical migration through ontogeny may also play a role in retaining larvae near adult habitat (e.g., Paris and Cowen, in press) without subjecting the larval patch to the shear dispersion effects of diel migrations (Fortier and Leggett 1982).

Accumulation of Larval Concentrations

Retention in recirculating or weak flows will retard the alongshore or offshore transport of larvae, but it will not increase the concentration of larvae. However, some retention zones may accumulate larvae, leading to an increase in concentration over time. For this to occur, there has to be an asymmetry in exchange, with import exceeding export. For bays where the alongshore flow separates from the coast, there is typically a fluid boundary that may be quite sharp (e.g., a front), so that small horizontal motions (e.g., swimming) can move a larva across this boundary and into a different flow regime, either into or out of the retention zone. Further, the simple vertical behavior of swimming toward the surface can retain larvae on the stratified side of a buoyant front, typically within a retention zone (e.g., Franks 1992; Graham and Largier 1997; Shanks et al. 2000; Bjorkstedt et al. 2002). Swimming (or buoyancy) allows larvae to break from the flow of water (i.e., to cross streamlines) and, thus, to accumulate in retention zones (i.e., larval concentration is no longer constrained to be conservative property of the flow). Through this selective interaction of larvae with patterns of water flow, local spawn can be retained (enhancing local recruitment) and remote spawn can be entrained (enhancing remote recruitment) at the same time. If this occurs then these bounded retention zones can exhibit greater larval concentrations than in adjacent waters (e.g., Graham et al. 1992; Wing et al. 1998), and these retention zones can be expected to play a disproportionate role in population-wide dispersal patterns.

Thus, these bounded retention zones are likely to contain sites with strong recruitment of both local and remote spawn, and this region should provide an effective anchor against "washout" by mean flow

(Gaines et al. 2003; Largier 2003). However, the population-wide relevance of this region depends also on the export of spawn that can successfully recruit elsewhere, thus providing a spillover effect. In wind-driven upwelling areas, retention sites exhibit time scales associated with the time-dependence of the wind forcing so that a retention phase is followed by an export phase as the retentive flow structure breaks down after a few days or a week. For example, after a few days of weak winds, the warm surface layer that developed in Monterey Bay during upwelling conditions is observed to dissipate, moving north along the coast and out of the Bay (Graham and Largier 1997). Along the California coast north of Point Reyes, pulses of settlement have been observed during relaxation periods, as the accumulated larvae in Drakes Bay retention zone are transported north past Point Reyes and up the coast in a coherent poleward coastal flow pattern (Send et al. 1987; Wing et al. 1995a, 1995b). This sequence of retention, accumulation, and export processes make these locations important anchors for the upstream edge of a population and prime candidates as MPAs.

In summary, "retention" is to be thought of as reduced dispersion rather than as denoting a closed, isolated region. While the focus here is on the importance of retention zones in limiting alongshore dispersal, coastal retention zones are also very important in limiting offshore dispersal, keeping larvae close to nearshore adult habitat. In the following section, I will explore the role of retention sites in alongshore dispersal.

Effect of Retention on Alongshore Larval Dispersal

The degree of connectivity between parts of a population due to larval dispersal is one of the key issues in understanding and modeling metapopulations (Botsford et al. 1994; Palumbi 2001). To evaluate the role of retention zones in alongshore dispersal, let us consider dispersal as a process with a quasi-Gaussian outcome and focus our attention on a line domain (e.g., a long coast like the U.S. West Coast [Botsford et al. 1998] or a string of islands or series of reef habitats as found along the Great Barrier Reef or in the Caribbean [Hughes et al. 1999; Cowen et al. 2000]). Following previous work, including Okubo (1994), Gaylord and Gaines (2000), Largier (2003), and Okubo and Levin (2001), this problem is generally expressed as an advection–diffusion problem:

$$d_t C = -d_y (vC - K_y d_y C) + \lambda, \quad (1)$$

where C is the larval concentration, t is time, y is alongshore distance, v is alongshore velocity, K_y is alongshore eddy diffusivity, and λ represents a local source (or loss, if $\lambda < 0$). A similar cross-shore advection–diffusion equation can be written, with parameters u and K_x. The diffusive parameters K_y and K_x represent the effect of small-scale variability in flows in the alongshore and cross-shore orientations, v' and u'. The one-dimensional Gaussian solution to this one-dimensional advection–diffusion equation (with $\lambda = 0$) is given by

$$C(y,t) = [C_0/(4\pi K_y t)]\exp(-y^2/4K_y t). \quad (2)$$

In spite of the inherent stochastic nature, it is generally accepted that one can obtain parameter values that characterize the dispersal of the population if one aggregates over appropriate time and space (alongshore) scales to represent the population; this means aggregating over the spawning season and over the spatial extent of the population. Typically one obtains two parameters: advection as a measure of mean transport, and diffusion as a measure of variability in individual transport and, thus, related to the spreading out of a concentration of larvae. Two parameters are inadequate to express the full complexity of real transport patterns and, thus, real dispersal kernels. But this approach does allow the debate to move beyond the definition of a single statistic (dispersal length scale) for a population (Botsford et al. 2001), or to invoke zero-diffusion advective transport (Roberts 1997). And using two parameters, one can recognize that a site may exhibit strong local recruitment at the same time as making important contributions to other sites. Although alongshore differences in dispersal are likely to be a key factor in structuring populations, for the sake of simplicity, they are not explicitly resolved here, nor in most studies to date.

Within this two-parameter Gaussian dispersal model, one can explore the different effects of advection and diffusion—or the interplay between these parameters, as in Largier (2003). The effect of increased advection is to offset the Gaussian curve farther from the origin and, thus, to reduce local recruitment, while the effect of increased diffusion is to broaden the Gaussian curve and lower the mode while increasing recruitment at distances away from the mode. In many circumstances, advection away from the origin can be mitigated by the effect of diffusion, resulting in enhanced local recruitment (Largier 2003).

Thus, it is important to obtain realistic parameter values, which means evaluating over appropriate space and time scales. This yields weaker advection and stronger diffusion than has been used in earlier models (e.g., Roughgarden et al. 1988; Possingham and Roughgarden 1990; and following papers). With realistic advection–diffusion values that recognize ocean circulation as having a "weak mean" and "strong variability" (Wunsch 1981; Swenson and Niiler 1996), local recruitment is often significant and may be enough to maintain the population over time (Gaines et al. 2003). From a review of PLD, length of spawning season, and behavior that makes use of shear in coastal currents, one comes to the conclusion that many populations exhibit strategies that favor widespread dispersal. In addition to allowing for distant connectivity (long-distance dispersal for some larvae), these strategies are also valuable in that they enhance local recruitment in an advective environment.

The effect of coastal retention zones can be explored as a "shear dispersion" (or "shear diffusion") problem where flow is slower near the coast and propagules are subject to this as they move toward and away from the coast. One may also view the effect of retention zones as a "hold-up dispersion" problem in which some propagules are held up in a bay for some time and then released again into the coastal flow, more fitting for semienclosed bays. The effect is similar in that some propagules get left behind and the Gaussian dispersal curve becomes asymmetrical (skewed) and exhibits long tails stretching back to the origin and far downstream. Given the immense variety of hold-up or retention scenarios, I will not explore specific details here but, rather, focus on the more general description of the effect as shear dispersion (Figure 4). This model is appropriate for the case of CBL or a series of retention zones but limited in the case of a single large bay, as there will be important alongshore differences in the dispersal kernel, and the specific problem should be explored in more detail.

The advection–diffusion approach can be adapted to account for shear dispersion, with an increased alongshore diffusion (K_y) due to propagules mixing across the shear in the alongshore flow ($d_x v$), with shear flow of width L_x. For shorter time scales ($t < L_x^2/K_x$), before mixing across the shear is complete, the effect of this shear dispersion can be scaled by $K_y = (d_x v)^2 K_x t^2$ (Okubo 1994). From this expression, one can see that increased shear ($d_x v$) and increased mixing across the shear (cross-shore diffusion, K_x) will result in greater alongshore diffusion (K_y). This is intuitive, as this will allow some

Figure 4.—Schematic of the effect of dispersing across a sheared flow, resulting in the along-stream spreading of the shaded patch (i.e., "shear dispersion").

propagules time in the slow flow near the coast while other propagules will continue in the main flow farther offshore. The greater the shear, the greater the difference in transport for different propagules. However, if cross-stream (cross-shore) mixing is very large, all propagules will be mixed rapidly across the shear, and all will exhibit the same alongshore transport (i.e., the shear dispersion contribution to K_y goes to zero, consistent with scaling $K_y = v_o^2 L_x^2/K_x$, where v_o is the maximum alongshore flow). On the other hand, if the cross-stream mixing is very weak, all propagules released near the coast will remain near the coast and exhibit similar mean transport; the shear in alongshore flow will not be experienced by the dispersing propagules and shear dispersion (K_y) will be negligible.

The effect of shear dispersion is best explored by way of "particle-tracking" models, as illustrated in Figure 5. In this example, particles are released at the coastal zero position at time zero and allowed to be carried by the sheared alongshore current. Cross-shore current has a zero mean, but there is some variable flow in both alongshore and cross-shore currents. This flow variability, and hence K_x and K_y, go to zero at the coast and attain a maximum offshore. The density of particles (propagules with no behavior) after 20 d is shown, with many particles still near the shore and exhibiting limited alongshore transport. The mode is 50 km downstream of the origin. Contrast this with obtaining a simple dispersal length scale from $u \times t =$ 173 km (using a 20-d period and offshore mean flow $u = 0.1$ m/s). Although those particles that have been diffused offshore have moved much farther alongshore, very few have been transported distances of 173 km or more. Meanwhile, some nearshore particles are found at the origin or even a bit upstream due to the effect of variability in the alongshore flow (diffusion effect). In this example, particles are seen up to 50 km offshore. The dispersal outcome (dispersal kernel for release from this origin) is shown in the lower

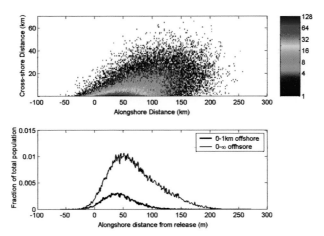

Figure 5.—Example of results of a particle-tracking model of dispersal. Particles (propagules) are released at position (0,0), and the distribution of particles is shown after 20 d (upper panel). The lower panel shows a cross-shore integration of particles, illustrating the alongshore distribution of larvae if all could find their own way to the shore: the bottom line indicates integration out to 1 km from the shore, and the top line integrates across the whole domain. A total of 40,000 particles are released. Alongshore flow is $v = 0.1$ m/s and decreases linearly from 0.1 m/s at 10 km offshore to 0.01 m/s at the shoreline. Flow variability is due to a tide-like 12-h variation in currents with random speeds between 0 and 0.2 m/s for v' (alongshore) and 0.1 m/s for u' (cross-shore).

panel, with curves illustrating the outcome for the case where all propagules find their way onshore to recruitment habitat and for the case where only propagules within 1 km of the shore find suitable recruitment habitat.

Comparing the dispersal curves in Figure 5 with Gaussian models, one immediately notes that these curves are skewed and that they have long tails. To express these two characteristics of the dispersal outcome, one would need an additional two parameters (skewness and kurtosis), indicating that the simple two-parameter advection–diffusion model has significant limits. One can see that these additional dispersal properties are important in quantifying spillover and the strength of long-distance dispersal on one hand and the strength of local recruitment on the other.

Clearly, the cross-shore distribution of propagules and their ability to move back onshore to coastal habitats are critical factors in alongshore dispersal (Figure 5). Depending on how effectively propagules are transported (or swim) onshore to recruitment habitat, one may obtain more local (Figure 5, bottom panel, bottom curve) or more dispersed recruitment (Figure 5,

bottom panel, top curve). Further, the cross-shore distribution of recruitment habitat is key, because a population that can settle in habitat further offshore is less reliant on onshore transport. In the plankton as well, the cross-shore structure of the pelagic community is important because differences in mortality may render either the more offshore or more nearshore dispersal trajectories more or less effective. Alternatively, the presence of both options allow bet-hedging and resilience during years when a widespread mortality event occurs either nearshore or offshore.

Swimming and Retention Zones

In the this section, I explore the importance of swimming in the dispersal of propagules and how swimming may influence this simple model of dispersal. Even weak swimming may be important in altering alongshore dispersal and in increasing the onshore movement of propagules immediately prior to and during competency. Specifically, swimming can play an enhanced role in the presence of coastal retention zones.

Larval swimming may have a major influence on dispersal patterns, although swimming speeds are typically slow compared with coastal currents. Many fish larvae and some invertebrate larvae display active swimming, with speeds of order 0.01–0.1 m/s (Shanks 1995a, 1995b; Leis and Stobutzki 1999), with coral reef fish developing even stronger swimming abilities (speed and persistence) at later larval stages. For example, Stobutzki and Bellwood (1997) observed larvae typically swimming 40 km unfed and without rest in laboratory chambers. Further, swimming may be directed in response to stimuli and navigation abilities (Armsworth 2000; Kingsford et al. 2002). Swimming can be represented as additional terms in the advection–diffusion model (equation 1), with the effect of mean directional swimming being represented by v_{swim} (an advective effect) and the effect of zero-mean variable swimming being represented by $K_{y\text{-swim}}$ (a diffusive effect). In this paper, I give attention to the effects of horizontal swimming. Vertical swimming may also play a key role, but it is not a primary focus of this paper as it has received much attention previously and, in particular, the retentive effect of diel vertical migration in vertically sheared flow is well recognized (Graham 1972; Peterson et al. 1979; Rothlisberg et al. 1983; Stephenson and Power 1988; Laprise and Dodson 1989). However, in addition to slowing advection, vertical swimming also greatly enhances K_x and K_y through shear dispersion effects (Rothlisberg et al. 1983; Okubo 1994; and see below).

Where larvae can navigate (Kingsford et al. 2002), horizontal swimming will exhibit a preferred direction and there will be a mean swimming effect (i.e., advection). Although likely to be weak (order of 0.01 m/s), it is comparable with the "weak mean" nature of ocean currents, and this may have an important impact on the number of propagules that recruit to adult habitat either by affecting the alongshore transport (e.g., see effect of small changes in advection in Largier 2003) or by affecting the onshore transport (e.g., Figure 5; onshore swimming of 0.01 m/s will clear all propagules within 1 km of the shore in 1 d, and 0.1 m/s will clear 10 km in 1 d). Swimming may be particularly important nearshore (e.g., within 1 km), where mean and low-frequency cross-shore flow is very weak due to the proximity of the coastal boundary. It may be that such onshore swimming at the time of competence is the most important direct effect of swimming.

In addition to the mean directional (advective) effect of swimming, there are also zero-mean nondirectional (diffusive) effects due to swimming in variable directions. This diffusive effect of swimming may contribute directly to alongshore or cross-shore dispersion in a similar way as zero-mean variability in water flow (i.e., one may add $K_{x\text{-swim}}$ and $K_{y\text{-swim}}$ terms into the advection–diffusion equation, or add v'_{swim} and u'_{swim} terms into the particle-dispersion model). As an example, the effect of increased alongshore diffusion is shown in the results of the particle-dispersion model (Figure 6): while modal recruitment decreases due to spreading out of the curve, the spatial extent of dispersion increases, and recruitment away from the mode is observed to increase. In particular, in this case, local recruitment (recruitment at the origin, $y = 0$ km) is enhanced by the additional diffusive effect of swimming. At the same time, recruitment at distances of more than 150 km downstream increases (and significant recruitment is now observed to distances of 250 km downstream instead of 200 km, and 80 km upstream instead of 20 km).

This increase in alongshore diffusion may be achieved either by variable alongshore swimming or through variable cross-shore swimming in the presence of shear $d_x v$. Under specific circumstances, enhanced alongshore diffusion is most effectively achieved through cross-shore swimming in the presence of shear in the alongshore flow and this may offer a more energy-efficient way to enhance dispersal. Following the shear dispersion expression in the previous section, this is achieved when $(d_x v)^2 t^2 > 1$ and $t < L_x^2/K_x$, conditions that are met in strongly sheared coastal flows where shear of order 0.1 m/s over 10 km is observed (e.g., Figure 1; D. M. Kaplan, J. L. Largier, and L. W. Botsford, unpublished). In terms of energetics, however, vertical swimming is likely to be most efficient as shear occurs over short distances, and variability in horizontal movement may be affected by moving less than 100 m between near-surface and near-bottom flows.

The effectiveness of nondirectional swimming in dispersion depends on the typical speed and persistence of swimming in a given direction (diffusivity K is scaled by the product of velocity and length $K_{swim} \sim u_{swim} l_{swim} \sim u^2_{swim} t_{swim}$). So, in evaluating the importance of swimming for the dispersion of a given population, one is interested in obtaining representative scales for the speed (u_{swim}) and persistence of swimming (t_{swim}). This is illustrated in Figure 7, where one can see that either faster swimming or more persistent swimming (swimming in one direction for a longer time before turning) will result in a stronger diffusive effect. As $K \sim u^2 t$, a threefold increase in swimming ability (u_{swim}) will result in an order of magni-

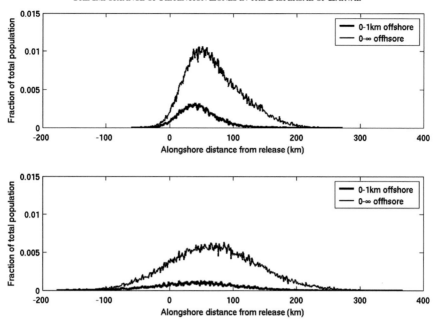

Figure 6.—Model results for alongshore particle distributions after 20 d: upper panel as in Figure 5; lower panel with increased alongshore diffusion due to additional diel flow variability (with varying amplitudes up to 0.2 m/s). This additional variability represents the effects of diel swimming behavior, either directly through alongshore swimming or, more likely, through shear dispersion effects. The effect of this increased diffusion for the top-line scenario is to increase local recruitment from ~0.002 to ~0.004 and to increase the spatial extent of recruitment exceeding 0.001 from a 180-km range ($-10 < y < 170$ km) to a 260-km range ($-60 < y < 200$ km).

tude increase in diffusion (K_{swim}), whereas the dependence on swimming persistence (t_{swim}) is linear.

Clearly, it is important to determine why larval fish swim, as this will affect the swimming time scale, direction, and speed. For example, if larvae swim to forage for food, then they are likely to describe a rather random motion with short-term variability. But if they are visual feeders or use a "solar compass" (Leis and Carson-Ewart 2003), then larvae may orient their swimming either into or away from the sun, resulting in diel variability in swimming and onshore or offshore directionality along meridional coasts, such as the U.S. West Coast. Diel vertical migrations may also result in a similar effect by moving larvae between two current directions above and below the thermocline. The strength of this diel current variability (of order 0.1 m/s), together with the daily time scale (of order 10^5 s), can account for diffusivities of order 1,000 m²/s (see Figure 7) and typically much larger than that achieved by short-term horizontal movements.

The above argues that even weak swimming of fish larvae can influence the pattern of dispersal and success of recruitment in a population. Further, it suggests that there are specific swimming behaviors that may be efficient in terms of the energy used versus the recruitment enhancement achieved – and this raises the question whether populations may even be able to select for these behaviors. Finally, the ability of late-stage larvae to navigate and to sustain swimming effort suggests that larvae may exhibit persistent "mean" velocity during critical phases that is stronger than that imposed by "weak mean" coastal currents.

Taking a more phenomenological approach, the importance of swimming can be well appreciated in the case of a retention zone such as a headland wake or bay, specifically where the retention zone is bounded by a narrow shear zone or buoyancy front. In Figure 8 (Drakes Bay), one can see how even weak swimming would result in larvae entering the retention zone much more frequently than passive drifters. As larger eddies are often absent from these shear fronts, current-induced cross-frontal exchange is very weak, and swimming-induced exchange may be critically important here. In other words, this swimming may control the rate at which propagules are entrained and released by the retention zone. Thus, swimming

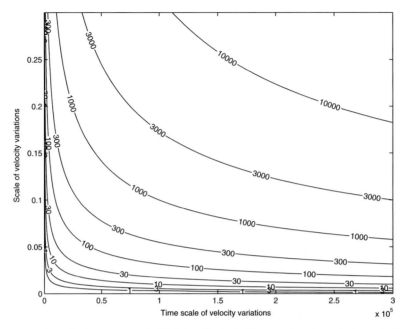

Figure 7.—The dependence of diffusivity $K \sim u^2 t$ (m²/s) on swimming velocity u (m/s) and the time scale of unidirectional swimming t (s).

may control the effect of this retention zone on both alongshore dispersal in general and on recruitment within this bay in particular. In addition to onshore swimming to get into the bay circulation (or offshore swimming to get out of the bay), nondirectional swimming will also increase the flux across the front allowing many propagules to have their time in the retention zone, but without increasing larval concentration in the retention zone.

Recognition of the potential importance of swimming in larval dispersal directs attention at questions about how and why larval fish swim. Existing research on swimming capability (speed and navigation) should also explore the persistence and directionality of larval swimming in natural environments. Further, study of the reasons why larval fish swim (e.g., Kingsford et al. 2002) and models of how different behaviors influence dispersal (e.g., James et al. 2002) will help to determine the links between larval swimming and larval dispersal patterns.

Retention Zones as Specific Habitat Zones

In addition to offering specific dispersal opportunities, retention zones also offer specific habitat oppor-

tunities for the meroplankton and associated adult populations. Significant differences can be expected between bays and adjacent coastal waters in terms of community structure, abiotic habitat characteristics, and human activity.

While detailed characterization must be bay-specific, one can note, for example, that bays in upwelling areas are warmer, more stratified, and more productive than those in surrounding regions (e.g., Graham and Largier 1997; Wing et al. 1998; Castilla et al. 2002). As can be seen in St Helena Bay, South Africa (Figures 2, 9), and as is shown in Wieters et al. (2003), Moita et al. (2003), and A. J. Vander Woude and A. J. Lucas (unpublished), bays in upwelling areas have higher primary production rates and higher phytoplankton biomass than do surrounding areas, thus offering a food-rich environment for planktotropohic larvae. At the same time, however, these bays may support different predator populations so that larval and postsettlement mortality may be quite different to that in open coastal waters.

Sheltered bay environments are more likely to be turbid, to have broad swaths of sandy bottom, to receive freshwater runoff, and to be warmer and more stratified (in summer) than are surrounding waters. These abiotic characteristics not only affect larvae, but they may result in there being more or less adult habi-

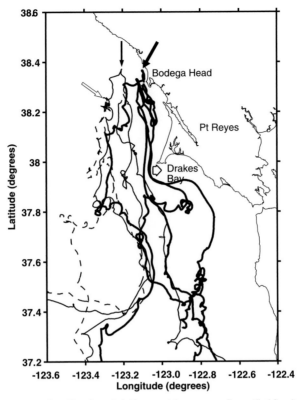

Figure 8.—Tracks of drifters with near-surface (5–10-m) drogues released off Bodega Head during the upwelling season (May 2001) show how water parcels that pass Pt Reyes close to the coast may be entrained into the retention zone. The outlined arrow shows the opportunity for even weak or diffusive onshore swimming to enhance the entrainment into Drakes Bay. Swimming of order 0.01 m/s will move a larva 1 km in a day, enough to cross the narrow shear zone observed off Pt Reyes. The outlined arrow is about 4 km. The solid arrows indicate the locations where the drifters were released.

tat and higher or lower adult population densities. Where these retention zones offer negligible adult habitat, they can still be important in reducing alongshore dispersal. Further, while enhancing local recruitment is of little value in the absence of much habitat, these bays may be important to enhancing local recruitment at adjacent sites. For example, a proportion of the propagules released along the coast north of Point Reyes may be expected to be entrained into Drakes Bay (Figure 8), where some of them can be retained (Wing et al. 1998), and then later transported back north to settle along the coast north of Point Reyes during a relaxation event (Wing et al. 1995a, 1995b). However, this ecological service will be supplied without the need to set the bay aside as an MPA.

Finally, bays tend to be more sheltered than do open coastal regions, and major metropolises have developed along their shores (e.g., San Francisco; Santa Cruz through Monterey; Cape Town, South Africa; Port Elizabeth, South Africa; Antofogasta, Chile; Valparaiso, Chile; Lisbon, Portugal; and so on). The associated threats of habitat loss, invasive species, water quality degradation, and accidental spills have to be recognized in a risk-based assessment of the importance of these environments for MPAs, even if their dispersal profile makes them prime candidates. The retentive characteristic that may render bays as good sites for MPAs also is a factor in water quality, as pollutants introduced to these retention zones may be accumulated and result in persistent concentrations

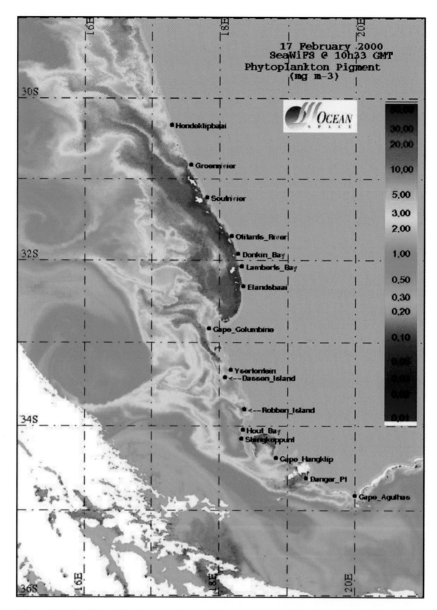

Figure 9.—Satellite-derived sea–surface chlorophyll concentration off the western coast of South Africa (17 February 2000), estimated from SeaWiFS satellite data. This image is concurrent with the thermal image in Figure 2. High levels of phytoplankton biomass are evident in upwelling shadows downstream of upwelling centers, specifically in St Helena Bay (within the bay, white indicates over-ranging on the color bar, i.e., >50 mg/m³). Data courtesy of Scarla Weeks (OceanSpace).

of pollutants there. Further, the association of harbors with bays, coupled with the tendency for retention of meroplankton and local recruitment, leads to these regions being susceptible to future invasions by exotic populations.

Conclusion

The dispersal of early life stages of sedentary adult populations is a critical question in determining the location, size, and spacing of aquatic protected areas.

In the case of coastal marine protected areas, the presence of retention zones is a key structure that has been under appreciated to date, specifically when one considers the potential importance of swimming in enhancing the value of these areas. Retention zones may provide any or all of three benefits that are likely to enhance local recruitment and recruitment levels in general: (1) limiting alongshore dispersal, (2) limiting offshore losses, and (3) offering habitat conducive to survival of larvae and juveniles. Thus, it makes sense to identify where such retention zones exist, how they relate to key adult habitat and spawning populations, and whether they do indeed offer these benefits. In systems where one can identify and validate the benefits of such an area, these sites would be prime candidates for protection—both from fishing effort and from negative external influences, such as pollution.

The swimming behavior of early life stages of fish is recognized here. But, the interaction of swimming and flow requires far more study before one can better assess how dispersal is influenced by the use of retention zones by larvae and juveniles. Even weak population-wide mean swimming may have major effects on the dispersal kernel and on the proportion of the dispersing propagules that return to coastal habitats and are then available for recruitment. Further, zero-mean diffusive swimming effects can also be very important in altering the dispersal kernel, specifically through shear dispersion. In this regard, it is not only the speed of swimming but also longer time scale variability in swimming behavior that will determine the importance of swimming in dispersal. Behavior that results in swimming variability on a diel time scale is likely to be important in population-wide dispersal outcomes.

While the problem can be characterized with a broad-brush shear-dispersion paradigm, evaluation of specific MPA options will require analysis of specific structures associated with retention zones. In particular, where the coastal flow separates from the coast at a headland, swimming may take on a major role in moving propagules across the narrow shear zone between the alongshore coastal flow and the retentive bay circulation. Further, the possibility of crossing the shear zone allows for the concurrent existence of two dispersal strategies—one part of the dispersing population being entrained into a retention zone and enhancing more local recruitment, while the other part of the dispersing population may move rapidly alongshore and result in distant recruitment. This helps one appreciate that spillover and distant recruitment can occur simultaneously with local recruitment, that it is

not an either–or question, and that it is seldom helpful (and often misleading) to identify a single statistic ("larval dispersal distance") for a given population.

Recognition that many marine populations have a long planktonic larval duration, exhibit a long spawning season, and are exposed to strong shear dispersion and recognition that these three features result in broad dispersal kernels lead one to wonder if there is a clear recruitment benefit that has allowed these populations to select for these features. If this selection has occurred, then it has happened through selecting for specific spawning and swimming behaviors. Future work in understanding these coastal marine populations must include an improved knowledge and better understanding of larval swimming behavior in the context of generic coastal flow structures.

Acknowledgments

This publication was supported in part by GLOBEC Grant OCE-0003254; CoOP Grant OCE 99-07884; the NOAA National Sea Grant College Program Grant NA06RG0142, project R/CZ-164, through the California Sea Grant College Program; and the California State Resources Agency. The views expressed herein do not necessarily reflect the views of any of those organizations. I express my thanks to Linden Clarke for generating the results from particle models and to Scarla Weeks at OceanSpace for providing the satellite imagery. Finally, I thank the American Fisheries Society for the invitation to prepare this paper.

References

Armsworth, P. R. 2000. Directed motion in the sea: efficient swimming by reef fish larvae. Journal of Theoretical Biology 210:1–11.

Armsworth, P. R. 2002. Recruitment limitation, population regulation, and larval connectivity in reef fish metapopulations. Ecology 83:1092–1104.

Austin, J. A., and J. A. Barth. 2002. Drifter behavior on the Oregon-Washington shelf during downwelling-favorable winds. Journal of Physical Oceanography 32:3132–3144.

Bakun, A. 1996. Patterns in the ocean: ocean processes and marine population dynamics. California Sea Grant College Program, La Jolla.

Bjorkstedt, E. P., L. K. Rosenfeld, B. A Grantham, Y. Shkedy, and J. Roughgarden. 2002. Distributions of larval rockfishes Sebastes spp. across nearshore fronts in a coastal upwelling region. Marine Ecology Progress Series 242:215–228.

Botsford, L. W., A. Hastings, and S. D. Gaines. 2001. Dependence of sustainability on the configuration of marine reserves and larval dispersal distances. Ecology Letters 4:144–150.

Botsford, L. W., C. L. Moloney, A. M. Hastings, J. L. Largier, T. M. Powell, K. Higgins, and J. F. Quinn. 1994. The influence of spatially and temporally varying oceanographic conditions on meroplanktonic metapopulations. Deep-Sea Research Part II Topical Studies in Oceanography 41:107–145.

Botsford, L. W., C. L. Moloney, J. L. Largier, and A. Hastings. 1998. Metapopulation dynamics of meroplanktonic invertebrates: the Dungeness crab (*Cancer magister*) as an example. Canadian Special Publication of Fisheries and Aquatic Sciences 125:295–306.

Castilla J. C., N. Lagos, R. Guinez, and J. L. Largier. 2002. Embayments and nearshore retention of plankton: the Antofogasta Bay and other examples. Pages 179–203 in J. C. Castilla and J. L. Largier, editors. The oceanography and ecology of the nearshore and bays in Chile. Ediciones Universidad Católica de Chile, Santiago.

Chadwick, D. B., and J. L. Largier. 1999. Tidal exchange at the bay-ocean boundary. Journal of Geophysical Research 104(C12):29901–29919.

Cowen, R. K., K. M. Lwiza, S. Sponaugle, C. B. Paris, and D. B. Olson. 2000. Connectivity of marine populations: open or closed? Science 287:857–859.

DiBacco, C., and D. B. Chadwick. 2001. Use of elemental fingerprinting to assess net flux and exchange of brachyuran larvae between regions of San Diego Bay, California and nearshore coastal habitats. Journal of Marine Research 59:1–27.

Escribano, R., V. Marin, P. Hidalgo, and G. Olivares. 2002. Physical–biological interactions in the pelagic ecosystem of the nearshore zone of the northern Humboldt Current System. Pages 145–175 in J. C. Castilla and J. L. Largier, editors. The oceanography and ecology of the nearshore and bays in Chile. Ediciones Universidad Católica de Chile, Santiago.

Fortier, L., and W. C. Leggett. 1982. Fickian transport and the dispersal of fish larvae in estuaries. Canadian Journal of Fisheries and Aquatic Sciences 39:1150–1163.

Franks, P. J. S. 1992. Sink or swim: accumulation of biomass at fronts. Marine Ecology Progress Series 82:1–12.

Gaines, S. D., B. Gaylord, and J. L. Largier. 2003. Avoiding current oversights in marine reserve design. Ecological Applications 13(1):S32–S46.

Gaylord, B., and S. D. Gaines. 2000. Temperature or transport? Range limits in marine species mediated solely by flow. American Naturalist 155:769–789.

Graham, J. J. 1972. Retention of larval herring within the Sheepscot Estuary of Maine. U.S. National Marine Fisheries Service Fishery Bulletin 70:299–305.

Graham, W. M., J. G. Field, and D. C. Potts. 1992. Persistent "upwelling shadows" and their influence on zooplankton distributions. Marine Biology 114:561–570.

Graham, W. M., and J. L. Largier. 1997. Upwelling shadows as nearshore retention sites: the example of northern Monterey Bay. Continental Shelf Research 17:509–532.

Hughes, T. P., A. H. Baird, E. A. Dinsdale, N. A. Moltschaniwsky, M. S. Pratchett, J. E. Tanner, and B. L. Willis. 1999. Patterns of recruitment and abundance of corals along the Great Barrier Reef. Nature 397:59–63.

Jackson, G. A. and C. D. Winant. 1983. Effect of a kelp forest on coastal currents. Continental Shelf Research 2(1):75–80.

James, M. K., P. R. Armsworth, L. B. Mason, and L. Bode. 2002. The structure of reef fish metapopulations: modelling larval dispersal and retention patterns. Proceedings of the Royal Society of London B 269:2079–2086.

Kimmerer, W. J., and A. D. McKinnon. 1987. Zooplankton in a marine bay. II. Vertical migration to maintain horizontal distributions. Marine Ecology Progress Series 41:53–60.

Kingsford, M. J., J. M. Leis, A. Shanks, K. C. Lindeman, S. G. Morgan, and J. Pineda. 2002. Sensory environments, larval abilities and local self-recruitment. Bulletin of Marine Science 70(1):309–340.

Laprise, R., and J. J. Dodson. 1989. Ontogeny and importance of tidal vertical migrations in the retention of larval smelt. *Osmerus mordax* in a well-mixed estuary. Marine Ecology Progress Series 55:101–111.

Largier, J., and A. Boyd. 2001. Drifter observations of surface water transport in the Benguela Current during winter 1999. South African Journal of Science 97(5/6):223–229.

Largier, J. L. 1996. Hydrodynamic exchange between San Francisco Bay and the ocean: the role of ocean circulation and stratification. Pages 69–104 in J. T. Hollibaugh, editor. San Francisco Bay: the ecosystem. Pacific Division, American Association for the Advancement of Science, Washington, D.C.

Largier, J. L. 2003. Considerations in estimating larval dispersal distances from oceanographic data. Ecological Applications 13(Supplement 1):S71–S89.

Largier, J. L., J. T. Hollibaugh, and S. V. Smith. 1997. Seasonally hypersaline estuaries in mediterranean-climate estuaries. Estuarine Coastal and Shelf Science 45:789–797.

Largier, J. L., B. A. Magnell, and C. D. Winant. 1993. Subtidal circulation over the northern California shelf. Journal of Geophysical Research 98:18147–18179.

Leis, J. M., and B. M. Carson-Ewart. 2003. Orientation of pelagic larvae of coral-reef fishes in the ocean. Marine Ecology Progress Series 252:239–253.

Leis, J. M., and I. C. Stobutzki. 1999. Swimming performance of late pelagic larvae of coral-reef fishes. Pages 575–583 *in* B. Seret and J. Y. Sire, editors. Proceedings of 5th Indo-Pacific Fisheries Conference. Société Française d'Ichtyologie and Institut de Recherche pour le Développement, Paris.

Lentz, S., R. T. Guza, S. Elgar, F. Feddersen, and T. H. C. Herbers. 1999. Momentum balances on the North Carolina inner shelf. Journal of Geophysical Research 104(C8):18205–18240.

Lutjeharms, J. R. E., H. R. Valentine, and R. C. Van Ballegooyen. 2000. The hydrography and water masses of the Natal Bight, South Africa. Continental Shelf Research 20:1907–1939.

Moita, M. T., P. B. Olliveira, J. C. Mendes, and A. S. Palma. 2003. Distribution of chlorophyll a and *Gymnodinium catenatum* associated with coastal upwelling plumes off central Portugal. Acta Oecologica 24:S125–S132.

Monteiro, P. M. S., and J. L. Largier. 1999. Thermal stratification in Saldanha Bay (South Africa) and subtidal, density-driven exchange with the coastal waters of the Benguela upwelling system. Estuarine Coastal and Shelf Science 49(6):877–890.

Nishimoto, M. M. 2000. Distributions of late-larval and pelagic juvenile rockfishes in relation to water masses around the Santa Barbara Channel Islands in early summer, 1996. *In* Fifth California islands symposium. OSC Study MMS 99-0038. Minerals Management Service, Santa Barbara, California.

Okubo, A. 1994. The role of diffusion and related physical processes in dispersal and recruitment of marine populations. Pages 5–32 *in* P. W. Sammarco and M. L. Heron, editors. The bio-physics of marine larval dispersal. American Geophysical Union, Washington, D.C.

Okubo, A., and S. A. Levin. 2001. Diffusion and ecological problems: modern perspectives. Springer-Verlag, New York.

Paduan, J. D., and L. K. Rosenfeld. 1996. Remotely sensed surface currents in Monterey Bay from shore-based HF radar (CODAR). Journal of Geophysical Research 101:20669–20686.

Palumbi, S. R. 2001. The ecology of marine protected areas. Pages 509–530 *in* M. Bertness, S. D. Gaines, and M. E. Hay, editors. Marine ecology: the new synthesis. Sinauer, Sunderland, Massachusetts.

Paris, C. B., and R. K. Cowen. In press. Direct evidence of a biophysical retention mechanism for coral reef fish larvae. Limnology and Oceanography.

Pattiaratchi, C., A. James, and M. Collins. 1987. Island wakes and headland eddies; a comparison between remotely-sensed data and laboratory experiments. Journal of Geophysical Research 92:783–794.

Penven, P., C. Roy, A. Colin de Verdiere, and J. Largier. 2000. Simulation of a coastal jet retention process using a barotropic model. Oceanologica Acta 23:615–634.

Peterson, W. T., C. B. Miller, and A. Hutchinson. 1979. Zonation and maintenance of copepod populations in the Oregon upwelling zone. Deep-Sea Research 26A:467–494.

Pettigrew, N. R., and S. P. Murray. 1986. The coastal boundary layer and inner shelf. Pges 95–108 *in* Baroclinic processes on continental shelves. C. N. K. Mooers, editor. American Geophysical Union, Washington D.C.

Pietrafesa, L. J., G. S. Janowitz, and P. A. Wittman. 1985. Physical oceanographic processes in the Carolina Capes. Pages 23–32 *in* L. P. Atkinson, D. W. Menzel, and K. A. Bush, editors. Oceanography of the southeastern U.S. continental shelf. American Geophysical Union, Washington D.C.

Possingham, H. P., and J. Roughgarden. 1990. Spatial population dynamics of a marine organism with a complex life cycle. Ecology 71:973–985.

Roberts, C. M. 1997. Connectivity and management of Caribbean coral reefs. Science 278:1454–1457.

Rothlisberg, P. C. 1982. Vertical migration and its effect on dispersal of penaeid shrimp larvae in the Gulf of Carpentaria, Australia. U.S. National Marine Fisheries Service Fishery Bulletin 80:541–554.

Rothlisberg, P. C., J. A. Church, and A. M. G. Forbes. 1983. Modeling the advection of vertically migrating shrimp larvae. Journal of Marine Research 41:511–538.

Roughgarden, J., S. D. Gaines, and H. P. Possingham. 1988. Recruitment dynamics in complex life cycles. Science 241:1460–1466.

Send, U., R. C. Beardsley, and C. D. Winant. 1987. Relaxation from upwelling in the coastal ocean dynamics experiment. Journal of Geophysical Research 92:1683–1698.

Shanks, A. L. 1995a. Orientated swimming by megalopae of several eastern North Pacific crab species and its potential role in their onshore migration. Journal of Experimental Marine Biology and Ecology 186:1–16.

Shanks, A. L. 1995b. Mechanisms of cross-shelf dispersal of larval invertebrates and fish. Pages 323–367 *in* L. R. McEdward, editor. Ecology of marine invertebrate larvae. CRC Press, Boca Raton, Florida.

Shanks, A. L., B. A. Grantham, and M. H. Carr. 2003. Propagule dispersal distance and the size and spacing of marine reserves. Ecological Applications 13(Supplement 1):S159–S169.

Shanks, A. L., J. L. Largier, L. Brink, J. Brubaker, and R. Hooff. 2000. Demonstration of the onshore transport of larval invertebrates by the shoreward movement of an upwelling front. Limnology and Oceanography 45(1):230–236.

Sinclair, M. 1988. Marine populations: an essay on population regulation and speciation. Washington Sea Grant Program, Seattle.

Stephenson, R. L., and M. J. Power. 1988. Semi-diel vertical movements in Atlantic herring *Clupea harengus* larvae: a mechanism for larval retention? Marine Ecology Progress Series 50:3–11.

Stobutzki, I. C., and D. R. Bellwood. 1997. Sustained swimming abilities of the late pelagic stages of coral reef fishes. Marine Ecology Progress Series 149:35–41.

Swearer, S. E., J. E. Caselle, D. W. Lea, and R. R. Warner. 1999. Larval retention and recruitment in an island population of a coral reef fish. Nature 402:799–802.

Swenson, M. S., and P. P. Niiler. 1996. Statistical analysis of the surface circulation of the California Current. Journal of Geophysical Research 101:22631–22645.

Warner, R. R., S. E. Swearer, and J. E. Caselle. 2000. Larval accumulation and retention: implications for the design of marine reserves and essential habitat. Bulletin of Marine Science 66:821–830.

Wieters, E. A., D. M. Kaplan, S. A. Navarrete, A. Sotomayor, J. Largier, K. J. Nielsen, and F. Veliz. 2003. Alongshore and temporal variability in chlorophyll *a* concentrations in Chilean nearshore waters. Marine Ecology Progress Series 249:93–105.

Wing, S. R., L. W. Botsford, J. L. Largier, and L. E. Morgan. 1995a. Spatial variability in the settlement of benthic invertebrates in a northern California upwelling system. Marine Ecology Progress Series 128:199–211.

Wing, S. R., L. W. Botsford, S. V. Ralston, and J. L. Largier. 1998. Meroplanktonic distribution and circulation in a coastal retention zone of the northern California upwelling system. Limnology and Oceanography 43:1710–1721.

Wing, S. R., J. L. Largier, L. W. Botsford, and J. F. Quinn. 1995b. Settlement and transport of benthic invertebrates in an intermittent upwelling region. Limnology and Oceanography 40:316–329.

Wolanski, E., D. Burrage, and B. King. 1989. Trapping and dispersion of coral eggs around Bowden Reef, Great Barrier Reef, following mass coral spawning. Continental Shelf Research 9:479–496.

Wolanski, E., and W. M. Hamner. 1988. Topographically controlled fronts in the ocean and their biological influence. Science 241:177–181.

Wolanski, E., J. Imberger, and M. L. Heron. 1984. Island wakes in shallow waters. Journal of Geophysical Research 89:10553–10569.

Wunsch, C. 1981. Low-frequency variability of the sea. Pages 342–347 *in* B. A. Warren and C. Wunsch, editors. Evolution of physical oceanography. Massachusetts Institute of Technology, Cambridge.

Part IV:
Evaluation

American Fisheries Society Symposium 42:125–131, 2004

Harvest Benefits: Marine Reserves or Traditional Fishery Management Tools

ROBERT L. SHIPP[1]

Department of Marine Sciences, University of South Alabama,
Mobile, Alabama 36688, USA

Abstract.—Increased harvest from fish stocks is often included among the many benefits attributable to marine reserves. Examination of the most targeted finfish stocks from the coasts of the United States finds that this is rarely the case. This is because the stocks are highly mobile or are not overfished, and overfishing is not occurring. Traditional fishery management tools such as seasonal closures, quotas, and bag limits are proving effective in management of most marine stocks and are preferred over marine reserves for purposes of maintaining or increasing finfish yield. Highly migratory stocks may receive temporary benefit as they pass through a reserve, but this is analogous to traditional methods. Overfished sedentary reef species from the U.S. South Atlantic and the Gulf of Mexico may receive limited benefits from a reserve; but in many instances, recently enacted catch quotas are enhancing stock recovery. For some of the severely stressed rockfishes off the Pacific Coast, stocks are so depleted that complete harvest moratoria are likely necessary for their recovery, and reserves would have to be of unrealistic size to substantially benefit these stocks. Reserves created for the benefit of very few species would negatively impact harvest of sympatric forms whose stocks are healthy. Marine reserves serve many other beneficial purposes in the marine environment.

Introduction

The decline of the world's fishery stocks has been well chronicled in recent years, especially for large highly migratory species (Myers and Worm 2003) and certain heavily exploited stocks such as Atlantic cod *Gadus morhua* (Myers et al. 1997). Sustainability of oceanic fisheries has been questioned unless drastic steps are taken (Pauly et al. 2002). Many of these articles describe the overcapitalization of fleets and improved technology that have led to increasing fishing pressures in most of the world's productive oceans and recommend various solutions, including establishment of marine reserves.

Thus, there has been increasing interest in the establishment of no-take marine protected areas, marine no-take areas, or marine sanctuaries (herein designated as "reserves") (NRC 2001; NRDC 2001; Halpern and Warner 2002; Palumbi 2002). Proponents of reserves have described the potential benefits as a

fishery management tool for sustaining and increasing yield as well as several other related advantages, specifically, conserving biodiversity, protecting (coastal) ecosystem integrity, preserving cultural heritage, providing educational and recreational opportunities, establishing sites for scientific research (NRC 2001), and especially as a tool for ecosystem-based management (Palumbi 2002). In this current era of global concern over the status of exploited stocks, these purported benefits have strong appeal.

The concept of reserves is initially attractive and will no doubt elicit a great deal of support and discussion among various groups interested in protecting marine habitats. However, the many offered benefits described above often overlap and become intertwined in the discussions which ensue. If reserves are to be considered as a management tool for sustained or increased yield, then that goal or objective needs to be clearly stated and distinguished from other, sometimes more theoretical goals. Although it is recognized that other, broader definitions exist and could include such goals as producing trophy fish, maintaining genetic integrity, etc., for purposes of this paper, a fishery

---

[1] E-mail: rshipp@jaguar1.usouthal.edu

management tool is defined as one that sustains and/ or increases through time the yield of a fish stock or several sympatric stocks of an ecosystem.

Methods

In this paper, I intend to compare the success or failure of traditional fishery management tools in three diverse regions of the United States. These conclusions are based on my experience as a fishery scientist (32 years) and fishery manager (9-year member of the Gulf of Mexico Fishery Management Council [GMFMC], twice its chair, currently serving on advisory panels for reef fish and red snapper *Lutjanus campechanus*).

Traditional management tools generally focus on controlling or reducing effort and protecting critical habitat. Effort reduction includes gear reduction, bag and size limits (sometimes including slot limits), quotas, seasonal or area closures, gear restrictions, and bycatch reduction. These have been applied with relative success for more than a century in freshwater and nearshore coastal environments (Ross 1997). Their use in offshore marine habitats has only become widespread in the United States in recent decades, especially since passage of the Fishery Conservation and Management Act in 1976. I will address finfish stocks under management by the South Atlantic Fishery Management Council (SAFMC), GMFMC, and Pacific Fishery Management Council (PFMC). In addition, I will include discussion of highly migratory species, managed directly by the National Marine Fisheries Service (NMFS). As a basis for addressing individual stocks, I am using the recently released National Oceanic and Atmospheric Administration (NOAA) Fisheries 2002 report to Congress (NMFS 2003). In particular, I will discuss those stocks whose status is known and are considered overfished or are experiencing overfishing. These terms have precise legal definitions. Briefly, "overfished" refers to a depleted stock status, and "overfishing" refers to a rate of fishing effort which will lead to a depleted stock. The purpose will be to determine whether traditional management tools are working or whether consideration of establishing marine reserves might be a better alternative to ensure a sustainable yield.

In considering whether the harvest of a finfish stock may benefit from establishment of a reserve, one must assume the stock is suffering from overfishing or is overfished. Otherwise, if a stock is healthy, not approaching an overfished state, and possesses a biomass at or above that to sustain a maximum sustainable yield, then establishment of a reserve will, at best, be yield neutral or, more likely, reduce yield or reduce harvest efficiency (e.g., DeMartini 1993; Holland and Brazee 1996). Therefore, in assessing potential yield benefits of a reserve, only stressed stocks (overfished or experiencing overfishing) will be considered.

The NMFS (2003) included 932 stocks (both finfishes and shellfishes) in its 2002 annual report to Congress. Of these, 259 were considered major stocks, and 673 were minor stocks. Of the major stocks, 41 were subject to overfishing, 129 were not subject to overfishing, 43 were overfished, and 117 were not overfished (some stocks were both subject to overfishing and were overfished). Of the minor stocks, 25 were subject to overfishing, 79 were not subject to overfishing, 43 were overfished, 33 were not overfished, and 1 was approaching the overfished condition. For 695 stocks, the fished status was unknown or undefined. Of this total, 86% were categorized as minor.

For purposes of this paper, in general, only finfish stocks designated by NMFS as major stocks whose stock status is known are considered. This is because the major stocks yield 99.9% of the landings (NMFS 2003). While reserves may have important benefits for the health of minor stocks, especially in terms of habitat protection from fishing gear since yield from these stocks is minimal, their inclusion in respect to yield is generally not considered relevant. However, for several "high profile" minor stocks (e.g., goliath grouper *Epinephelus itajara*), discussion of various management strategies is included. In addition, certain major stocks whose status is unknown but for which traditional management tools may not be appropriate (e.g., deepwater groupers), are also discussed.

From the most recent NOAA Fisheries report to Congress (NMFS 2003), I have developed tables of overfished–overfishing finfish stocks under management jurisdiction of the councils listed above, as well as highly migratory stocks managed directly by NMFS. These are listed in Tables 1–4.

U.S. South Atlantic Stocks

Eight major stocks under management of the SAFMC are considered overfished and are experiencing overfishing (Table 1). All these are reef species, and most are relatively sedentary. Two species (snowy grouper and tilefish) are found in relatively deep water, one (yellowtail snapper) is only in the extreme southern

Table 1.—Major finfish stocks managed by the South Atlantic Fishery Management Council that are overfished and current year and duration of rebuilding plan (from NMFS 2003). Y = yes; N = no.

Species	Overfishing	Overfished	Current year	Total duration
Vermilion snapper *Rhomboplites aurorubens*	Y	Y	4	10
Red snapper *Lutjanus campechanus*	Y	Y	12	15
Snowy grouper *Epinephelus niveatus*	Y	Y	12	15
Tilefish *Lopholatilus chamaeleonticeps*	Y	Y	11	15
Yellowtail snapper *Ocyurus chrysurus*	Y	Y	11	10
Red grouper *E. morio*	Y	Y	12	15
Black sea bass *Centropristis striata*	Y	Y	3	10
Gag *Mycteroperca microlepis*	Y	Y	12	15

portion of the council's jurisdiction, while the other species are found in shallower to middle depths, throughout the council's jurisdiction. Currently, all species are subject to rebuilding plans, most more than halfway toward completion, using traditional management measures. The SAFMC is confident that all these stocks will be rebuilt using traditional management measures, and, in fact, gag was recently removed from the overfished list, and black sea bass and vermilion snapper are expected to be removed after the next assessment (G. Waugh, SAFMC, personal communication).

While these eight stocks are officially listed as currently experiencing overfishing, this designation cannot be removed until future assessments are completed, but the current measures are likely preventing overfishing.

Goliath grouper, a minor stock, has been a species of great concern for more than a decade. In fact, a total harvest prohibition was placed on this species in 1990. Although no formal stock assessment has been performed, at a recent SEDAR (southeastern data assessment and review) meeting (2003),

[D]ivers pointed out that there were aggregations of goliath grouper off the southeastern coast of Florida, near Jupiter, in the 1950s. These aggregations were fished-out soon after discovery, and the Goliath grouper had not been reported from that area for several decades. However, in 2002, an apparent aggregation of 50 individuals was observed in that same area. Reports of fish in the northeastern Gulf of Mexico and northeastern coast of Florida are beginning to come in through the state of Florida tagging hotline.

Many commercial and recreational fishermen have expressed concern that this species' predatory behavior may negatively impact populations of sympatric reef species, especially spiny lobsters *Panulirus*

argus (also known as Caribbean spiny lobsters). At the recent (January 2003) meeting of the Reef Fish Advisory Panel of the GMFMC, several members noted that these stocks have rebounded so strongly and are impacting their prey species so heavily that the panel voted unanimously to request that the council consider a controlled harvest to determine the status of the stocks. Thus, the total moratorium (a traditional management tool) appears to be leading to a recovery of the species.

Nassau grouper *E. striatus*, also a minor stock, are found only in the most extreme southern United States, primarily the Florida Keys (Sadovy and Eklund 2000). The status of their stocks has also been of great concern, especially because of their well documented spawning aggregations (Colin 1992) which make them vulnerable to intense harvest at that time. For this reason, protection of these sites during spawning is certainly a positive function of a reserve. Whether these sites should be so designated permanently would require additional studies to determine if habitat requirements were threatened by harvest activities during other times. In addition, designation of areas other than the spawning sites as reserves for protection of Nassau grouper would not be beneficial, since they would leave those areas during spawning and, thus, become vulnerable to capture (Bolden 2000).

Therefore, establishment of reserves in the U.S. South Atlantic is likely unnecessary for purposes of maintaining or increasing harvest for these eight major and two minor stocks and, in any case, would likely be impractical given the diverse and expansive range of these species. However, the SAFMC is concerned about several deepwater groupers (e.g., misty grouper *E. mystacinus*, yellowedge grouper *E. flavolimbatus*, and yellowfin grouper *Mycteroperca venenosa*) whose status is considered unknown. As a precautionary measure, it might be appropriate to designate identified important habitat for these species as a reserve until their status can be determined.

Gulf of Mexico Stocks

Six major Gulf of Mexico stocks are overfished or are subject to overfishing (Table 2).

King mackerel has just been removed from the overfishing designation, and stocks are close to not being overfished. Recent data on mixing with the healthy U.S. South Atlantic stocks, based on otolith shape, may improve the knowledge of this stock. But in any case, this is a very migratory species and, in my opinion, is not likely to benefit from a reserve.

Of the five additional stocks, four are reef species. Red snapper is likely the most stressed, although stocks have continued to recover during the last decade under relatively stringent quotas and size and bag limits (R. Leard, GMFMC, personal communication). However, high mortality of 0-year-class and 1-year-class juveniles in the Gulf shrimp bycatch has been cited as the major factor in the stressed status of this stock (Schirripa and Legault 1999). Recent implementation of bycatch reduction devices, mandated by the GMFMC, has greatly reduced this mortality. In light of these management measures now in place,

recovery of this wide-ranging form would likely receive very little additional benefit by establishment of a reserve. The other three reef species are of questionable status, but recent revisions of their stock status indicates they may be close to recovery using traditional methods.

Red drum is protected in federal waters of the Gulf of Mexico and under varying but stringent harvest requirements in state territorial seas. All Gulf states have been experiencing recovery of the stock. And given that the broodstock of this species is highly mobile, it would appear that creation of a reserve would be of little or no benefit.

Pacific Species

Stocks of Pacific groundfish (Table 3) have declined sharply in recent years due to heavy overexploitation combined with extremely long life cycles of many of these species. However, recent action by the PFMC has imposed stringent measures on the most stressed of these stocks (e.g., bocaccio, canary rockfish, and darkblotched rockfish), resulting in harvest levels that are near moratoria for several of the most stressed species (NMFS 2003). Thus, while these stocks are currently not subject to overfishing, they are severely overfished, and recovery time will require many decades.

Table 2.—Major finfish stocks managed by the Gulf of Mexico Fishery Management Council that are overfished and current year and duration of rebuilding plan (from NMFS 2003). Y = yes; N = no; ns = not yet submitted; nd = not defined.

Species	Overfishing	Overfished	Current year	Total duration
King mackerel *Scomberomorus cavalla* (Gulf group)	N	Y	17	nd
Red snapper	Y	Y	12	29
Red grouper	Y	Y	ns	ns
Greater amberjack *Seriola dumerili*	N	Y	ns	ns
Vermilion snapper	Y	unknown	ns	ns
Red drum *Sciaenops ocellatus*	Y	Y	12	nd

Table 3.—Major finfish stocks managed by the Pacific Fishery Management Council are overfished and current year and duration of rebuilding plan (from NMFS 2003). Y = yes; N = no.

Species	Overfishing	Overfished	Current year	Total duration
Lingcod *Ophiodon elongatus*	N	Y	3	10
Pacific ocean perch *Sebastes alutus*	N	Y	3	42
Bocaccio *Sebastes paucispinis*	N	Y	3	110
Canary rockfish *Sebastes pinniger*	N	Y	2	76
Darkblotched rockfish *Sebastes crameri*	N	Y	1	47
Widow rockfish *Sebastes entomelas*	N	Y	1	38
Pacific hake *Merluccius productus* (also known as Pacific whiting)	Y	Y	To be developed	To be developed

While establishment of reserves, as has been done recently for part of the Channel Islands, California, may aid the process, this is likely to be of minimal impact given the extensive range, both in depth and latitude, of these species. Of greater import is to guarantee that only minimal bycatch mortality is experienced by these species from harvest of sympatric species of healthy stocks.

Highly Migratory Species

Highly migratory species (Table 4) are among the most overexploited finfish species in the world and, thus, have received prominent attention recently (Myers and Worm 2003). Management and subsequent recovery of these forms is made especially difficult due to the multiple jurisdictions through which they travel. While the most stringent management measures are necessary to affect recovery of these stocks, reserves are of minimal value due to the migratory nature of these stocks. However, areas identified as serving as spawning aggregations for any of these species should definitely be designated at least as temporary reserves.

Discussion

When one examines the finfish stocks found in territorial seas and the Exclusive Economic Zone of the United States, there appear to be very few stocks for which permanent reserves would provide a significant benefit toward rebuilding the stocks from an overfished or overfishing level to one that is not overfished or where overfishing is not occurring. But when intended to serve as a fishery management tool, there are several situations for which they may be extremely beneficial and many others for which more traditional methods are much preferred. These are discussed in Shipp (2003) and reviewed briefly as follows.

Benefits of Reserves as Mnagement Tools

Reserves are beneficial during periods of active spawning by aggregations, when species may be especially vulnerable to harvest, and when certain components of the stock (e.g., dominant males) may be especially vulnerable to capture. The utility of these is likely to be seasonal and normally would not require year-round catch restrictions.

For stocks severely overfished and subject to little or no management, a reserve can be used along with other measures to more rapidly replenish populations. This is especially true in isolated, insular populations (e.g., Roberts et al. 2001, for St. Lucia) which are not strongly connected to proximal populations for replenishment.

Where habitats are damaged by fishing practices, establishment of reserves may help ensure habitat recovery. Oftentimes, however, gear restrictions can be enacted to lessen the social impact that would result in declaration of a total no-take zone.

Reserves may also be beneficial where ecosystem management is employed in fisheries (primarily of near-sedentary species) where bycatch of non-targeted species has become excessive, or conversely, where a protected species has reached population levels which increase natural mortality rates of targeted species, preventing a reasonable harvest

A reserve will allow some version of dynamic equilibrium to return.

Liabilities of Reserves as Management Tools

When establishment of a reserve is intended as a near proxy for a virgin stock, several factors need to be kept in mind. First, by definition, a virgin stock provides no yield. Therefore, a perfect proxy would be a negative in terms of management goals to produce a maximum sustainable yield or optimal yield.

Table 4.—Major highly migratory finfish stocks managed by the National Marine Fisheries Service that are overfished and current year and duration of rebuilding plan (from NMFS 2003). Y = yes; N = no; ns = not yet submitted; nd = not defined.

Species	Overfishing	Overfished	Current year	Total duration
Bigeye tuna *Thunnus obesus*	Y	Y	Implemented	nd
Albacore *Thunnus alalunga*	Y	Y	ns	ns
Bluefin tuna *Thunnus thynnus* (West Atlantic)	Y	Y	5	20
Swordfish *Xiphias gladius* (North Atlantic)	Y	Y	4	10
Sandbar shark *Carcharhinus plumbeus*	Y	Y	39	nd
Blacktip shark *Carcharhinus limbatus*	Y	Y	30	nd
Bull shark *Carcharhinus leucas*	Y	Y	30	nd
Finetooth shark *Carcharhinus isodon*	Y	N	ns	ns

And the postulated "spillover effect" of harvestable adults to adjacent areas will always be less than that of a properly managed stock, which generates the optimal yield per recruit (YPR), again, by definition. These YPR models are discussed in numerous classical and modern texts (e.g., Rounsefell 1975; Iverson 1996).

Another claim is that larvae from a reserve will be a significant addition to the overall stocks. This may be beneficial, but only for a very seriously depleted stock. In other cases, larval production, always in excess of the carrying capacity of the habitat, does not normally relate to year-class strength. Rather, density-dependent factors usually control ultimate recruitment to the harvestable stock. While this principle has been the subject of scores of books and probably thousands of publications, it was espoused nearly 150 years ago by Darwin and restated frequently in most every fishery text (e.g., Gulland 1977; Rothschild 1986).

Stocks within a Reserve

There are numerous examples in the literature of stock increases within a reserve (e.g., Johnson et al. 1999; Roberts et al. 2001). However, one must not forget what the point is here, in regard to yield. While effective reserves may support a stock with relatively great biomass, perhaps larger individuals, and a high spawning potential ratio, this portion of the stock has been removed from harvest. Therefore, unless there is an increase in effort, the overall yield is reduced by whatever fraction could be contributed to overall harvest from this protected stock and mitigated only by the possibility of spillover or larval contribution, as discussed above.

Pragmatic Perspective

Examination of the major species of finfishes from the U.S. South Atlantic, Gulf of Mexico, and U.S. Pacific coasts, as well as highly migratory species, reveals that few species are known to be both overfished and experiencing overfishing and are sedentary. Those candidates that are in both categories and may possibly benefit from a reserve are found in widely differing geographic ranges, with optimal potential reserves sites far apart (e.g., snowy grouper and yellowtail snapper in the U.S. South Atlantic). To establish a reserve for the benefit of those few species would remove harvest potential of the scores of sympatric forms, most of which are not overfished.

And while this may not reduce the overall harvest of these species, it would definitely reduce efficiency and increase fishing effort in other, adjacent areas.

Far better would be to impose more traditional methods to restore the overfished stocks, as has been done for many species. This becomes more and more successful as we adopt more precautionary harvest levels and improve our methods of stock assessment and methods for studying stock–recruit relationships and life history information.

References

Bolden, S. K. 2000. Long-distance movement of Nassau grouper (*Epinephelus striatus*) to a spawning aggregation in the central Bahamas. U.S. National Marine Fisheries Service Fishery Bulletin 98:642–645.

Colin, P. L. 1992. Reproduction of the Nassau grouper, *Epinephelus striatus*, (Pisces: Serranidae) and its relationship to environmental conditions. Environmental Biology of Fishes 34:357–377.

DeMartini, E. E., 1993. Modeling the potential of fishery reserves for managing Pacific coral reef fishes. U.S. National Marine Fisheries Service Fishery Bulletin 91:414–427.

Gulland, J. A. 1977. Fish population dynamics. Wiley, New York.

Halpern, B. S., and R. R. Warner. 2002. Marine reserves have rapid and lasting effects. Ecology Letters 5:361–366.

Holland, D. S., and R. J. Brazee. 1996. Marine reserves for fisheries management. Marine Resource Economics 11:157–171.

Iverson, E. S. 1996. Living marine resources, their utilization and management. Chapman and Hall, New York.

Johnson, D. R., N. A. Funicelli, and J. A. Bohnsack. 1999. Effectiveness of an existing estuarine no-take fish sanctuary within the Kennedy Space Center, Florida. North American Journal of Fisheries Management 19:436–453.

Myers, R. A., J. A. Hutchings, and N. J. Barrowman. 1997. Why do fish stocks collapse? The example of cod in Atlantic Canada. Ecological Applications 7:91–106.

Myers, R. A., and B. Worm. 2003. Rapid worldwide depletion of predatory fish communities. Nature (London) 423:280–283.

NMFS (National Marine Fisheries Service). 2003. NOAA Fisheries 2002 report to Congress. NMFS, Silver Spring, Maryland.

NRC (National Research Council). 2001. Marine protected areas, tools for sustaining ocean ecosystems. National Academy of Sciences, Washington, D.C.

NRDC (National Resources Defense Council). 2001. Keeping oceans wild. NRDC, April 2001 report, New York.

Palumbi, S. R. 2002. Marine reserves: a tool for ecosystem management and conservation. Pew Oceans Commission, Arlington, Virginia.

Pauly, D., V. Christensen, S. Guenette, T. J. Pitcher, U. R. Sumaila, C. J. Walters, A R. Watson, and D. Zeller. 2002. Towards sustainability in world fisheries. Nature (London) 418:689–695.

Roberts, C. M., J. A. Bohnsack, F. Gell, J. P. Hawkins, and R. Goodridge. 2001. Effects of marine reserves on adjacent fisheries. Science 294:1920–1923.

Ross, M. R. 1997. Fisheries conservation and management. Prentice Hall, New Jersey.

Rothschild, B. J. 1986. Dynamics of marine fish populations. Harvard University Press, Cambridge, Massachusetts.

Rounsefell, G. A. 1975. Ecology, utilization, and management of marine fisheries. C.V. Mosby, St. Louis.

Sadovy, Y., and A. E. Eklund. 2000. Synopsis of biological data on the Nassau grouper, *Epinephelus striatus* (Bloch, 1792) and the jewfish, *E. itajara* (Lichtenstein, 1822). NOAA Technical Report NMFS 146, FAO Fisheries Synopsis 157.

Schirripa, M. J, and C. M. Legault. 1999. Status of the red snapper stock in U.S. waters of the Gulf of Mexico: updated through 1998. National Oceanic and Atmospheric Administration, National Marine Fisheries Service, SFD-99/00-75, Miami.

Shipp, R. L. 2003. A perspective on marine reserves as a fishery management tool. Fisheries 28(12):10–21.

American Fisheries Society Symposium 42:133–154, 2004

Effects of Marine Protected Areas on the Assessment of Marine Fisheries

ANDRÉ E. PUNT[1]

School of Aquatic and Fishery Sciences, University of Washington,
Seattle, Washington 98195-5020, USA

RICHARD D. METHOT[2]

Northwest Fisheries Science Center, National Oceanic and Atmospheric Administration,
Fisheries, 2725 Montlake Boulevard East, Seattle, Washington 98112, USA

Abstract.—Fishery stock assessments are designed to estimate the current status of fished marine resources relative to target and limit reference levels and to provide advice on the implications of future harvest rates and other management actions. Most fisheries stock assessments are based on the assumption that the fishery or the fish population is distributed homogeneously or freely mixes across the region being assessed. Any local patterns in density, age structure, or mortality are assumed to be ephemeral and to diffuse quickly throughout the population. The introduction of no-fishing zones (here termed marine protected areas, MPAs) that are effective in preserving the part of the population found in the MPA is, therefore, likely to provide unique challenges, both positive and negative, for fisheries stock assessment. In principle, MPAs provide a new opportunity to estimate quantities, such as natural mortality and growth, needed when defining management reference points. However, MPAs, if effective, will further distort any spatial variation in the density and age structure of the population, thereby further increasing the extent to which the homogeneous distribution assumption underlying most stock assessments is violated. Simulations based on two West Coast groundfish species (widow rockfish *Sebastes entomelas* and lingcod *Ophiodon elongatus*) are used to assess the likely impact of the latter factor on the performance of the age-structured, unit-stock models used for the assessments of most West Coast groundfish populations. The results suggest that assessments can be substantially in error if the data are aggregated across the MPA and the area subject to fishing. However, the negative impact of an MPA on assessment ability is relatively slight if account is taken of the impact of the MPA on the spatial pattern in density and age structure by using spatially structured assessment models. The ability to improve estimates of parameters using data from the MPA is found to be relatively slight for the quantities and scenarios examined in the simulations.

Introduction

Although often incompletely articulated, the management goals for most of the world's marine fish species are to select management strategies and tactics to achieve the maximum long-term average yield without substantially negatively impacting the marine ecosystems on which the fisheries are based and the fish-ing communities which depend on them. Within the United States, the management goals are framed by the 10 national standards of the Sustainable Fisheries Act (SFA) of 1996. In particular, national standard 1 of the SFA states that "conservation and management measures shall prevent overfishing while achieving, on a continuing basis, the optimum yield from each fishery for the United States industry."

The need to satisfy national standard 1 has led, inter alia, to the requirement for the eight regional fishery management councils in the United States to develop control rules that are used to assess whether overfishing is occurring or a stock is in an overfished

---

[1] E-mail: aepunt@u.washington.edu
[2] E-mail: Richard.Methot@noaa.gov

state (e.g., Restrepo and Powers 1999). In addition, the SFA specifies that a rebuilding plan has to be developed for any fish stocks that are designated to be overfished. (In this paper, and consistent with usage by the Pacific Fishery Management Council [PFMC], "overfishing" means that the level of fishing mortality exceeds the maximum fishing mortality threshold, which is currently set at the rate associated with maximum sustainable yield (MSY), and "being in an overfished state" means that the current spawning output is less than a minimum spawning stock threshold.)

The fishery management plan for the groundfish fishery off the U.S. West Coast includes 82 species. The management regulations for these species established by the PFMC are based on a harvest control rule that aims to maintain the population near the proxy for the biomass at which MSY is achieved (B_{MSY}) of 40% of the estimate of the pre-fishery spawning output (i.e., $0.4B_0$) and which defines $0.25B_0$ as the level at which a stock is declared to be overfished. Spawning output is the sum over age of the product of egg production at age and numbers at age.

Nine groundfish species are currently designated as overfished. Management regulations recommended by the PFMC and implemented by the National Marine Fisheries Service for the U.S. West Coast groundfish fishery include reduced trip limits and large areas on the continental shelf with greatly reduced fishing opportunities. These regulations are designed to reduce the fishing mortality on the overfished species to allow them to recover to the proxy for B_{MSY} of $0.4B_0$. The closed areas, which prohibit all bottom trawling and most nontrawl gears, are somewhat of a marine protected area (MPA). In this paper, MPAs are assumed to be areas in which no harvest-related removals are permitted. However, long-term use of MPAs currently is not part of the management system since regulations regarding the closed areas are implemented through annual specifications and can be eliminated depending on the status of the stocks they were designed for. The strategic plan for the Pacific Fishery Management Council (PFMC 2003) includes the future possibility of implementing MPAs as part of a long-term fishery management plan for West Coast groundfish.

Closed areas are currently seen as ways to protect and recover overfished species. However, some modeling work (e.g., Polacheck 1990; Sladek Nowlis and Roberts 1998; Apostolaki et al. 2002) has suggested that marine reserves or no-take MPAs could lead to higher sustainable yields than can conven-

tional management approaches such as effort limitations and trip limits.

There is, however, a gap (Holland 2002) between the types of models used to evaluate the benefits of no-take MPAs (e.g., Man et al. 1995; Lauck et al. 1998; Sladek Nowlis and Roberts 1998; Guénette and Pitcher 1999; Hastings and Botsford 1999; Pezzey et al. 2000; Apostolaki et al. 2002; Beattie et al. 2002; Sumaila 2002) and those on which stock assessments and, hence, the scientific management advice provided to the U.S. regional fishery management councils are typically based. The models used to evaluate MPA benefits typically emphasize spatial differences in population structure, while those used for stock assessment purposes are more focused on (1) changes in the age and length structure of the entire population, (2) the abundance of the resource relative to target and limit levels, and (3) the link between the parameters of the model and the data available for assessment purposes. Several analytical methods are used to conduct stock assessments in the United States (e.g., see NRC 1998). However, the bulk of these assessments are conducted using two basic approaches: ADAPT (Gavaris 1988) and integrated analysis (Fournier and Archibald 1982; Methot 1993, 2000). Integrated analysis is currently the "method of choice" for assessments of West Coast groundfish species.

The quantities on which traditional stock assessment models depend include: (1) the rate of natural mortality (usually assumed to be independent of age and time), (2) fishery selectivity (often for a number of fishery types and occasionally time dependent), (3) the selectivity and catchability of the fishery-independent survey gears used to sample the population, (4) the length, weight, and fecundity of fishes of different ages, and (5) the sizes of the cohorts that have recruited into the population for some of the years since the start of substantial exploitation. Including MPAs in fisheries management will require that the models used for assessment purposes be extended to account for spatial structure and, hence, include parameters related to movement and mixing. Unfortunately, the ability to estimate such parameters, even in the medium-term to long-term, is limited unless research focuses more on the estimation of mixing, movement, and dispersal rates (Holland 2002). This led Botsford et al. (2001, 2003) to conclude that although MPAs decrease some types of uncertainty, they also increase uncertainty because they require information on movement. Therefore, for the foreseeable future, MPAs and stock assessment are likely to be unhappily married. Even in the longer term, it is questionable whether accurate estimates of movement rates for adults and larvae are possible (Holland 2002).

One impact of a large closed area would be to establish a persistent spatial pattern in the density and age structure of the population, thereby further increasing the extent to which the homogeneous distribution assumption underlying most stock assessment methods is violated. Therefore, it is possible that implementing an MPA might actually lead to poorer estimates of the quantities on which management is currently based for the more data-rich target species. This issue has not been considered in much detail to date. For example, in their review of marine reserves, Guénette et al. (1998) did not consider how the imposition of a marine reserve would impact the ability to conduct assessments using typical assessment methods. One possible reason for this is that many of the studies on the impact of introducing MPAs have focused on data-poor situations in which management is either not present or not linked closely to the results of any stock assessment. However, the issue of the impact of MPAs on the accuracy of stock assessment methods is important for those species that are managed using the results of stock assessments, such as the U.S. West Coast groundfish fishery.

Holland (2002) examines the impact of MPAs on stock assessments in terms of how current stock assessment models will need to be modified and how data collection protocols will need to be changed to tailor stock assessment methods for use in fisheries where there are large MPAs. We consider the alternative question: What is the ability of current assessment methods to estimate parameters of management interest if data collection and model structure are not changed?

Methods

The most common method used to determine how well a stock assessment method is likely to perform is Monte Carlo simulation (e.g., de la Mare 1986; Patterson and Kirkwood 1995; Sampson and Yin 1998; Punt et al. 2002). Monte Carlo simulation involves the following steps:

(1) definition of a model of the system to be assessed; this model (often referred to as the operating model) will represent the truth for the simulations;

(2) generation by the operating model of data that will be used by the assessment methods;

(3) application of a number of alternative stock assessment methods to the generated data sets; and

(4) comparison of the estimates provided by the stock assessment methods with the true values from the operating model.

The Operating Model

The operating model (Appendix 1) is a spatially-explicit age-structured and sex-structured population dynamics model. The region to be managed is divided into 10 spatial cells of equal size where the possibility exists for movement of larvae and age-1 and older animals among cells. Ten variants of the operating model (scenarios A–J in Table 1) are considered to examine the sensitivity of the results to the extent of adult movement, how density dependence operates, whether the probability of movement of age-1 and older animals increases or decreases with age, the ex-

Table 1.—The scenarios considered in this paper. The features of the operating model for each scenario that differ from those for the baseline scenario (A) are indicated in bold italics. The spatially disaggregated assessment method is applied to cells 1–2, 3–6, and 7–10 for Scenario J.

Scenario	Asymptotic movement rate	Movement increases with age	Stock–recruitment relationship	Variation in selectivity	Surveys in the closed area	Natural mortality	MPA
A	0.01	Yes	3a	0	Yes	Known exactly	Cells 1–4
B	*0.05*	Yes	3a	0	Yes	Known exactly	Cells 1–4
C	*0.3*	Yes	3a	0	Yes	Known exactly	Cells 1–4
D	0.01	Yes	*3b*	0	Yes	Known exactly	Cells 1–4
E	0.01	Yes	*3c*	0	Yes	Known exactly	Cells 1–4
F	0.01	Yes	3a	*1.0*	Yes	Known exactly	Cells 1–4
G	0.3	*No*	3a	0	Yes	Known exactly	Cells 1–4
H	0.01	Yes	3a	0	*No*	Known exactly	Cells 1–4
I	0.01	Yes	3a	0	Yes	*Estimated*	Cells 1–4
J	0.01	Yes	3a	0	Yes	Known exactly	*Cells 1–2*

tent of interannual variation in selectivity, whether survey data are available for the MPA, whether natural mortality is estimated or assumed to be known exactly, and the spatial extent of the MPA relative to the area open to fishing.

Three scenarios (A, D, and E; stock–recruitment relationships 3a, 3b, and 3c in Table 1 and Appendix A) examine the sensitivity to how larval distribution and density dependence operate:

(1) Density dependence acts at the level of the total population (i.e., density dependence is post-dispersal) and the number of larvae recruiting to a cell is independent of the spawning output in that cell (equation A3a; baseline assumption);

(2) Density dependence acts at the cell level (i.e., density dependence is pre-dispersal) and the number recruiting to a cell depends only on the spawning output in the cell (equation A3b); and

(3) Density dependence acts at the cell level, but all larvae then mix, and the number of larvae recruiting to a cell is independent of the spawning output in the cell (equation A3c).

Although none of these three scenarios may reflect reality, they should bound the likely behavior of systems that do not have clear source–sink dynamics.

The parameters of the operating model are based on the actual situation of assessing widow rockfish *Sebastes entomelas* and lingcod *Ophiodon elongatus* off the West Coast of the United States. These two species were chosen because they have both been declared to be overfished and because they differ in terms of the apparent relationship between spawning biomass and subsequent recruitment (Figure 1); recruitment for widow rockfish declines with declining spawning biomass whereas recruitment is almost independent of spawning stock size over the observed range of lingcod spawning stock abundance. The biological parameters for lingcod have been set equal to those for the southern area (Jagielo et al. 2000).

The model is projected for 110 years. The catches for the first 40 years for widow rockfish and 50 years for lingcod are based on the actual catches for the two stocks (Figure 2) and occur in all ten cells. The catches, thereafter, are based on fixing the fishing mortality rate in the cells open to fishing. Cells 1–4 are declared to be an MPA after the period of historical catches. The stock is assumed to be uniformly distributed among the spatial cells prior to exploitation; the status of the population in each cell at the time of the imposition of the MPA depends on how catchability differs spatially (see equation A2) and on spatial variation in recruitment. The size of the spawning output (all 10 cells) when the MPA is declared is set to 20% of the virgin spawning output. This specification is implemented for each simulated replicate by generating the recruitment residuals (ε_y and η_y^c; see equation A3) and selecting the value for the total number of recruits in the absence of exploitation (R_0^T) so that when the model is projected from year 0 to the year when the MPA is declared, the spawning output in the latter year equals 20% of the virgin spawning output.

The fishing mortality in the area open to fishing after the imposition of the MPA is constant and is chosen so that recovery of the total (across cells open and closed to fishing) spawning output to $0.4B_0$ occurs within a prespecified time. Future fishing mortality is prespecified to avoid confounding the performance of the stock assessment method with the level of fishing mortality. In actuality, the fishing mortality would

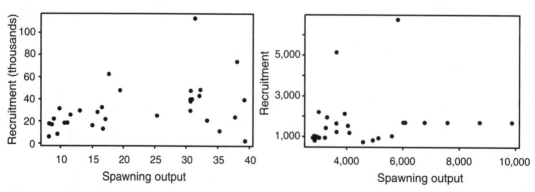

Figure 1.—Recruitment versus spawning output for widow rockfish (left panel; Williams et al. 2000) and lingcod south of 40°10N (right panel; Jagielo et al. 2000).

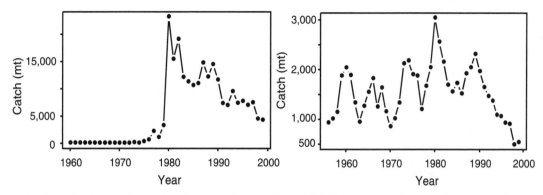

Figure 2.—Historical catches (in metric tons; mt) of widow rockfish (left panel; Williams et al. 2000) and lingcod south of 40°10N (right panel; Jagielo et al. 1997, 2000).

depend on the results of the stock assessment because the stock assessment would be used to determine the level of fishing mortality consistent with recovery.

Movement of animals age 1 and older is assumed to depend on age. The bulk of the scenarios are based on the assumption that the probably of migrating from a cell increases with age (from 50% of the maximum rate of movement at age 4 to 95% of this rate by age 8), although scenario G examines the impact of the probability of moving decreasing with age. Movement from a cell occurs to all other cells, and the probability of an animal moving from one cell to another depends on the distance between the cells so that the probability of movement between adjacent cells is greater than between nonadjacent cells (see equation A8). The highest value for the asymptotic annual movement rate (0.3) implies that there is a probability of more than 99% that an animal moves from its current cell over a 50-year period while the lowest movement rate (0.01) corresponds to a probability of 60% of movement from a cell over a 50-year period.

The data available to the assessment methods are assumed to be survey estimates of abundance, the age-composition of the survey catches, and the age-composition of the fishery catches. Although the actual assessments of lingcod and widow rockfish (Jagielo et al. 2000; Williams et al. 2000) are based on additional information, the analyses of this paper are restricted to the types of data commonly available for assessment purposes. Surveys are assumed to be conducted every third year starting in simulation year 13, and the surveys are assumed to provide an index of abundance (with an associated coefficient of variation) and age-composition data within each cell. Fishery age-composition data are assumed to be available for a cell when it

is open to fishing. The bulk of the calculations of this paper are predicated on the assumption that surveys would continue to operate in closed areas, although scenario H examines the implications of there being no surveys in the MPA. The extent of variation in survey abundance across all cells (σ_p) is set to 0.25 while the total number of animals aged is taken to be 100 (under the assumption of multinomial sampling).

The remaining parameters of the operating model (Table 2; Figure 3) are the same for all of the simulations. All of the scenarios are conducted when cells 1–4 are an MPA and when there is no MPA at all so that it is possible to detect the impact of implementing an MPA on assessment performance. Results are not, however, shown for all combinations of ways of implementing the stock assessment method and options for specifying the operating model to reduce the volume of results presented.

The Assessment Models

The assessment model (Appendix 2) is based on an age-structured and sex-structured population dynamics model. Unlike the operating model, it assumes that fishery selectivity is time-invariant, and the catch is taken in a pulse in the middle of the year. The first assumption was made because attempting to estimate time-dependence in selectivity did not improve estimation performance even when fishery selectivity changed over time. Assuming that the catch is taken in the middle of the year rather than continuously throughout the year reduces the time needed to fit the model to the data but has little quantitative impact on the results because the fishing mortality rates are seldom more than 20% per year.

Table 2.—The fixed parameters of the operating model.

Parameter	Value
Plus-group age, x	20 years
Natural mortality, M	0.15/year (widow rockfish); 0.18/year (lingcod)
Variation in attractiveness, $\sigma_{\bar{q}}$	0.4
Steepness, h	0.4 (widow rockfish); 0.9 (lingcod)
Recruitment variation, σ_T	0.424
Recruitment variation, σ_R	0.424
Correlation in selectivity ρ_S	0
Age at 50% movement, a_{50}	4
Change in movement with age, δ	1.358 (baseline)
Variation in movement, σ_X	2
Variation in selectivity, σ_S	0 (baseline)

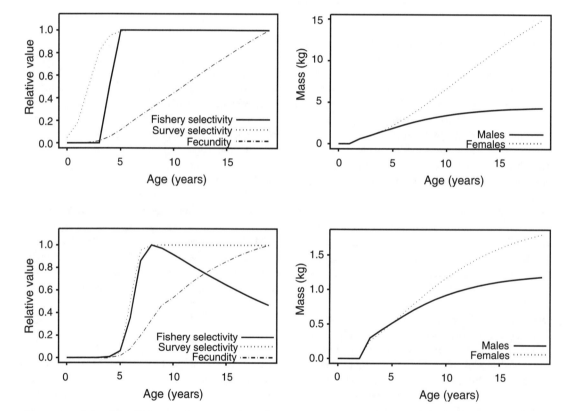

Figure 3.—Selectivity (fishery and survey), fecundity at age, and mass at age for lingcod (upper panels) and widow rockfish (lower panels).

The assessment model can be applied to a number of "assessment cells" simultaneously. Each assessment cell is one or more of the cells in the operating model. The population in each assessment cell is assumed to be homogeneously distributed across the area being assessed and, apart from correlation in recruitment among assessment cells, the populations in each

assessment cell are assumed to be independent (i.e., in the assessment it is assumed that there is no movement of age-1 and older animals among assessment cells).

The parameters of the assessment model (four fishery selectivity parameters, two survey selectivity parameters, survey catchability for each assessment cell, mean recruitment for each assessment cell, and a recruitment residual for each year for each assessment cell) are estimated by minimizing an objective function. This function includes contributions from the survey index information, the age-composition data from the surveys, the age-composition for the fishery catches, and a constraint on the values for the recruitment residuals. This latter constraint prevents the recruitment residuals being set to unrealistic values to fit some "quirk" in the data and includes the assumption that recruitment for the various assessment cells are correlated a priori (which is true for the bulk of the simulations). For simplicity, the effective sample sizes needed to calculate the contribution of the age-composition data to the objective function are taken to be the actual number of fish aged.

Two variants of the assessment model are examined: (1) a case when all ten cells are treated as a single assessment cell, and (2) a case in which cell groups 1–4, 5–6, and 7–10 are each treated as assessment cells. Assessment variant 1 has much fewer estimable parameters than assessment variant 2 so should be more precise, but it ignores the impact of spatial differences in population structure so may be more biased. The survey and fishery selectivity patterns for case 2 are assumed to be the same for each of the three assessment cells (fishery and survey selectivity are, of course, estimated separately) to reduce the number of estimable parameters.

Summary Statistics

Each simulation trial involved 100 replicates. The results from a simulation study are always voluminous, and this study is no exception. For each trial, the actual values and the assessment model predictions are compared when the MPA is first implemented and 20, 40, and 60 years afterward. The quantities examined are the annual spawning outputs (aggregated over all cells), spawning output expressed relative to B_0, fishing mortality in the area open to fishing, and recruitment. The statistic used to measure the differences is the relative error, expressed as a percentage:

$$100 \times \frac{X^{est} - X^{true}}{X^{true}}, \qquad (1)$$

where X^{est} is the estimate of X and X^{true} is the true (i.e., operating model) value of X. To further summarize the results, the output from the simulations is summarized by the median of the absolute values of the relative errors. This statistic captures both bias and variance and is less sensitive to outlying estimates than, for example, the mean square error.

Results

Prior to examining the results, it is necessary to determine that the operating model behaves as expected. Figures 4 and 5 show time trajectories of spawning output (expressed relative to B_0) for a single simulation for cells 1–10 combined and for cells 1–4 (the closed area), 5–6 (an area of mixing), and 7–10 for operating model scenarios A–D. The extent to which the time trajectories differ among cells 1–4, 5–6, and 7–10 is determined by the extent of movement of adults among areas and how larval dispersal operates. In particular, these differences are largest when the extent of adult migration equals its lowest value (0.01; scenario A) and when density dependence is local (equation A3b; scenario D).

The time trajectory of spawning output for cells 1–4 is more similar to that for cells 5–6 than to that for cells 7–10 because, by design (see equation A8), adult movement from the MPA is greatest to the cells open to fishing closest to the closed area. There are notable qualitative differences between the population trajectories for lingcod (Figure 5) and those for widow rockfish (Figure 4). The impact of the MPA is explored further for lingcod in Figure 6, which shows the population age structure in cells 1–4, 5–6, and 7–10 when the MPA is declared (year 40 for widow rockfish and year 50 for lingcod), after an additional 40 years (when the total spawning output has recovered to $0.4B_0$), and after an additional 20 years (the last year of the simulation period considered in this paper). The population age structure in the MPA is virtually pristine again after 60 years of protection. The impact of movement of animals out of the MPA is evident by the small number of older animals in cells 5–6; such animals are almost absent from cells 7–10. This is because fishing mortality is chosen so that the total stock (which is dominated by animals in the MPA) recovers to $0.4B_0$, thus allowing high fishing mortality outside of the MPA.

Estimation Error for a Single Scenario

Figures 7–10 show time trajectories of relative error for widow rockfish and lingcod for the baseline scenario A. Figures 7 and 8 show results for the four management-related quantities after 20 years of the implementation of the MPA, while Figures 9 and 10

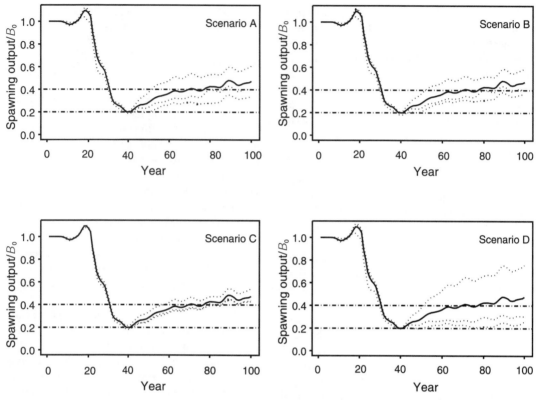

Figure 4.—Time trajectories of spawning output (expressed relative to B_0) for widow rockfish for a single simulation for operating model variants A–D (see Table 1). The solid lines are the time trajectories for the whole population and the dotted lines are the time trajectories for the populations in cells 1–4 (upper dotted line), 5–6 (middle dotted line), and 7–10 (lower dotted line). The horizontal dashed lines indicate the proxy for B_{MSY} and the size of the resource when the MPA is implemented.

show results for one management quantity (the ratio of the current spawning output to B_0) when the MPA is declared and 20, 40, and 60 years thereafter. The results are summarized by the median and 90% intervals for the relative error, and separate results are shown when the assessment method is applied to data pooled over all cells (solid lines) and when the data are pooled into three assessment cells (corresponding to cells 1–4, cells 5–6, and cells 7–10 of the operating model).

Some differences occur between conducting a spatially-aggregated assessment for widow rockfish and assessing the total area using three assessment cells; the spatially-aggregated assessments tend to lead to wider 90% intervals for relative error. The spatially-aggregated assessment fails to predict fishing mortality in the open area (Figure 7, lower right panel), although this is not surprising given that the spatially-aggregated assessment has no real notion

of "open area." Therefore, in some sense, this is mainly a definitional problem. The extent of this bias is reduced when the extent of movement is increased from 0.01 per year to 0.3 per year and is negligible if there is no MPA (results not shown). The diverging results over time in Figure 7 are probably predominately due to the spatially-aggregated model not being aware of the large difference in population age-composition between the MPA and the fished areas.

In contrast to the situation for widow rockfish, a spatially aggregated assessment performs much more poorly for lingcod than a spatially disaggregated assessment (Figures 8, 10). This result is, perhaps, not surprising given that the differences among cells in population trajectories is much greater for lingcod than for widow rockfish (see Figures 4, 5). This may be due to the rapid growth of female lingcod which results in a large accumulation of spawning biomass in the MPA.

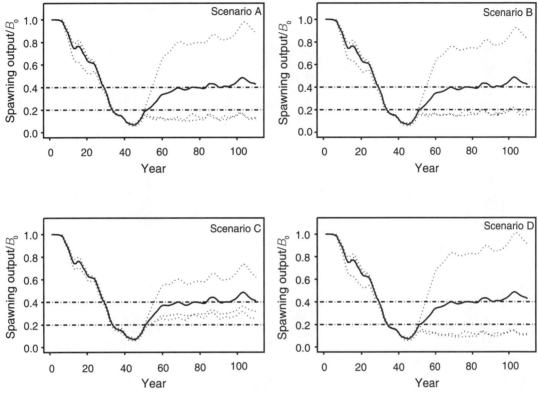

Figure 5.—Time trajectories of spawning output (expressed relative to B_0) for lingcod for a single simulation for operating model variants A–D (see Table 1). The solid lines are the time trajectories for the whole population and the dotted lines are the time trajectories for the populations in cells 1–4 (upper dotted line), 5–6 (middle dotted line), and 7–10 (lower dotted line). The horizontal dashed lines indicate the proxy for B_{MSY} and the size of the resource when the MPA is implemented.

The patterns in relative error for spawning output expressed relative to B_0 are similar for assessments conducted 20, 40, and 60 years after the MPA is implemented (Figures 9, 10), suggesting that it is primarily the transient effects of implementing the MPA that lead to the severe deterioration in the performance of the spatially aggregated assessment method.

The estimates of annual recruitment and spawning output are least precise for the earliest and most recent years. This is hardly unexpected given that few data are available for the early years and the cohorts that constitute the population in the last years of the assessment period have only been indexed by surveys and the fishery for a few years. Although the estimates for widow rockfish and (particularly) lingcod are imprecise (wide 90% intervals), the estimates tend to be close to unbiased (in median terms) when the MPA and non-MPA data are kept separate. This is encouraging and implies that assessment results will not be

seriously impacted when an MPA is implemented if appropriate account of this is taken when conducting assessments.

The remaining calculations in this paper are based on the variant of the assessment method that fits separate population dynamics models to cells 1–4, cells 5–6, and cells 7–10, as this approach to assessment is clearly superior to the spatially aggregated assessment method (particularly for lingcod).

The Implications of Having an MPA

Figures 11 and 12 show medians and 90% intervals for relative error for four management-related quantities (widow rockfish and lingcod, respectively) for scenario A for cases in which there is an MPA in cells 1–4 and in which there is no MPA at all. In general, there are no major differences between the results for the two cases, although there is some evidence for

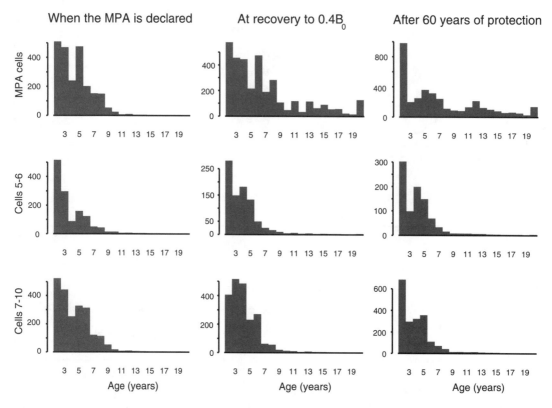

Figure 6.—Population age structure for lingcod after 50 years (when the MPA is declared), 90 years (when the resource recovers to $0.4B_0$) and 110 years (after 60 years of protection) for a single simulation for scenario A. Results are shown for the MPA (cells 1–4), the area of mixing (cells 5–6), and cells 7–10.

slightly greater precision when there is no MPA (fishing mortality in the open area for both species, and generally for lingcod).

The MPA should include more large and old animals (e.g., Figure 6). Therefore, it could be expected that having data from an MPA will lead to improved estimation of some key management-related parameters even if the estimates of overall biomass are not markedly impacted. Sladek Nowlis and Roberts (1998) argue that MPAs may offer the potential to generate more accurate estimates of natural mortality because the confounding impact of fishing mortality is eliminated. This assumes, of course, that the MPAs are sited randomly with respect to all ecosystem factors that may affect natural mortality for the species of interest. Two parameters that may be estimated more adequately when data from an MPA are available are the instantaneous rate of natural mortality (M) and the selectivity on the oldest age-class. The latter is particularly important because declining selectivity with age implies a "refuge" from fishing.

Figure 13 shows the relative error distributions for the selectivity of widow rockfish at age 20 based on data generated for years 1–60 (i.e., until 20 years after the implementation of an MPA) using the baseline operating model for cases in which there is and is not an MPA in cells 1–4. There is no discernable improvement in estimation performance if there is an MPA. Figure 14 shows the estimates of M for widow rockfish after 100 years (i.e., 60 years after the implementation of an MPA) for scenario I. Although the estimates based on the MPA are closer to the true value (0.15/year), the improvement in estimation ability is negligible.

Sensitivity Analyses

The results in this section are restricted to the median absolute relative error for the spawning output at the start of year 60 (widow rockfish) and year 70 (lingcod) and the median absolute relative error for spawning output relative to B_0 at these points in time. These

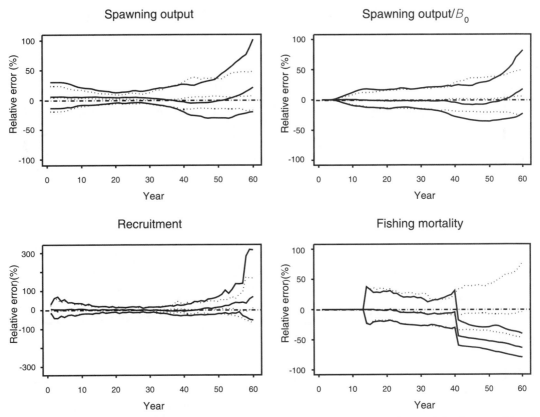

Figure 7.—Median and 90% intervals for the annual relative errors (expressed as percentages) for four quantities of interest for widow rockfish (operating model scenario A). Results are shown when the stock assessment method is applied to all ten cells simultaneously (solid lines) and when it is applied to the data separated into three assessment cells defined by operating model cells 1–4, 5–6, and 7–10 (dotted lines). The MPA is implemented in year 40.

two management-related quantities were chosen because management decisions for U.S. West Coast groundfish species depend critically on them; whether a stock is declared overfished or not depends on the size of the current spawning output relative to B_0, and the annual optimum yield depends on the absolute size of the biomass.

Table 3 shows the values for these statistics for the spatially disaggregated stock assessment method for the scenarios in Table 1. No results are shown for scenario F for lingcod because the selectivity pattern for lingcod is asymptotic rather than being dome-shaped (Figure 3). There are some general features in Table 3 related to the impact of MPAs on the performance of spatially disaggregated assessment methods.

(1) The extent of error tends (with one exception) to increase as the extent of movement increases (compare the results for scenario B ($X_{max} = 0.05$ in Table 3), and particularly C

($X_{max} = 0.3$), with those for scenario A. Increased movement will tend to further violate the assumption implicit when applying the assessment method that the area being assessed contains a closed population that is distributed homogenously over the area being assessed.

(2) Performance is worse if the probability of movement decreases with age. This is probably due to scenario G (movement increases with age; Table 3), increasing the effective number of animals moving among cells by reducing the first age at which they can move, so scenario G behaves rather like a scenario in which X_{max} is increased slightly.

(3) Performance is not very sensitive to how larval dispersal and density dependence are modeled (scenarios A, D, and E), probably because the number of recruits entering the population

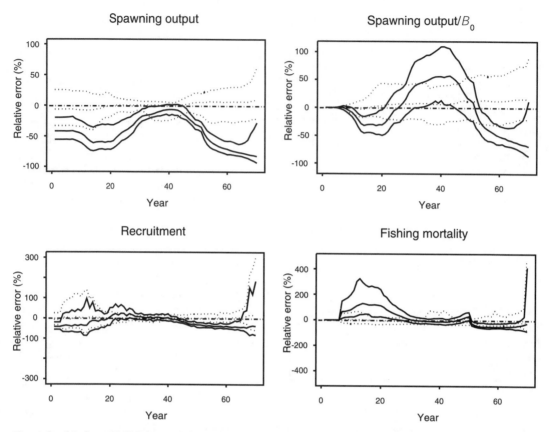

Figure 8.—Median and 90% intervals for the annual relative errors (expressed as percentages) for four quantities of interest for lingcod (operating model scenario A). Results are shown when the stock assessment method is applied to all ten cells simultaneously (solid lines) and when it is applied to the data separated into three assessment cells defined by operating model cells 1–4, 5–6, and 7–10 (dotted lines). The MPA is implemented in year 50.

each year are treated as estimable parameters and not related to the size of the spawning output when conducting stock assessments.

(4) Except in one case (spawning output for lingcod) having no survey (or fishery) data for the MPA (scenario H) leads, as expected, to poorer estimation performance. Note that the extent of deterioration in estimation performance is not very large, possibly because information on recruitment in the MPA can be obtained from the estimates of recruitment for open areas and because the total sample size for the survey age-composition does not change (i.e., the sample sizes for the area open to fishing are increased), and the coefficients of variation for the surveys in the open area are reduced compared to the baseline scenario.

(5) Estimating M rather then using the true value for this parameter (scenario I in Table 3) leads, as expected, to poorer estimation performance, and this is also the case when selectivity varies over time but the assessment assumes that selectivity is time invariant.

(6) Performance is generally poorer when the MPA is restricted to cells 1–2 only (scenario J in Table 3).

(7) The impact of MPAs on assessment performance is case specific. For example, the extent to which performance deteriorates when X_{max} is increased from 0.05 to 0.3 is much greater for lingcod than for widow rockfish.

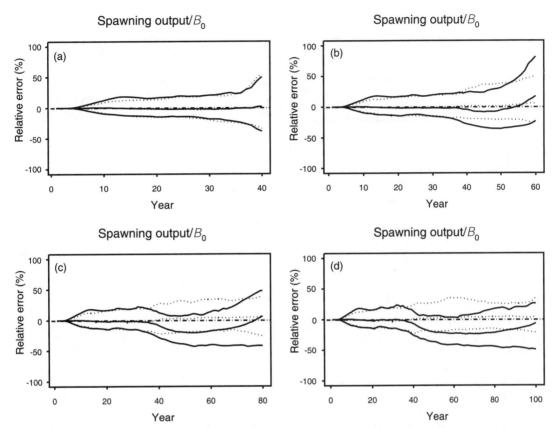

Figure 9.—Median and 90% intervals for the annual relative errors (expressed as percentages) for spawning output expressed relative to B_0 for widow rockfish (operating model scenario A) for assessments when the MPA is first implemented (year 40; a), and after the MPA has been in place for 20 (b), 40 (c), and 60 (d) years. Results are shown when the stock assessment method is applied to all 10 cells simultaneously (solid lines) and when it is applied to the data separated into three assessment cells defined by operating model cells 1–4, 5–6, and 7–10 (dotted lines). The MPA is implemented in year 40.

Discussion

Performance of Assessment Methods in the Presence of MPAs

The results of the study suggest that the (negative) impact of MPAs on assessment ability is relatively slight if appropriate account is taken of differences in population structure due to movement when conducting assessments. The assessment approach considered in this paper to deal with MPAs was to divide the area to be assessed into several assessment cells and to fit population models to the data for each assessment cell that ignore movement of adults among assessment cells and that assume that recruitment is unrelated to spawning output but is correlated among the assessment cells. It may be possible to further improve estimation per-

formance if data on tagging and larval movement patterns are available. However, testing of such estimators is beyond the scope of the present study.

Although having an MPA could improve the ability to estimate some quantities of management interest (such as natural mortality), the results obtained in this study suggest that the extent of such improvement (if any) will be slight. However, this result is almost certainly case specific. For example, Punt et al. (2001) found that having data for unfished populations of coral trout *Plectropomus leopardus* substantially improved the ability to estimate M. However, the simulations of Punt et al. (2001) were based on the assumption of no movement of adults.

The estimators were relatively successful in this study, in contrast to studies based on operating models with complicated movement patterns (e.g., Punt et

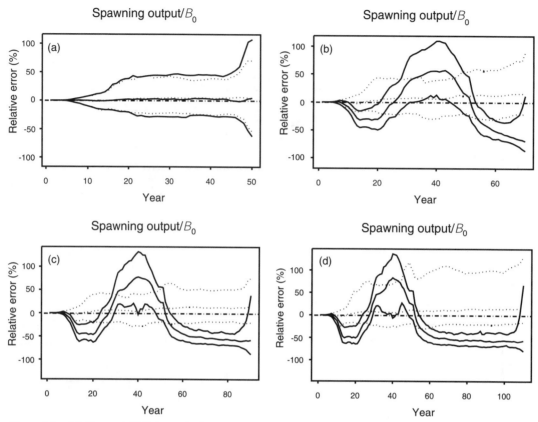

Figure 10.—Median and 90% intervals for the annual relative errors (expressed as percentages) for spawning output expressed relative to B_0 for lingcod (operating model scenario A) for assessments when the MPA is first implemented (year 50;a), and after the MPA has been in place for 20 (b), 40 (c), and 60 (d) years. Results are shown when the stock assessment method is applied to all ten cells simultaneously (solid lines) and when it is applied to the data separated into three assessment cells defined by operating model cells 1–4, 5–6, and 7–10 (dotted lines). The MPA is implemented in year 50.

al. 2002). However, unlike the operating model of Punt et al. (2002), the operating model of this study did not allow for very realistic fleet dynamics and stochastic or density-dependent movement. Incorporation of these factors will lead to a deterioration in estimation performance, but whether this will be compounded by having an MPA is unknown.

Caveats and Future Work

The simulations described above were based on the assumption that mass at age and fecundity at age were known and time invariant. Although beyond the scope of the present paper, MPAs may provide the opportunity to better characterize growth and fecundity of large individuals. While improved information on the size and fecundity of large individuals is unlikely to im-

prove estimates of the current biomass of stocks that are presently highly depleted (such stocks will have few large individuals), it will improve estimates of management reference points that are defined to be a prespecified percentage of B_0, such as B_{MSY}, because the population will consist of more large, old individuals at these levels.

Another assumption not examined in this paper, and on which the benefits to stock assessment of MPAs critically depends, is that the values for population dynamics processes such as growth and natural mortality are not different in the fished area and in the MPA. If these processes are density dependent, as might be expected and as has been shown in some cases (Trippel 1995; Helser and Brodziak 1998), the ability to extrapolate estimates based on data collected within the MPA to the fished area will be limited.

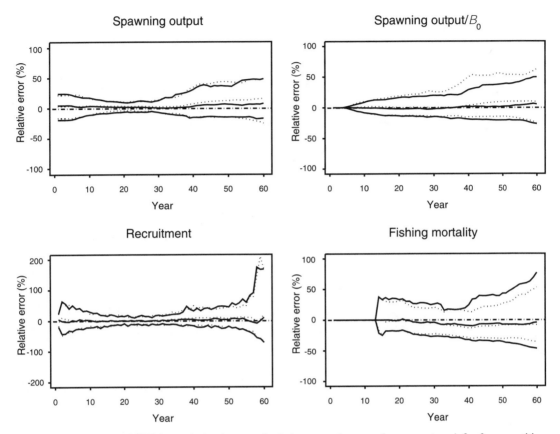

Figure 11.—Median and 90% intervals for the annual relative errors (expressed as percentages) for four quantities of interest for widow rockfish (operating model scenario A) when the stock assessment method is applied to the data separated into three assessment cells. Results are shown when there is an MPA (solid lines) and when there is no MPA (dotted lines).

The purpose of this paper was to examine the impact of imposing an MPA on the ability to estimate the quantities on which fisheries management advice is usually based. Several studies (e.g., Polacheck 1990; Sladek Nowlis and Roberts 1998; Apostolaki et al. 2002) have argued that higher yields can be achieved through the imposition of MPAs than through control of fishing effort. Cases in which this could occur include, for example, when the MPA reduces growth overfishing by protecting juveniles (Guénette et al. 1998). However, the bulk of these studies have been predicated on the assumptions that fishing mortality is currently not controlled, that future management is based on setting a target level of fishing mortality, and that the managers have perfect information about stock size as well as the biological parameters of the population. Exceptions to these assumptions exist. For ex-

ample, Lauck et al. (1998) based their projections on the assumption that catch (rather than effort) limits are set. No account has been taken in previous studies of the impact of having an MPA on the ability to estimate the quantities of interest to management.

The present paper represents the first attempt to examine the impact of MPAs on the performance of stock assessment methods. Although it seems likely that the general conclusions of this study are robust to adjustments to many of the features of the operating model, future work needs to examine the impacts on estimation performance of, inter alia, different movement patterns, including whether movement is density dependent, fleet dynamics, different density dependence assumptions (such as that density dependence acts on larval survival at the cell level and that the number of larvae recruiting to a cell depends

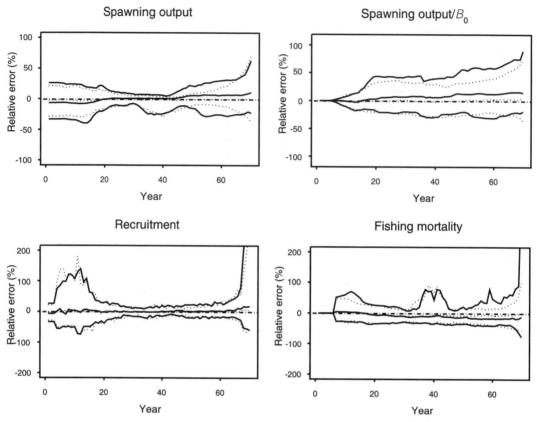

Figure 12.—Median and 90% intervals for the annual relative errors (expressed as percentages) for four quantities of interest for lingcod (operating model scenario A) when the stock assessment method is applied to the data separated into three assessment cells. Results are shown when there is an MPA (solid lines) and when there is no MPA (dotted lines).

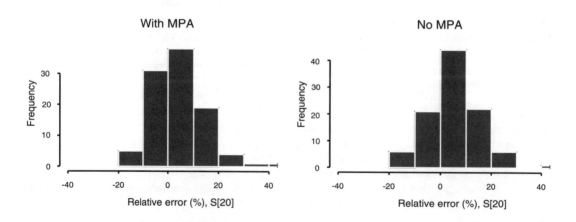

Figure 13.—Distributions of relative error for the selectivity on widow rockfish aged 20 years for the baseline (scenario A) operating model. Results are shown for cases in which there is an MPA and in which there is no MPA.

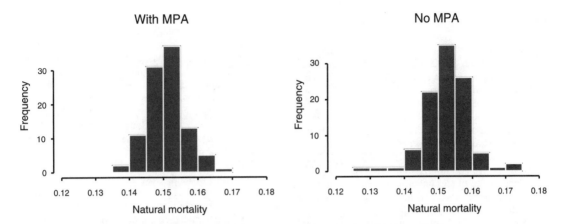

Figure 14.—Distributions for the estimate of M for widow rockfish for the operating model in which the biological assumptions are the same as those for the baseline operating model but in which M is treated as an estimable parameter (scenario I). Results are shown for cases in which there is an MPA and in which there is no MPA.

Table 3.—Median absolute relative errors for two management-related quantities (current spawning output [B] and current spawning output expressed relative to B_0) for widow rockfish and lingcod, for assessments conducted 20 years after an MPA is implemented. Results are shown for ten operating model scenarios (see Table 1 for more details) and for spatially disaggregated assessment models.

Scenario: Quantity	Widow rockfish		Lingcod	
	B	B/B_0	B	B/B_0
A: Baseline	12.47	12.89	15.43	18.58
B: X_{max} = 0.05	12.30	14.07	19.01	24.88
C: X_{max} = 0.3	17.01	17.15	56.07	58.19
D: Equation 3b	12.24	10.31	14.64	17.65
E: Equation 3c	12.68	13.12	12.25	22.40
F: σ_s = 1	52.21	43.97		
G: Movement increases with age	14.48	15.76	26.02	32.30
H: No survey in MPA	14.04	16.97	13.03	18.54
I: M estimated	12.72	19.25	25.34	36.48
J: MPA in cells 1-2 only	16.06	24.12	16.11	21.02

on cell-specific density dependence), and having a network of MPAs rather than just a single MPA. Only two scenarios regarding the spatial structure of the resource and the management system were examined in this paper; further work should examine the consequences of having several rather than only one MPA.

The set of quantities examined in this paper is not the full set that would be of interest to the decision makers. For example, there would be interest in how well it is possible to estimate the fraction of the population in the MPA. Finally, this study has been predicated on a data-moderate to data-rich situation. In contrast, MPAs are being implemented in data-poor situa-tions where the method of stock assessment, if any, is likely to be quite different from that examined in this paper. Future work needs to examine the performance of data-poor assessment methods in the presence of MPAs.

Acknowledgments

AEP acknowledges funding through National Marine Fisheries Service grant NA07FE0473. Loo Botsford, Kevin Piner, Pamela Mace, Lisa Wooninck, and an anonymous reviewer are thanked for their comments on an earlier version of this paper.

References

Apostolaki, P., E. J. Milner-Gulland, G. Kirkwood, and M. McAllister. 2002. Modelling the effects of establishing a marine reserve for mobile fish species. Canadian Journal of Fisheries and Aquatic Sciences 59:405–415.

Beattie, A., U. R. Sumaila, V. Christensen, and D. Pauly. 2002. A model for the bioeconomic evaluation of marine protected area size and placement in the North Sea. Natural Resource Modeling 15:413–437.

Botsford, L. W., A. Hastings, and S. D. Gaines. 2001. Dependence of sustainability on the configuration of marine reserves and larval dispersal distance. Ecology Letters 4:144–150.

Botsford, L. W., F. Micheli, and A. Hastings. 2003. Principles for the design of marine reserves. Ecological Applications 13(supplement):S25–S31.

Butterworth, D. S., G. S. Hughes, and F. Strumpfer. 1990. VPA with ad hoc tuning: implementation for disaggregated fleet data, variance estimation, and application to the horse mackerel stock in ICSEAF divisions 1.3+1.4+1.5. South African Journal of Marine Science 9:327–357.

de la Mare, W. K. 1986. Fitting population models to time series of abundance data. Report of the International Whaling Commission 36:399–418.

Fournier, D., and C. P. Archibald. 1982. A general theory for analyzing catch at age data. Canadian Journal of Fisheries and Aquatic Sciences 39:1195–1207.

Gavaris, S., 1988. An adaptive framework for the estimation of population size. Canadian Atlantic Fisheries Scientific Advisory Committee (CAFSAC), Research Document 88/29, Dartmouth, Nova Scotia.

Guénette, S., T. Lauck, and C. Clark. 1998. Marine reserves: from Beverton and Holt to the present. Reviews in Fish Biology and Fisheries 8:251–272.

Guénette, S., and T. J. Pitcher. 1999. An age-structured model showing the benefits of marine reserves in controlling overexploitation. Fisheries Research 39:295–303.

Hastings, A., and L. W. Botsford. 1999. Equivalence in yield from marine reserves and traditional fisheries management. Science 284:1537–1538.

Helser, T. E., and J. K. T. Brodziak. 1998. Impacts of density-dependent growth and maturation on assessment advice to rebuild depleted U.S. silver hake (Merluccius bilinearis) stocks. Canadian Journal of Fisheries and Aquatic Sciences 55:882–892.

Holland, D. S. 2002. Integrating marine protected areas into models for fishery assessment and management. Natural Resource Modeling 15:369–386.

Jagielo, T., P. Admas, M. Peoples, S. Rosenfield, K. Silberberg, and T. Laidig. 1997. Assessment of lingcod in 1997. In Appendix to status of Pacific

Coast groundfish fishery through 1997 and recommended acceptable biological catches for 1998. Stock assessment and fishery evaluation. Pacific Fishery Management Council, Portland, Oregon.

Jagielo, T., D. Wilson-Vandenberg, J. Sneva, S. Rosenfield, and F. Wallace. 2000. Assessment of lingcod (Ophiodon elongates) for the Pacific Fishery Management Council in 2000. In Appendix to status of Pacific Coast groundfish fishery through 2000 and recommended acceptable biological catches for 2001. Stock assessment and fishery evaluation. Pacific Fishery Management Council, Portland, Oregon.

Lauck, T., C. W. Clark, M. Mangel, and G. M. Munro. 1998. Implementing the precautionary principle in fisheries management through marine reserves. Ecological Applications 8(supplement):S72–S78.

Man, A., R. Law, and N. V. C. Polunin. 1995. Role of marine reserves in recruitment to reef fisheries: a metapopulation model. Biological Conservation 71:197–204.

Methot, R. D. 1993. Synthesis model: an adaptable framework for analysis of diverse stock assessment data. International North Pacific Fisheries Commission Bulletin 50:259–277.

Methot, R. D. 2000. Technical description of the stock synthesis assessment program. NOAA Technical Memorandum NMFS-NWFSC-43.

NRC (National Research Council). 1998. Improving fish stock assessments. National Academy Press, Washington, D.C.

Patterson, K. R., and G. P. Kirkwood. 1995. Comparative performance of ADAPT and Laurec-Shepherd methods for estimating fish population parameters and in stock management. ICES Journal of Marine Science 52:183–196.

Pezzey, J. C. V., C. M. Roberts, and B. T. Urdal. 2000. A simple bioeconomic model of a marine reserve. Ecological Economics 33:77–91.

PFMC (Pacific Fishery Management Council). 2003. Marine reserves. Available: www.pcouncil.org/reserves/reservesback.html (October 2003).

Polacheck, T. 1990. Year around closed areas as a management tool. Natural Resource Modeling 4:327–354.

Punt, A. E., A. D. M. Smith, and G. Cui. 2002. Evaluation of management tools for Australia's South East Fishery. 2. How well do commonly-used stock assessment methods perform? Marine and Freshwater Research 53:631–644.

Punt, A. E., A. D. Smith, A. J. Davidson, B. M. Mapstone, and C. R. Davies. 2001. Evaluating the scientific benefits of spatially explicit experimental manipulations of common coral trout (Plectropomus leopardus) on the Great Barrier Reef, Australia. Pages 67–103 in G. H. Kruse, N. Bez, A. Booth, M. W. Dorn, S. Hills, R. N. Lipcius, D. Pelletier, C. Roy, S. J. Smith, and D. Withrell, editors. Spatial processes and management of marine populations. University of Alaska Sea Grant Report AK-SG-01-02.

Restrepo, V. R., and J. E. Powers. 1999. Precautionary control rules in US fisheries management: specification and performance. ICES Journal of Marine Science 56:846–852.

Sampson, D. B., and Y. Yin. 1998. A Monte Carlo evaluation of the stock synthesis assessment program. Pages 315–338 in F. Funk, T. J. Quinn, II, J. Heifetz, J. N. Ianelli, J. E. Powers, J. F. Schweigert, P. J. Sullivan, and C. I. Zhang, editors. Fishery stock assessment models. University of Alaska Sea Grant Report AK-SG-98-01.

Sladek Nowlis, J., and C. M. Roberts. 1998. Fisheries benefits and optimal design of marine reserves.

U.S. National Marine Fisheries Service Fishery Bulletin 97:604–616.

Sumaila, U. R. 2002. Marine protected area performance in a model of the fishery. Natural Resource Modeling 15:439–451.

Trippel, E. A. 1995. Age at maturity as a stress indicator in fisheries. BioScience 45:759–771.

Williams, E. H., A. D. MacCall, S. V. Ralston, and D. E. Pearson. 2000. Status of the widow rockfish resource in Y2K. In Appendix to status of Pacific Coast groundfish fishery through 2000 and recommended acceptable biological catches for 2001. Stock assessment and fishery evaluation. Pacific Fishery Management Council, Portland, Oregon.

Appendix A: The Operating Model

Basic Dynamics

The dynamics of the population are governed by the equation

$$N_{y+1,a}^{g,c} = \begin{cases} 0.5R_{y+1}^c & \text{if } a = 0 \\ \sum_{c'} X_{a-1}^{c',c} N_{y,a-1}^{g,c'} e^{-(M+S_{y,a-1}F_y^{c'})} & \text{if } 0 < a < x \\ \sum_{c'} X_{x-1}^{c',c} N_{y,x-1}^{g,c'} e^{-(M+S_{y,x-1}F_y^{c'})} \\ \quad + \sum_{c'} X_x^{c',c} N_{y,x}^{g,c'} e^{-(M+S_{y,x}F_y^{c'})} & \text{if } a = x \end{cases},$$

(A1)

where $N_{y,a}^{g,c}$ is the number of animals of sex g (f = female, m = male) and age a in cell c (the total area considered in the operating model consists of $n = 10$ cells) at the start of year y,

$X_a^{c',c}$ is the fraction of animals of age a in cell c' that move to cell c,

M is the instantaneous rate of natural mortality (assumed to be independent of age, sex, and location),

$S_{y,a}$ is the selectivity of the fishing gear on animals of age a during year y, and

F_y^c is the fully-selected fishing mortality in cell c during year y:

$$F_y^c = \tilde{q}^c \chi_y^c \tilde{F}_y,$$

(A2)

where χ_y^c is 1 if cell c is open to fishing during year y and 0 otherwise,

$\tilde{F}_y$ is the fully selected fishing mortality in cells open to fishing during year y,

$\tilde{q}^c$ is a measure of the attractiveness of cell c to fishing, $\log_e \tilde{q}^c \sim N(0; \sigma_{\tilde{q}}^2)$,

$\sigma_{\tilde{q}}^2$ is the variability in the relative attractiveness of different cells to the fishery,

R_y^c is the recruitment to cell c during year y, and

x is the maximum age, taken to be a plus group.

The age-structure, sex-structure, and spatial-structure of the population at the start of the first year are taken to be those of a population at its deterministic unfished equilibrium.

Recruitment

The number of 0-year-olds in cell c at the start of year y is related to the spawner stock size by means of a Beverton–Holt stock–recruitment relationship:

$$R_y^c = \begin{cases} \lambda^c R_0^T \dfrac{B_y^T / B_0^T}{\alpha + \beta(B_y^T / B_0^T)} e^{\varepsilon_y - \sigma_T^2/2} e^{\eta_y^c - \sigma_R^2/2} & \text{(A3a)} \\[3mm] \lambda^c R_0^T \dfrac{B_y^c / B_0^c}{\alpha + \beta(B_y^c / B_0^c)} e^{\varepsilon_y - \sigma_T^2/2} e^{\eta_y^c - \sigma_R^2/2} & \text{(A3b)} \\[3mm] \lambda^c R_0^T \sum_{c'} \dfrac{\lambda^{c'} B_y^{c'} / B_0^{c'}}{\alpha + \beta(B_y^{c'} / B_0^{c'})} e^{\varepsilon_y - \sigma_T^2/2} e^{\eta_y^c - \sigma_R^2/2} & \text{(A3c)} \end{cases}$$

where B_y^c is the spawning output in cell c at the start of year y (a subscript of 0 indicates the spawning output in the pre-exploitation equilibrium state):

$$B_y^c = \sum_{a>0} f_a N_{y,a}^{f,c} , \qquad (A4)$$

f_a is the fecundity of a female of age a,

B_y^T is the total (summed over all cells) spawning output at the start of year y,

R_0^T is the total number of 0-year-olds in a pre-exploitation equilibrium state,

λ^c is the fraction of the total recruitment that recruits to cell c (set to $1/n$ for the calculations of this paper),

σ_T^2 is the variation in recruitment that is common across cells, $\varepsilon_y \sim N(0;\sigma_T^2)$,

σ_R^2 is the variation in recruitment that is cell-specific, $\eta_y^c \sim N(0;\sigma_R^2)$, and

α and β are the parameters of the stock–recruitment relationship (parameterized in terms of the steepness of the stock–recruitment relationship and R_0^T).

Equation A3a is based on the assumption that density dependence is a function of the total spawning output and that the expected fraction of the total recruitment that recruits to each cell is independent of the size of the spawning output in that cell. Equation A3b is based on the assumption that the recruitment to a cell depends only on the amount of spawning output in that cell, and equation A3c assumes that density dependence acts at the cell level but that the larvae which survive density-dependent mortality are distributed randomly among the n cells.

Catches

The catch (in mass) during year y from cell c is given by

$$C_y^c = \sum_g \sum_{a=0}^x w_a^g \frac{S_{y,a} F_y^c}{M + S_{y,a} F_y^c} N_{y,a}^{g,c} [1 - e^{-(M+S_{y,a}F_y^c)}] \quad , (A5)$$

where w_a^g is the mass of an animal of sex g and age a.

Selectivity

Fishery selectivity is modeled by a double-logistic equation that allows the age at 50% selectivity to vary over time:

$$S_{y,a} = S'_{y,a} / \max_{a'} S'_{y,a'};$$

$$S'_{y,a} = \frac{1}{1+e^{-\delta_1(a-a_{50}^1+\gamma_y)}} \times \frac{1}{1+e^{-\delta_2(a_{50}^2-a)}} \quad , (A6)$$

where a_{50}^1, a_{50}^2, δ_1, and δ_2 are the parameters of the double-logistic equation,

γ_y is the deviation from the average selectivity pattern during year y:

$$\gamma_y = \rho_S \gamma_{y-1} + \varepsilon_y^S \qquad \varepsilon_y^S \sim N(0;\sigma_S^2) \quad , \quad (A7)$$

ρ_S is the interannual correlation in the deviations from average selectivity, and

σ_S is a measure of the standard deviation of the interannual deviations in average selectivity.

Movement

The fraction of animals of age a in cell c' that move to cell c ($X_a^{c',c}$) is given by

$$X_a^{c',c} = \begin{cases} 1 - \dfrac{X_{max}}{1+e^{-(a-a_{50})/\delta}} & \text{if } c = c' \\[2em] (1-X_a^{c',c'}) \dfrac{\exp\left[-\dfrac{(L_c - L_{c'})^2}{2\sigma_X^2}\right]}{\displaystyle\sum_{c''\neq c'} \exp\left[-\dfrac{(L_{c''} - L_{c'})^2}{2\sigma_X^2}\right]} & \text{otherwise,} \end{cases}$$

$$(A8)$$

where X_{max} is the asymptotic movement rate,

a_{50} and δ are the parameters that determine the fraction of animals that leave a cell, and

σ_X determines the extent of movement.

Data Generation

Individual Cell

The data available for assessment purposes are survey indices of abundance, survey age-composition data, and fishery age-composition data. The survey estimates of abundance are assumed to be lognormally distributed about a survey-selected biomass:

$$P_y^c = \bar{P}_y^c e^{\phi_y^c - (\sigma_P^2)/2} \qquad \phi_y^c \sim N(0;\sigma_P^2) \quad , \quad (A9)$$

where P_y^c is the survey estimate for cell c and year y,

$\tilde{P}_y^c$ is the actual survey-selected biomass for cell c and year y:

$$\tilde{P}_y^c = \sum_g \sum_{a=0}^x S_a^g w_a^g N_{y,a}^{g,c} , \qquad (A10)$$

S_a^g is the selectivity of the survey gear on fish of sex g and age a,

σ_P^2 is (approximately) the square of the coefficient of variation of P_y^c taken to be $n\tilde{\sigma}_P^2$, and

$\tilde{\sigma}_P^2$ is (approximately) the square of the coefficient of variation of a survey of the whole area.

The survey age-composition data for cell c and year y are assumed to be a multinomial sample of size N_y^c from the survey-selected numbers for cell c and year y. The number of animals aged for a given cell is assumed to be proportional to the survey abundance estimate for that cell (i.e., $N_y^c = N_y P_y^c / \sum_{c'} P_y^{c'}$). The fishery age-composition data are generated analogously to the survey age-composition data except that these data are generated from the fishery selectivity-selected numbers and the number of animals aged is

assumed to be proportional to the catch in mass for cell c (i.e., $N_y^c = N_y C_y^c / \sum_{c'} C_y^{c'}$).

Groups of Cells

The survey estimate of abundance for a given year y and group of cells C (P_y^c) and its coefficient of variation are computed by treating each of the cells of which group C consists as a survey stratum:

$$\sum_{c\in C} P_y^c; \quad \sqrt{\sum_{c\in C} (\sigma_P P_y^c)^2} / \sum_{c\in C} P_y^c . \qquad (A11)$$

The survey age-composition data for a group of cells (C) is obtained by weighting the age-composition data for each of its constituent cells by the survey estimate of abundance for that cell:

$$\rho_{y,a}^C = \frac{1}{P_y^C} \sum_{c\in C} \rho_{y,a}^c P_y^c , \qquad (A12)$$

where $\rho_{y,a}^c$ is fraction of the survey catch during year y in cell c that is aged to be age a.

The fishery age-composition data are generated analogously to the survey age-composition data except that the samples for a given cell are weighted by the catch in mass for that cell.

Appendix B: The Assessment Method

The Population Dynamics Model

The dynamics of the population are governed by the equation

$$N_{y+1,a}^{g,c} = \begin{cases} 0.5\bar{R}^c e^{\varepsilon_{y+1}^c} & \text{if } a = 0 \\ N_{y,a-1}^{g,c} e^{-M}(1-S_{a-1}F_y^c) & \text{if } 0 < a < x \\ N_{y,x-1}^{g,c} e^{-M}(1-S_{x-1}F_y^c) & \text{if } a = x \\ \quad + N_{y,x}^{g,c} e^{-M}(1-S_x F_y^c) \end{cases}$$

(B1)

where $N_{y,a}^{g,c}$ is the number of animals of sex g and age a in assessment cell c (the total area consists of m assessment cells, where $m \leq n$) at the start of year y,

M is the instantaneous rate of natural mortality,

S_a is the selectivity of the fishing gear on animals of age a (assumed to be independent of time),

F_y^c is the exploitation rate on fully selected animals in assessment cell c during year y,

$\bar{R}^c$ is the mean recruitment for assessment cell c,

ε_y^c is the recruitment residual for assessment cell c and year y, and

x is the maximum age, taken to be a plus group.

The age structure at the start of the first year for which catches are available is taken to be that corresponding to deterministic unfished equilibrium.

The exploitation rate (F_y^c) is computed using the equation

$$F_y^c = C_y^c / V_y^c , \qquad (B2)$$

where C_y^c is the catch (in mass) during year y from assessment cell c, and

V_y^c is the exploitable biomass in assessment cell c in the middle of year y (fishing is assumed to occur in a pulse in the middle of the year after half of natural mortality):

$$V_y^c = \sum_g \sum_{a=0}^x w_a^g S_a N_{y,a}^{g,c} e^{-M/2} , \qquad (B3)$$

and w_a^g is the mass of an animal of sex g and age a.

Fishery selectivity is modeled by a double-logistic equation:

$$S_a = S_a' / \max_{a'} S_{a'}';$$

$$S_a' = \frac{1}{1+e^{-\delta_1(a-a_{50}^1)}} \times \frac{1}{1+e^{-\delta_2(a_{50}^2-a)}} \quad , \quad (B4)$$

where $a_{50}^1, a_{50}^2, \delta_1,$ and δ_2 are the parameters of the double-logistic equation.

Parameter Estimation

The parameters of the assessment model are those that define fishery selectivity, survey selectivity, natural mortality, growth, fecundity, and recruitment. Fecundity and mass at age are assumed known from previous studies and, for the bulk of the analyses of this paper, natural mortality is also assumed to be known exactly. The values for the remaining parameters are determined by maximizing an objective function that includes a weak constraint on the extent to which recruitment can vary about its mean and contributions for the survey index data, the age-composition data for the surveys, and the age-composition data from the fishery catches.

The constraint on the recruitment residuals is

$$\frac{1}{8}\sum_{c'}\sum_{c''}\sum_{y}\varepsilon_y^{c'} X^{c',c''}\varepsilon_y^{c''} \quad , \quad (B5)$$

where X is the inverse of a correlation matrix with 1 on the diagonal entries and 0.7 in the off-diagonal entries. The division by 8 implies that very low emphasis is placed on this constraint (equivalent to a coefficient of variation for recruitment of 2).

The survey indices are assumed to be lognormally distributed about the corresponding model quantity. The contribution of the survey indices to the objective function is, therefore,

$$\sum_{c}\sum_{y}\frac{(\log_e P_y^c - \log_e \hat{P}_y^c)^2}{2\sigma_P^2} \quad , \quad (B6)$$

where $\hat{P}_y^c$ is the model estimate corresponding to the survey index for assessment cell c and year y:

$$\hat{P}_y^c = q^c \sum_g \sum_a w_a^g S_a^s N_{y,a}^{g,c}(1-S_a F_y^c/2)e^{-M/2}, \quad (B7)$$

S_a^s is the survey selectivity for fish of age a:

$$S_a^s = \frac{1}{\left(1+\exp\left[-\log_e 19 \dfrac{a-a_{50}^s}{a_{95}^s - a_{50}^s}\right]\right)} \quad , \quad (B8)$$

a_{50}^s and a_{95}^s are the ages at which 50% and 95%, respectively, of animals are vulnerable to the survey gear, and

q^c is the catchability coefficient for the survey in assessment cell c.

The value for the parameter q^c is determined analytically by differentiating equation B6 with respect to q^c, setting the resultant equation to 0, and solving for q^c.

The contribution of the survey age-composition data to the objective function is

$$\sum_{c}\sum_{y}\sum_{g}Q_y^{g,c}\sum_{a}\rho_{y,a}^{g,c}\log_e \hat{\rho}_{y,a}^{g,c} \quad , \quad (B9)$$

where $Q_y^{g,c}$ is the effective sample size for animals of sex g for the survey conducted during year y in assessment cell c,

$\rho_{y,a}^{g,c}$ is the observed fraction of the survey catch during year y in assessment cell c of fish of sex g that is aged to be age a, and

$\hat{\rho}_{y,a}^{g,c}$ is the model-estimate of the fraction of the survey catch during year y in assessment cell c of fish of sex g that is aged to be age a:

$$\hat{\rho}_{y,a}^{g,c} = \frac{S_a^s N_{y,a}^{g,c}(1-S_a F_y^c/2)}{\sum_{a'}S_{a'}^s N_{y,a'}^{g,c}(1-S_{a'}F_y^c/2)} \quad , \quad (B10)$$

The contribution of the catch age-composition data to the objective function follows equation B9 except that the data relate to the fishery age-composition data and the model estimate is given by

$$\hat{\rho}_{y,a}^{g,c} = \frac{S_a N_{y,a}^{g,c}}{\sum_{a'}S_{a'}N_{y,a'}^{g,c}} \cdot \quad (B11)$$

The recruitment estimates for the most recent few years of a stock assessment are frequently very imprecise (Butterworth et al. 1990). Therefore, in this study, instead of attempting to estimate the last three recruitment residuals (years z, z-1, and z-2), the values for these parameters are instead predicted from the results of a regression of the values of the recruitment residuals for the 10-year period z-12 to z-3 on time.

American Fisheries Society Symposium 42:155–164, 2004

Marine Protected Areas as Biological Successes and Social Failures in Southeast Asia

PATRICK CHRISTIE[1]

School of Marine Affairs and Henry M. Jackson School of International Studies,
University of Washington, 3707 Brooklyn Avenue Northeast,
Seattle, Washington 98105-6715, USA

Abstract.—Marine protected areas (MPAs) are of growing interest globally. They are principally studied from a biological perspective, with some cases documenting improved environmental conditions and increased fish yields. The MPAs that meet narrowly defined biological goals are generally presented as "successes." However, these same MPAs may, in fact, be social "failures" when social evaluation criteria are applied. A review of four MPAs in the Philippines and Indonesia demonstrates this scenario. The cases are reviewed using standard measures of biological and social success. Their historic and present management structures are reviewed. It is suggested that a strong linkage exists between social and biological success, with social considerations determining long-term biological success. This finding implies that standards for measuring both biological and social success should be applied equally and that MPAs should be designed to meet multiple social and biological goals. The evaluation and portrayal of MPAs has implications for the management of a particular MPA and the broader discourse surrounding marine environmental management.

Introduction

The marine protected area (MPA) literature to date is mainly comprised of studies considering the biological significance of this management approach. The so-called "spill-over effect," connectivity, appropriate dimensions, and habitat representation are some of the most active areas of inquiry (e.g., Russ and Alcala 1996; Salm et al. 2000; NRC 2001; Roberts et al. 2001). As highlighted in a recent essay by seventeen social scientists, MPA research and the resultant literature is generally lacking detailed accounts of the social implications of MPAs and the activities associated with them such as fishing, recreational diving, tourism, and research (Christie et al. 2003c). This paper grew out of a conference sponsored by the National Oceanic and Atmospheric Administration (NOAA) in 2002 as an attempt to fill this notable gap in MPA research and published literature (NOAA 2002). There are a few notable exceptions to this characterization (e.g., Trist 1999; Sandersen and Koester 2000; Pollnac et al. 2001),

and it is clear that MPAs, and protected areas in general, are beginning to attract considerable attention by those mainly interested in the human dimensions of environmental management.

The lack of social research on MPAs has led to at least two unfortunate conditions: an incomplete understanding of how to most effectively utilize this popular management tool and omissions from the scientific literature of potentially fascinating accounts of human responses to MPAs (Christie et al. 2003c; Mascia et al. 2003). One example of an omission is the general underrepresentation of conflict surrounding MPA establishment and implementation in the MPA literature. This paper will demonstrate that, in the tropics, conflict often stems from the marginalization of artisanal fisheries by other forms of resource utilization such as dive tourism. While this conflict (and its reporting) may be disconcerting to some environmentalists and scientists advocating MPAs, a careful consideration of the receptivity of fishing communities to MPAs is fundamental for their long-term success (Agardy et al. 2003).

If the measure of MPA success is mainly based on biological metrics, then it is plausible that some MPAs, at least in the short term, may be considered

---

[1] E-mail: patrickc@u.washington.edu

biological successes while simultaneously causing social harm such as conflict and economic and social dislocation for disadvantaged communities (such as artisanal fishing communities near MPAs). In response, the marginalized community may either strongly resist the imposition of the MPA or initially support the MPA but then lose interest. Field research presented in this paper and other accounts demonstrates that this scenario is not uncommon and has a strong destabilizing effect on any MPA (Trist 1999; Sandersen and Koester 2000; Christie et al. 2003a, 2003b; Oracion 2003).

Based on the experiences of four failing or vulnerable MPAs, this study comments on the implications of ignoring social complexities associated with MPAs. In conclusion, a case is made to improve our understanding of the complex and mixed results of MPAs thus far. The intent is to help ensure their biological and social success and to improve the likelihood that they will provide tangible benefits such as increased biodiversity and improved fisheries and tourism management.

Methods

This study is based on a comparative analysis of four MPAs in Southeast Asia—San Salvador Island (Philippines), Twin Rocks (Philippines), Balicasag Island (Philippines), and Bunaken National Park (BNP, Indonesia). All locations are coral reefs. The author was directly involved for 3 years in the establishment of the MPA on San Salvador Island. The first three MPAs are small (between 3-ha and 125-ha no-take areas), while Bunaken National Park consists of a large, zoned space (89,056 ha of land and sea area) that includes small no-take areas (up to approximately 25 ha) for dive tourism.

All four MPAs have both conservation and economic development goals. The Philippine management processes—commonly characterized as "community based" or "comanagement"—were particularly attentive to issues of social equity and grassroots participation from the inception (White and Savina 1987; White et al. 1994; Christie et al. 2002, 2003a, 2003d). These MPAs were intended to both improve coral reef conditions and spur community-level sustainable development. Bunaken National Park was established to simultaneously meet conservation and economic development needs at a number of levels (Merrill 1998; Salm et al. 2000). The MPAs, therefore, overlap in both environmental and social goals,

but, as highlighted in the subsequent analysis, utilize different approaches; operate at different scales; and have different histories. All management regimes aspired to some form of comanagement but with differing degrees of resident, private sector, and governmental–institutional influence (Christie and White 1997).

The analysis draws from published accounts and recent biological and social field research. To consistently evaluate these four MPAs, commonly utilized measures of biological and social success are applied to each case. An evaluative matrix was developed that includes biological and social variables. For this study, measures of biological success for an MPA include increased fish abundance, fish diversity, and living coral cover. Other data on coral substrate are available in cited studies.

The social indicators of success are drawn from the following sources: the National Research Council report on MPAs (NRC 2001), the recently developed social research agenda derived from a NOAA-sponsored workshop involving over 100 social scientists (NOAA 2002; Christie et al. 2003c), a recently published guide to the socioeconomic dimensions of coral reefs (Bunce et al. 2000), and the few social research studies of MPAs in the tropics (e.g., Trist 1999; Pollnac et al. 2001). While there are numerous possible measures of social impact, the following measures of success are applied since they have been shown to be critical in the Southeast Asia context: broad stakeholder participation, equitable sharing of economic benefits, and the presence of conflict–resolution mechanisms (White et al. 1994; Pollnac et al. 2001, 2003; Pomeroy et al. 2003).

Fish Species Richness and Density

The author recorded the diversity and abundance of fish in a 500-m^2 area demarcated by a 50-m transect line (laid at approximately 7 m deep, parallel to the reef crest) serving as the upper boundary. The observers swam 10 m along the line, then down the slope and 10 m parallel to the line, and then back to the line in this pattern until reaching the transect end. This procedure was repeated in the opposite pattern back to the beginning of the transect line. The number of individuals per species was noted, employing logarithmic categories for those species with large numbers of individuals. The families surveyed were surgeonfishes (acanthurids), rabbitfish (siganids; also known as spinefoots), sea basses (serranids; also known as groupers), snappers (lutjanids), grunts (haemulids; also known as sweetlips), emperors (lethrinids), jacks (carangids), fusiliers (caesionids), breams (nemipterids), goatfishes (mullids), parrotfishes (scarids), sea chubs (kyphosids; also known as rudderfish), triggerfish (balistids; also known as leatherjackets), butterflyfishes (chaetodontids), angelfishes (pomacanthids), wrasses (labrids), and damselfishes (pomacentrids). Anthids (family Serranidae) and *Zanclus cornutus* (known as the moorish idol) were also counted. The first twelve fish listed are commonly targeted by fishers due to higher market values and recorded as "target species." A range of 4–8 transects, each covering 500 m², was completed for each monitoring site, and confidence in mean estimates are represented by 95% confidence intervals on each figure. Dr. Alan White collected most of the pre-1998 data in Anilao and Balicasag (both in the Philippines). Mr. Jonathan Apurado was also involved in collecting recent data in these two sites.

Interviews

In addition to biological assessments, in-depth, semi-structured interviews were conducted with key informants to investigate, among other topics, local opinions of MPAs, management systems and rules, perceived benefits and costs, and implementation challenges. The 73 informants were dive resort owners, fishers, MPA advocates, and scientists. Eighteen informants in Twin Rocks, 15 informants in Balicasag, and 40 informants in Bunaken were interviewed. The author lived on San Salvador Island for 3 years (1987–1990) and periodically visits to conduct research. Dozens of management documents and published accounts were reviewed in order to identify management goals and issues. Interview data for Twin Rocks and Bunaken were analyzed using Atlas.ti software (Scientific Software Development, Berlin, *www.atlasti.de*) that allows for systematic analysis of qualitative information.

This analysis consisted of identifying and labeling relevant themes within interviews (e.g., employment, perception of the MPA, etc.). Once interviews were coded, search commands (using code labels) were used to scan the interviews for quotes meeting two or more criteria (e.g., "diver resort owner" and "perception that the environment is improving"). As trends emerged (or predicted ones did not), theoretical memos were affixed to each code label. These analytic memos served as starting points to relate findings to the relevant literature and biological findings.

This approach allowed the researcher to create an "analytic trail" that demonstrates how conclusions were reached.

Results and Discussion

Of the four well-documented MPAs in the Philippines and Indonesia that were chosen, all met standard biological criteria of success more consistently than standard criteria of social success (Table 1). In general, initially successful management processes at San Salvador Island, Twin Rocks, and Balicasag Island MPAs have deteriorated over time without consistent and long-term support of governmental agencies and nongovernmental organizations that initially established them. Poorly managed controversy and conflict are derailing these MPAs. In Bunaken National Park, mandated by the Indonesian national government and supported with external aid, management and the enforcement of no-take areas is proceeding but in a manner that does not necessarily reflect the interests of many local fishing communities. Based on lessons from the other sites, this represents an unstable situation that likely requires corrective measures. To highlight gross similarities and differences, one point was assigned whenever a site effectively met a criterion of success (even in the most lenient sense).

San Salvador: Initial Success Eroded by Interpersonal Conflict

Each MPA has a unique and interesting history that helps explain the above characterizations. On San Salvador Island, initial success in community-based management (Christie and White 1994; Christie et al. 1994; White et al. 1994; Katon et al. 1999) has given way to intense interpersonal conflicts that have arisen between long-standing rivals within the community (Christie et al. 2003a). The MPA management process has become an opportunity through which such conflict, ongoing between key community leaders for more than 40 years, has expanded. While seemingly trivial (and underreported), such interpersonal conflict can have a strong detrimental impact considering the community-based nature of the management system.

Established in July 1989, San Salvador Island's 125-ha no-take MPA continues to be protected by a few committed advocates from the community and a supportive local mayor. Therefore, at least until 1999, environmental conditions were improving or staying constant while the management process become in-

Table 1.—Evaluation matrix of four MPAs in Southeast Asia: San Salvador Island, Philippines; Twin Rocks, Philippines; Balicasag Island, Philippines; and Bunaken National Park no-take tourism zones, Indonesia.

Criteria	San Salvador Island	Twin Rocks	Balicasag Island	Bunaken National Park[a]	Criteria summation
		Biological			
Increased fish abundance	Yes	Yes	Initially, no longer	Likely	3
Increased fish biodiversity	Yes	Yes	Initially, no longer	Likely	3
Improved habitat (coral substrate)	Yes	Yes	Yes	Likely	4
		Social			
Broad stakeholder participation	Initially, no longer	Initially, no longer	Initially, no longer	Yes	1
Broad sharing of economic benefits	Possibly (due to increased fishery yields)	No	No	No	1
Presence of conflict-resolution mechanisms	Initially, no longer	Initially, no longer	Initially, no longer	Yes, only sporadically utilized	1

[a] There are no baseline data available for Bunaken National Park. Therefore, the characterization of increases in fish and improved coral cover as "likely" is based on interviews and comparisons between sites within the park.

creasingly tenuous (Figure 1). Species richness has increased from 126 species belonging to 19 families in 1988 to 138 species belonging to 28 families in 1998 (Christie and White 1994; Christie et al. 2003a).

Some former supporters complain that advocates are unwilling to share responsibility and are heavy handed in their methods of enforcement. As a result, the island community, which formerly appeared to be unified behind the MPA (Christie et al. 1994; Katon et al. 1999), is now clearly divided (Christie et al. 2003a). While enforcement is an important ingredient for successful programs, it is important for long-term sustainability that wide stakeholder support exists (Peluso 1992; Brechin et al. 2002; Lowe 2003). Without considerable conflict–resolution interventions, the likelihood that management will continue for another

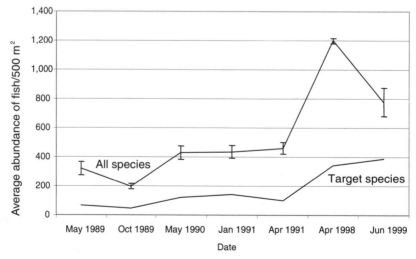

Figure 1.—Average abundance of fish over time, San Salvador Island, Philippines. From Christie et al. (2003a).

decade is unlikely based on recent interviews and comparative research on design principles for such management systems (Ostrom 1992; Pollnac et al. 2001).

Twin Rocks: Initial Success Usurped by Influential Stakeholder

Twin Rocks is the site of a destabilizing conflict between dive resort owners and local fishing communities (Christie et al. 2003d; Oracion 2003). In the most superficial sense, this conflict stems from disagreements over whether recreational diving—a practice formally banned by the MPA's regulations but broadly ignored by dive shop owners—should be allowed in the small no-take area (approximately 3 ha). The inter-stakeholder conflict is grounded in class distinctions and perceptions of environmental management, a phenomenon apparent in other Philippine contexts (Nazarea et al. 1998). The involved dive shop owners are generally from the capital city, much more affluent than local fishers, and politically well connected with local officials (partly as a result of election campaign contributions). As a result, these elites are able to wield greater influence over MPA management practices and have usurped control from the founding community (Peluso 1992; Trist 1999; Sandersen and Koester 2000; Lowe 2003; Oracion 2003). As of 1999, the resort owners purchased the nearshore lands and are the main enforcers of the MPA. In July 2001, one owner was particularly committed and vigilant:

> But what I'm telling the people in this community is, for the reef…we take care of it. [I spent] many sleepless nights [protecting the sanctuary]. I have to bear the burden of getting the ire of these people. That's okay. I don't care. As long as the fish are there. We will have to bribe people. I will resort to anything that will prevent any direct negative impact [on the sanctuary]…

Predictably, fishers, who initially voluntarily protected the no-take area (over a period of 7 years) as part of the community-based management regime, are either losing interest in the MPA or are plotting how to stop diving inside the reserve and reassert their influence. When asked why they are losing interest, informants expressed a general sense of mistrust of the dive industry and concern that MPA management is no longer fair. There is a struggle for ownership over this MPA and the resort owners are perceived as having violated the tenets of community-based resource management (White et al. 1994; Pollnac et al. 2001; Oracion 2003). One community leader who dedicated years of voluntary effort has now distanced herself from this work:

> Now, since the resort was established they [resort owners] are the ones who guard and protect the sanctuary. But I think they already took over the sanctuary and that's the problem now… Umm, they might hear my interview. They'll be angry with me…

Asked why this control was a problem as long as the sanctuary was protected, she replied, "it's the same, but the only thing is that sanctuary is for the community, now they [the resort owners] are already taking it over it." This MPA management process is suffering the fate of its own success in one community leader's opinion: "If there's good management, our coastal resources bloom. That's when divers came in. Resorts came in. But community-based management has also vanished…"

Figure 2 displays data from three locations—Twin Rocks (the enforced MPA), Arthur's Rock (a nonenforced MPA due to past conflicts), and a nearby non-MPA reef. Increase in fish abundance for target species for all sites has been marginally significant since 1995 (two-way analysis of variance [ANOVA], time, $P = 0.065$). There is a significant difference between sites (two-way ANOVA, site, $P = 0.033$), with Twin Rocks being significantly different from non-MPA sites (Scheffé's test, $P < 0.01$) but not significantly different from Arthur's Rock (Scheffé's test, $P = 0.195$). Twin Rocks target fish abundance in 2001 was 280.9 (± 134) individuals per 500 m^2. Target fish abundance has remained constant for the nearby non-MPA sites since 1995. This is an indication that any "spillover" from the MPAs is likely being caught by local fishers—a condition consistent with other MPAs in the Philippines (Christie et al. 2002). The greatest increase in target fish abundance for Twin Rocks took place between sampling in 1997 and 2001. A plausible conclusion is that once local resort owners took over management and enforcement of the no-take area there were immediate beneficial biological impacts.

From an exclusively biological perspective, conditions at Twin Rocks are only improving. From a social perspective, such disregard for the community-based regime represents a failure.

The scenario of inter-stakeholder tensions, particularly between tourist brokers and resource users,

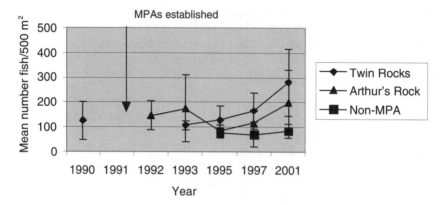

Figure 2.—Target fish abundance change over time (mean ± 95% confidence interval). Two-way ANOVA for 1995–2001: time, $P = 0.065$; site, $P < 0.05$; time × site, not significant. $N > 5$ per site. From Christie et al. (2003d).

is common (West et al. 2003). In fact, it is apparent (although again underreported by some advocates) in other conservation–tourism destinations such as Soufrière, St Lucia (Trist 1999; Sandersen and Koester 2000; Roberts et al. 2001) and Bunaken National Park (Salm et al. 2000; Christie et al. 2003b; Lowe 2003).

Some MPA advocates and scientists seemingly have determined that tourism is the most effective economic engine to propel the conservation agenda forward (Nichols 1999; Trist 1999; Lowe 2003). In such cases, it appears that enforcement systems are more common than incentive-based or self-monitoring management systems based on compliance rather than enforcement (Peluso 1992; Brechin et al. 2002). While most would argue that enforcement is necessary for MPA success, the growing tendencies toward coercive mechanisms that are not compatible with other cooperative options represents a break from the early successes of community-based and comanagement regimes in the Philippines and elsewhere (White and Savina 1987; White et al. 1994; Christie and White 1997; Brechin et al. 2002).

Balicasag Island: Lost Community Control Has Negative Biological Impacts

The insertion of central government agency control over a community-based MPA has the potential to undermine community support on Balicasag Island (Christie et al. 2002). Historically, the Philippine national government had formal control over fisheries resources. The passage of decentralization laws in the 1990s allowed community-based reserves to flourish in that context (White et al. 2002). However, the Phil-

ippines National Tourism Authority (NTA), which is an arm of the central government, has effectively laid claim to the Balicasag Island MPA. The NTA built a resort at the shores of the no-take area and now captures, along with offshore dive businesses, the majority of revenues generated by this MPA. Local residents are relegated to selling shells and t-shirts to visitors. While the NTA has stationed an armed guard at the MPA, he is unable to monitor the area effectively. Formerly supportive community members are now likely poachers, as manifested by declining fish populations inside and outside the no-take area (Christie et al. 2002).

Poorly managed social dynamics have real consequences for biological resources. Fish abundances (of target species within families such as Serranidae, Lutjanidae, Lethrinidae, and Carangidae) within the no-take area have declined 291% from a peak in 1986 (one year after MPA implementation) to a low in 1999 (Christie et al. 2002). Fish abundance on the reef near the no-take area have also severely declined from 1986 (1,642 ± 223 individuals/500 m²) to 1999 (230 ± 65 individuals/500 m²). There is no longer any significant difference in fish abundance when comparing fishing areas on Balicasag (within 1 km of the no-take area) with nearby control sites where fishing is allowed, but without nearby MPAs. Even if Balicasag's MPA were effectively managed, it is likely that isolated MPAs will have a declining effect without a wider policy of fishing effort reduction in the Philippines (White et al. 2002). Initial success stories probably become magnets for increasingly desperate fishers from other areas in the Philippines. The Balicasag case provides evidence of the biological consequences of poorly functioning social management systems and may suggest reorientations relevant to Bunaken National Park where fisheries and tourism coexist.

Bunaken National Park: Will Emerging Tensions Be Adequately Managed?

The Bunaken case illustrates the perennial issues that emerge when trying to simultaneously meet sometimes conflicting management goals for conservation, fishery enhancement, and tourism development in a complex context (Merrill 1998; Agardy et al. 2003). Bunaken National Park is unique from the other MPAs in this analysis since it represents an example of a relatively large national park established through national government decree in a context other than the Philippines (Salm et al. 2000). Nonetheless, lessons from the Philippines are relevant.

Since 1994, its management has been the focus of two successive projects funded by the U.S. Agency for International Development that helped with management plan development and implementation. Recently, management has focused on rezonation, improving the enforcement of park zones, and addressing the impacts of dive tourism. As a result, biological conditions, especially the abundance of large target fish species, appear to have improved within tourism dive zones and now are significantly greater than in nearby zones with very similar physical and oceanographic conditions (Christie et al. 2003b). Baseline data are not available to assess this impact definitively; however, key informant interviews and comparisons of fish and coral data (between zones within the park) strongly suggest that the strict no-take management system is having environmentally favorable effects within tourism zones.

The management process in Bunaken has not been a smooth one and has generated considerable controversy (Merrill 1998; Salm et al. 2000; Lowe 2003). Initially, Indonesian central government agencies considered dislocating local fishing communities within the park. Local communities and an Indonesian nongovernmental organization effectively resisted that proposal. More recently, park managers have engaged local communities in a consultative process (hence the notation of broad community participation in Table 1) (Erdmann et al. 2004). While consultation is ongoing, some social scientists and local Indonesians have expressed concerns that the park management system is too hierarchical and implicitly favors the dive industry and particular ethnic or religious groups (Lowe 2003).

In short, some stakeholders do not feel that economic benefits are being equitably shared, a fundamental condition for long-term success of MPAs (Pollnac et al. 2001, 2003; Pomeroy et al. 2003). While dive tourism has been active in the area since the 1970s, the number of visitors has greatly increased since 1993. Approximately 13,361 Indonesian tourists and 7,213 foreign tourists visited BNP between March 2001 and March 2002. This is a dramatic increase from 2,248 visitors in 1985. These numbers are based on entrance fee collection figures collected by the local management board. In a 1999 survey, it was found that out of 368 jobs in BNP's tourism industry, only 24.5% went to native Bunaken National Park residents (V. Lee, University of Waterloo, unpublished report, 1999). As demonstrated in the Twin Rocks MPA, disparities in income and the perception that benefits are not shared equally is a potent scenario that can quickly undermine what popular support may exist. Neither these economic impacts nor hoped-for improvements in fish yields are adequately documented.

The new zonation scheme is also controversial. Interviews demonstrated that some fishers feel that the current zonation scheme is unfair on the grounds that it protects the best fishing areas exclusively for diving and does not allow for necessary seasonal relocation of fishing around the islands (Merrill 1998; Christie et al. 2003b). A Western dive shop owner in BNP in July 2002 stated, "I haven't given up any primary diving sites from zonation... The fishers probably feel that they gave up some of their primary fishing areas..."

These issues are being discussed through a complex consultation process, but it is unclear if park managers are willing to make major changes to management practices and zonation schemes. Consultative participation is generally considered to be the lowest form of participation in management (Christie and White 1997; Kay and Alder 1999). The terms of discussion have largely been set by the general principles outlined by Indonesian law regarding national parks and the conservation agendas of environmental organizations.

Conclusion: A Clear Understanding is Fundamental

The intent of this review is not to be dismissive of MPAs or the dedicated efforts of managers, but to highlight the complexities of establishing ambitious conservation areas in impoverished and socially stratified contexts. It also suggests the need for a monitoring system that matches this biological and social complexity. The means by which MPAs are evaluated is not trivial and can lead to biased assessments. This study demonstrates that, while particular MPAs may meet biological goals, they may be, at least in the short term, considerably less effective in attaining basic measures of social success. If these social considerations—which will largely determine the fate of these MPAs—are ignored, MPAs are likely to continue to have high failure rates and eventually may fall out of favor as a management tool.

Perhaps the most troubling aspect of this review is that none of these MPAs have formal conflict–resolution mechanisms that operate impartially and represent all stakeholder interests equally. Rather, conflicts emerge and are generally addressed on an ad hoc basis or ignored until they reach a crisis stage. At this point, entrenched opinions make it difficult to diffuse the conflict (McCreary et al. 2001). While the creation of dependent relationships between local communities and external agents may be undesirable, there may be no other choice, considering the challenges facing these MPAs and local communities. Again, the provision of comprehensive assessments and third-party broker systems are fundamental to address entrenched conflict.

If large-scale social dislocation and strife are considered the necessarily price for environmental improvements, it may be tempting to consider coercive mechanisms as a means to force acquiescence to management regimes. Social theory strongly suggests that this strategy is likely to fail in the long term (Ostrom 1992; Peluso 1992; Brechin et al. 2002). Social monitoring has the potential to identify stakeholder opinions of management regimes and to moderate one-sided agendas.

In short, to begin to unravel the puzzle of how to meet both biological and social goals with MPAs, greater attention needs to be paid to social research as a complement to the already extensive biological research agenda (Agardy et al. 2003; Christie et al. 2003c; Mascia et al. 2003). Since MPA establishment and management is a complex undertaking with frequently contentious outcomes, particular attention should be paid to comparative studies and those that explore the likely inter-group and intra-group differences (Nazarea et al. 1998). A clear understanding of how different constituencies value marine resources and MPAs is a logical first step toward improving management practices. Based on site-specific studies, MPA management plans and monitor-

ing protocols should be designed to address local conditions.

Acknowledgments

The author takes sole responsibility for any errors in this analysis. Much of this research was made possible with the support of the David and Lucile Packard Foundation (grant number 2000-14654) and the National Science Foundation (grant number DGE-0132132). Field research relied on the ongoing collaboration with Alan White, Bogor Agricultural University and Silliman University. The comments of two anonymous reviewers were very helpful.

References

Agardy, T., P. Bridgewater, M. P. Crosby, J. Day, P. K. Dayton, R. Kenchington, D. Laffoley, P. McConney, P. A. Murray, J. E. Parks, and L. Peau. 2003. Dangerous targets? Unresolved issues and ideological clashes around marine protected areas. Aquatic Conservation: Marine and Freshwater Ecosystems 13:1–15.

Brechin, S., P. Wilshusen, C. Fortwangler, and P. West. 2002. Beyond the square wheel: toward a more comprehensive understanding of biodiversity conservation as social and political process. Society and Natural Resources 15:41–64.

Bunce, L., P. Townsley, R. Pomeroy, and R. B. Pollnac. 2000. Socioecomic manual for coral reef management. National Ocean Service, National Oceanic and Atmospheric Administration, Silver Spring, Maryland.

Christie, P., D. Buhat, L. R. Garces, and A. T. White. 2003a. The challenges and rewards of community-based coastal resources management. Pages 231–249 in S. R. Brechin, P. R. Wilshusen, C. L. Fortwangler, and P. C. West, editors. Contested nature, promoting international biodiversity with social justice in the twenty-first century. State University of New York Press, Albany.

Christie, P., D. Makapedua, and Ir. L. T. X. Lalamentik. 2003b. Bio-physical impacts and links to integrated coastal management sustainability in Bunaken National Park, Indonesia. Indonesian Journal of Coastal and Marine Resources Special Edition 1:1–22.

Christie, P., B. J. McCay, M. L. Miller, C. Lowe, A. T. White, R. Stoffle, D. L. Fluharty, L. Talaue McManus, R. Chuenpagdee, C. Pomeroy, D. O. Suman, B. G. Blount, D. Huppert, R. L. Villahermosa Eisma, E. Oracion, K. Lowry, and R. B. Pollnac. 2003c. Toward developing a complete understanding: a social science research agenda for marine protected areas. Fisheries 28(12):22–26.

Christie, P., and A. T. White. 1994. Reef fish yield and reef condition for San Salvador Island, Luzon, Philippines. Asian Fisheries Science 7:135–148.

Christie, P., and A. T. White. 1997. Trends in development of coastal area management in tropical countries: from central to community orientation. Coastal Management 25:155–181.

Christie, P., A. T. White, and D. Buhat. 1994. Community-based resource management on San Salvador Island, the Philippines. Society and Natural Resources 7:103–117.

Christie, P., A. T. White, and E. Deguit. 2002. Starting point or solution? Community-based marine protected areas in the Philippines. Journal of Environmental Management 66:441–454.

Christie, P., A. T. White, B. Stockwell, and C. R. Jadloc. 2003d. Factors influencing integrated coastal management sustainability: focus on environmental conditions in two locations in the Philippines. Silliman Journal 44(1):286–323.

Erdmann, M. V., P. R. Merrill, M. Mongdong, I. Arsyad, Z. Harahap, R. Pangalila, R. Elverawati, and P. Baworo. 2004. Building effective co-management systems for decentralized protected areas management in Indonesia: Bunaken National Park case study. United States Agency for International Development, Natural Resources Management Program, Jakarta, Indonesia.

Katon, B. M., R. S. Pomeroy, L. R. Garces, and A. M. Salamanca. 1999. Fisheries management of San Salvador Island, Philippines: a shared responsibility. Society and Natural Resources 12:777–796.

Kay, R., and J. Alder. 1999. Coastal planning and management. E&FN Spon, New York.

Lowe, C. 2003. Sustainability and the question of "enforcement" in integrated coastal management: the case of Nain Island, Bunaken National Park. Indonesian Journal of Coastal and Marine Resources Special Edition 1:49–63.

Mascia, M. B., J. P. Brosius, T. A. Dobson, B. C. Forbes, L. Horowitz, M. A. McKean, and N. J. Turner. 2003. Conservation and the social sciences. Conservation Biology 17:649–650.

McCreary, S. J. Gamman, B. Brooks, L. Whitman, R. Bryson, B. Fuller, A. McInerny, and R. Glazer. 2001. Applying a mediated negotiation framework to integrated coastal zone management. Coastal Management 29:183–216.

Merrill, R. 1998. The NRMP experience in Bunaken and Bukit-Bukit Raya National Parks: lessons learned from PAM in Indonesia. Natural Resources Management Program. Available: www.nrm.or.id/Content/Resources/Bibliography.asp (April 2003).

Nazarea, V., R. Rhoades, E. Bontoyan, and G. Flora. 1998. Defining indicators that make sense to local people: intra-cultural variation in perceptions of natural resources. Human Organization 57:159–170.

Nichols, K. 1999. Coming to terms with "integrated coastal management": problems of meaning and method in a new arena of resource regulation. Professional Geographer 51:388–405.

NOAA (National Oceanic and Atmospheric Administration). 2002. Marine protected areas social science workshop notes from breakout groups. NOAA, National MPA Center, Santa Cruz. Available: www.mpa.gov (April 2003).

NRC (National Research Council). 2001. Marine protected areas: tools for sustaining ocean ecosystems. National Academy Press, Washington, D.C.

Oracion, E. 2003. The dynamics of stakeholder participation in marine protected area development: a case study in Batangas, Philippines. Silliman Journal 44(1):95–137.

Ostrom, E. 1992. Governing the commons, the evolution of institutions for collective action. Cambridge University Press, New York.

Peluso, N. 1992. Coercing conservation: the politics of state resource control. Global Environmental Change 4:199–218.

Pomeroy, R., E. Oracion, D. A. Caballes, and R. B. Pollnac. 2003. Economic benefits and integrated coastal management sustainability. Silliman Journal 44(1):75–94.

Pollnac, R. B., B. R., Crawford, and M. L. G. Gorospe. 2001. Discovering factors that influence the success of community-based marine protected areas in the Visayas, Philippines. Ocean and Coastal Management 44: 683–710.

Pollnac, R. B., R. Pomeroy, and L. Bunce. 2003. Factors influencing the sustainability of integrated coastal management projects in Central Java and North Sulawesi, Indonesia. Indonesian Journal of Coastal and Marine Resources Special Edition 1:24–33.

Roberts, C. M., J. A. Bohnsack, F. Gell, J. P. Hawkins, and R. Goodridge. 2001. Effects of marine reserves on adjacent fisheries. Science 294:1920–1923.

Russ, G. R., and A. C. Alcala. 1996. Do marine reserves export adult fish biomass? Evidence from Apo Island, central Philippines. Marine Ecology Progress Series 132:1–9.

Salm, R. V., J. Clark, and E. Siirila. 2000. Marine and coastal protected areas: a guide for planners and managers. International Union for Conservation of Nature and Natural Resources, Washington, D.C.

Sandersen, H. T., and S. Koester. 2000. Co-management of tropical coastal zones: the case of the Soufrière Marine Management Area, St. Lucia, WI. Coastal Management 28:87–97.

Trist, C. 1999. Recreating ocean space: recreational consumption and representation of the Caribbean marine environment. Professional Geographer 51:376–387.

West, P. C., C. L. Fortwangler, V. Agbo, M. Simsik, and N. Sokpon. 2003. The political economy of ecotourism—Pendjari National Park and ecotourism concentration in Northern Benin. Pages 103–115 *in* S. R. Brechin, P. R. Wilshusen, C. L. Fortwangler, and P.C. West, editors. Contested nature—promoting international biodiversity conservation with social justice in the twenty-first century. State University of New York Press, Albany.

White, A. T., L. Z. Hale, Y. Renard, and L. Cortesi, editors. 1994. Collaborative and community-based management of coral reefs: lessons from experience. Kumarian Press, West Hartford, Connecticut.

White, A. T., A. Salamanca, and C. A. Courtney. 2002. Experience with marine protected area planning and management in the Philippines. Coastal Management 30:1–26.

White, A. T., and G. Savina. 1987. Community-based marine reserves: a Philippine first. Pages 2022–2036 *in* Proceedings of coastal zone '87. American Society of Civil Engineers, Seattle.

Part V:
Case Studies

American Fisheries Society Symposium 42:167–184, 2004

Spillover Effects from Temperate Marine Protected Areas

STEVEN MURAWSKI, PAUL RAGO, AND MICHAEL FOGARTY

National Marine Fisheries Service, Woods Hole, Massachusetts 02543, USA

Abstract.—Economic benefits of permanently closed areas can accrue to fisheries in two ways. Export of reproductive products can increase recruitment in open areas, while movement of harvestable-sized animals provides benefits in the form of "spillover" from the refuge, resulting in elevated catch rates near closed area boundaries. Here, we evaluate potential spillover effects from four large marine protected areas in temperate New England waters, closed beginning in 1994. True spillover, as differentiated from seasonal, ontogenetic, or environmentally driven movements, requires differential densities within and adjacent to the closed areas. Density-related spillover, thus, is typified by a biomass or abundance gradient beginning at the boundary and declining as a function of increasing distance. Moderate rates of dispersion are required to establish the density gradient, which may be enhanced by differential distribution of fishing intensity. We tested for density gradients as a function of distance from the closed areas using otter trawl tow-by-tow data collected by scientific observers aboard commercial fishing vessels. Data were adjusted for tow duration (e.g., kg/h towed). A total of 51 species–area combinations were evaluated for the presence of density gradients consistent with implied spillover effects. Of this total, five species–area combinations exhibited statistically significant declines in catch rates with distance. These significant combinations were generally consistent with research vessel surveying information showing year-round catches higher inside the closed area as compared to adjacent open areas after they were closed (e.g., "reserve effects"). Combined groundfish species catches did not show significant declining trends as functions of distances from the four closed areas, nor did the numbers of species caught per haul. Haddock *Melanogrammus aeglefinus* associated with Closed Area I demonstrated the most pronounced apparent spillover, but the pattern of relative fish density at the reserve boundary exhibited a seasonal cycle associated with spawning. We conclude that spillover effects are not a universal consequence of siting marine protected areas in temperate waters but are related to the specifics of the degree of random and directional movements, the fishing intensity field in the adjoining open areas, seasonal migration patterns, and optimal habitat preferences of individual species in relation to the placement of reserve boundaries, all of which may confound the interpretation of spillover.

Introduction

Marine protected areas (MPAs) are increasingly advocated as a tool in fishery management and for the protection and enhancement of oceanic biodiversity (Allison et al. 1998; Fogarty et al. 2000; NRC 2001; Shipley 2004). As a fishery management measure, permanently closed MPAs have the potential for reducing fishing mortality rates on exploited stocks and for providing enhanced economic benefits to fisher-

ies (Botsford et al. 2003; Halpern and Warner 2003). These potential economic benefits are realized either through "export" of reproductive products from the reserve to open areas, where they eventually contribute to improved fishery recruitment, or as a source of large (adult) animals in the form of "spillover" to open areas (see Ward et al. 2001 for a comprehensive review of reserve, export, and spillover effects and their definition and requirements).

A number of studies have evaluated "reserve" effects (e.g., differential abundance and demographics between open and closed areas) and spillover from coral reef or tropical lagoon reserves, generally docu-

---

[1] Corresponding author: Steve.Murawski@noaa.gov

168   MURAWSKI ET AL.

menting increased densities and demographic shifts consistent with reduced total mortality rates on resource species (McClanahan and Mangi 2000; Roberts et al. 2001; Ward et al. 2001; Halpern 2003; Russ et al. 2003; Zeller et al. 2003). For some tropical reserve situations, density gradients have been identified using scientifically designed sampling schemes (McClanahan and Mangi 2000; Zeller at al. 2003) or by interpreting patterns in fishery catch per unit effort (CPUE; Russ et al. 2003) and size compositions of fishery catches near reserve boundaries (Roberts et al. 2001). Large-scale, year-round protected areas are less common in temperate and boreal waters than in the tropics, and the demonstration of reserve and spillover effects have been more variable (Horwood et al. 1998; Piet and Rijnsdorp 1998; Millar and Willis 1999; Frank et al. 2000; Murawski et al. 2000; Fisher and Frank 2002). Here, we evaluate evidence for spillover from a network of year-round closed areas off the New England coast (Fogarty and Murawski 1998; Fogarty 1999; Murawski et al. 2000; Link et al., in press). These MPAs—closed beginning in 1994 and totaling over 20,000 km² (Figure 1)—are among the largest and longest duration anywhere in the

world's oceans. Extensive preclosure and postclosure monitoring surveys and routine fishery data collections allow interpretation for consistency with hypothesized reserve effects and spillover. We utilize routine bottom trawl survey data (Azarovitz 1981) collected in and near the closed areas to evaluate density differentials inside and outside the closures. Observer sampling aboard commercial otter trawl fishing vessels (Murawski 1996) is used to evaluate relative biomass (CPUE) as a function of distance from the four closed areas as a potential indicator of spillover. A series of dispersion models is fit to catch information, and ancillary environmental data are evaluated as covariates. We also evaluate the degree of aggregation of catch and fishing effort as a function of distance from closed area boundaries (Wilcox and Pomeroy 2003).

Background—Interpreting the Effects of Marine Reserves from Data

Evaluating hypotheses of reserve and spillover effects as they relate to MPAs presents a difficult problem of sampling and data interpretation (Willis et al. 2003).

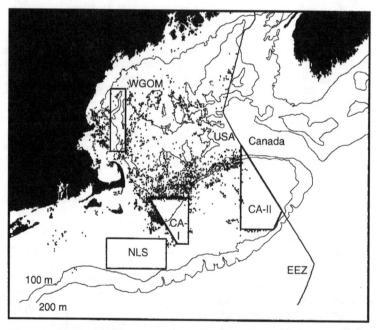

Figure 1.—Four year-round protected areas off the northeastern USA: NLS = Nantucket Lightship (6,275 km²), CA-I = Closed Area I (3,960 km²) , CA-II = Closed Area II (6,927 km²), and WGOM = Western Gulf of Maine (3,025 km²). EEZ is the exclusive economic zone. Points are the locations of observed otter trawl tows sampled in 2001–2002 (n = 4,050 tows).

In particular, hypothesis testing is confounded by issues of replication (both in sampling and reserve siting), spatial heterogeneity of data in and around MPAs (especially when interpreting CPUE statistics, since fishing effort is rarely homogeneously distributed), and movements of animals across reserve boundaries that may occur for reasons other than density differentials. The challenge is particularly difficult for temperate marine ecosystems where animals tend to be distributed along environmental gradients (Swartzman et al. 1992; Perry and Smith 1994) and obligate habitats (such as reefs) may be rare.

Interpreting effects of reserves is generally undertaken by evaluating preclosure and postclosure trends in abundance and demographics or by evaluating spatial variation in these variables in and around the closures. A research strategy based on preclosure and postclosure evaluation assumes that the effect of the reserve is the only or dominant "treatment" effect influencing species of interest. While this may be true, there are so many uncontrolled variables affecting marine ecosystems (e.g., climate change, fishery regulations in open areas) that this assumption can never be completely met in nature. A particularly vexing problem is the fact that even where multiple reserves may be coincidently adopted, no two will be alike in habitat characteristics, species community representation, or dynamics in and outside the reserves. Thus, the first two tenants of hypothesis testing—replication and adequacy of controls—may be seriously violated (Willis et al. 2003).

An alternative approach to classical experiment-based hypothesis testing is to evaluate lines of evidence for their consistency with hypotheses and necessary corollaries of reserve effects and spillover. Density-driven spillover, as differentiated from seasonal, ontogenetic, or environmentally induced movements, requires differential densities within and adjacent to the closed areas. Thus, a logical test to be applied to the reserve is an evaluation of the density (biomass or numbers) within the reserve as opposed to outside the reserve. Such tests are potentially influenced by the characteristics of the specific reserve being evaluated. If the reserve is sited in the core habitat of a particular animal, and the environmental conditions outside the reserve are less conducive, then a density differential may exist, even in the absence of the establishment of a closed area. In this case, a compelling line of evidence in support of a reserve effect would be the ratio of resource abundance inside to outside the closed area, spanning the time sequence before and after the closure. Evidence for reserve effects, in this case, does not require a change in ratio from 1:1 since even in the absence of the reserve, the area inside or outside the closure may be the preferential habitat. Rather, the change in ratio to an increasing differential inside the closure would be of most importance. Similarly, if fishing is a significant source of mortality, and dispersion rates are low to moderate, then there will be a demographic response when comparing populations inside and outside the closures. Such an effect was documented by Murawski et al. (2000) for Atlantic sea scallop *Placopecten magellanicus* associated with Closed Areas I and II (CA-I and CA-II) on Georges Bank (Figure 1). In this case, there was a significant change in the ratio of biomasses inside and outside, with a marked increase in average size within the closed areas. Such responses are also well documented in reef and other tropical ecosystems, where movements from the preferred habitats are limited (Fogarty et al. 2000; Ward et al. 2001; Halpern 2003).

In some circumstances, there may be a change in the ratio of animals inside and outside the area after establishment, but such a differential only occurs seasonally. This may happen, for example, if a particular MPA is seasonally used as a spawning habitat by animals. Density-induced spillover will be confounded with spawning or postspawning migratory behavior and, thus, difficult to distinguish. Other types of distribution patterns that may confound the interpretation of spillover include ontogenetic movements. For example, Overholtz (1985) showed progressive northward (deeper) shifts in distribution of strong (1975, 1978) year-classes of haddock *Melanogrammus aeglefinus* on Georges Bank as a function of age. Hypothetically, siting reserves in locations that are "ontogenetic sinks" could be erroneously interpreted as a reserve effect, especially since there might be a consistent demographic cline. In this case, the apparent reserve effect may be a function of "spill-in" rather than the reverse. Finally, it is possible that apparent spillover effects could be influenced by environmentally induced movements. For example, wind-driven water displacements across reserve boundaries could make fish vulnerable to capture outside reserve boundaries even though directed migrations or ontogenetic movements do not exist. In these circumstances, elevated catch rates at the boundaries would be associated with some environmental condition and be unrelated to spawning season or other biological phenomenon.

Density-related spillover, thus, is typified by a biomass or abundance gradient beginning at the boundary and declining as a function of increasing distance. Botsford et al. (2003) note that for spillover to contribute significant fishery benefits, instantaneous rates of dispersion (*D*) should be sufficiently low so that a reserve effect is established and not so high that the animals spend too much time outside the reserve to provide a refuge. This is particularly important since the probability of capture may increase if fishing effort is "attracted" to the boundary of the reserve and open areas. However, if *D* is trivial, then the net benefits to fishery productivity become negligible.

Interpreting density gradients as evidence for spillover requires sufficient understanding of seasonal and ontogenetic movements, the influences of distribution in relation to environmental gradients, and short-term, environmentally driven forces. While a pattern of declining fish density, as a function of distance from a closed area boundary, is a necessary condition indicative of spillover, it is by no means sufficient evidence for concluding that such an effect is a consequence of the presence of the reserve. These considerations form the framework of our research strategy for evaluating the potential for spillover effects for the groundfish assemblages in and around four New England MPAs (Figure 1).

Study Area and Data Sources

Year-round closed areas have been used as a fishery management tool in New England waters since 1994, and seasonal closures have been used since 1970 (Murawski et al. 2000; Figure 1). The four year-round closures (totaling about 20,000 km²) were originally sited to protect and help restore overfished groundfish resources. The three southern areas (CA-I, CA-II, and the Nantucket Lightship Area [NLS]) were closed year-round to all fishing gears capable of retaining groundfishes beginning in December 1994. The Western Gulf of Maine closure (WGOM) was added in 1996. An additional year-round closed area is located in the central part of the Gulf of Maine (Cashes Ledge) but is not analyzed herein. Since closure, the only gears that have been allowed in the reserves include lobster traps, midwater trawls (for Atlantic herring *Clupea harengus*), and some limited sea scallop dredge fishing. Some scientific sampling of the groundfish resources therein has also been allowed (Link et al., in press).

There has been a limited amount of statistically designed sampling focused on the effects of closed areas on finfish biomass in and around the New England MPAs (Link et al., in press). Rather, most of

what can be inferred about reserve, spillover, and export effects from these reserves is based on data from opportunistic sampling during routine bottom trawl and sea scallop resource surveys. An additional source of opportunistically derived data is sea sampling (using scientifically trained observers) aboard commercial otter trawl, gill-net, and hook vessels (Murawski 1996). The intensity of sea sampling (number of days at sea sampled per year) was increased abruptly in 2002 due to ongoing litigation regarding conservation efforts for New England groundfish. The sea sampling program now directly samples at least 5% of the fishery catch and effort at sea, and these samples are an approximate random sample of the fishery.

Bottom Trawl Survey Data

Data from routine trawl surveys have the great advantage of being temporally synoptic. The surveys provide an unbroken series of abundance indices from 1963 (autumn) and 1968 (spring). The data are obtained from a stratified random sampling scheme, with station numbers per stratum allocated proportional to stratum areas. The survey sampling design was not modified when the year-round closures were instituted; thus, interpreting reserve effects involves poststratification of individual stations to determine their membership in sets inside or outside the closed area boundaries. We evaluated the ratio of catches inside versus outside closed area boundaries by computing the average catch rates (kg/tow) for those survey strata that intersected each of the closure areas. Data were averaged separately for the subset of tows inside and outside each closed area. These data were used to evaluate hypothesized reserve effects (e.g., changes in ratios of catch rates inside and outside and before and after closures).

While bottom trawl surveys provide time series of abundance indices, they are far to sparse in spatial coverage on individual cruises for interpretation of the effects of distance from closed areas versus catch rates. In fact, only a few tows occur each survey in each of the four MPAs, and thus, the indices of catch rates inside versus outside the areas provide an unbiased but imprecise estimate of the relative resource abundance therein. Accordingly, we used time series smoothing to interpret trends in the ratios of catches inside closures versus the adjacent areas outside. The spring and autumn catch data also provide two snapshots per year, separated by 6 months (March versus October), allowing inferences regarding the seasonality of catch ratios, to evaluate if closed areas result in year-round reserve effects.

Sea Sampling Data

Whereas bottom trawl survey data provide unbiased preclosure and postclosure abundance indices, albeit at low levels of annual precision, sea sampling provides intensive and extensive data regarding catch rates and relative fishery performance in any particular year (Figure 1). We used sampling data from large mesh otter trawl fisheries to evaluate the potential for spillover by relating catch rates to distances from each of the closed areas (Figures 1–4; Tables 1, 2). Data were screened to eliminate information from trips targeted at species not considered large-mesh target species (e.g., we used vessel trips targeted to the species indicated in Table 1, except for silver hake). Additionally, we eliminated gill-net and hook-and-line gears from our initial analyses because of the limited number of target species sought by these fisheries. Using these criteria, we obtained information from 4,050

individual otter trawl tows sampled during the latter part of 2001 and 2002, that provided complete records of catch, effort, discards, depth, and precise location (Figure 1). From these records, we constructed a database of catch and CPUE (catch per hour towed) by species, with ancillary information suitable for our purposes. Catches of each species included both kept and discarded portions of the trawl contents. We used ArcView (ESRI, Inc., version 3.3, Redlands, California) geographic information system (GIS) software to compute the distances (km) of each trawl location (starting points) to the nearest boundary of each of the four year-round closed areas (Figure 2). Results of tow-by-tow sampling were highly variable, reflecting the spatial and seasonal heterogeneity of the fishery (Figure 1). The average catch rate was 279.24 kg/h towed (SD = 568.84), average time per tow was 3.61 h (SD = 1.55), and the average numbers of species caught per haul was 9.64 (SD = 2.92).

The data were standardized to catch per hour towed, but no additional standardization was undertaken (e.g., for vessel size, predominant species in the

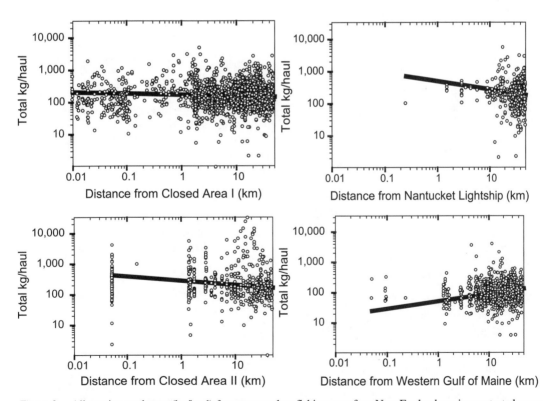

Figure 2.—All species catch rate (kg/haul) for otter trawlers fishing near four New England marine protected areas, 2001–2002.

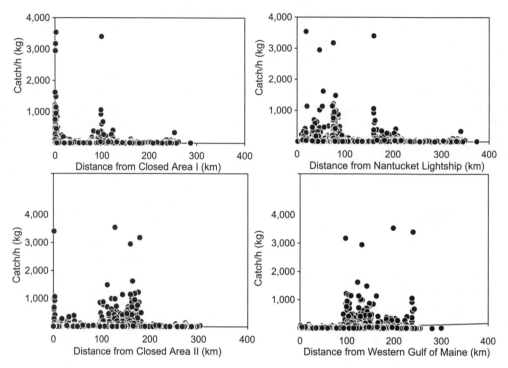

Figure 3.—Catch rates (kg/h) of haddock by otter trawlers fishing in the vicinity of four New England marine protected areas, 2001-2002. The four panels represent the catch rates versus the distance (km) from the designated closed area.

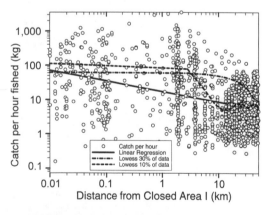

Figure 4.—Catch rate (kg/h fished) of haddock by otter trawlers versus distance from Closed Area I. Three models of catch rate versus distance are plotted: (1) solid line indicates $\log_e$(CPUE) versus $\log_e$(distance); (2) short dashed line indicates a LOWESS smooth using a window including the adjacent 10% of data; and (3) LOWESS smooth using the adjacent 30% of data. Models include all data irrespective of distance from the closed area (Figure 3).

catch, or other factors). We chose not to standardize the data for these factors since this was not the specific aim of the study and because previous evaluations of these factors indicated that mesh size and primary species (or species group) sought were the most significant factors determining catch rate in New England trawl fisheries (Murawski 1996). Since the data were preselected to include only groundfish as target species and using large mesh trawls, these factors were partially controlled in our analyses. Our analyses of trawl survey data were undertaken at the tow level (i.e., assuming independence between tows and summarizing them as if each represented a random sample of relative resource abundance at a particular location and time). This assumption of independence may not have been totally met since there is likely some covariance of individual tows within a trip (e.g., on individual fishing trips there may be several to dozens of tows made).

We selected a subset of the species captured by the otter trawl fleet for analyses of potential spillover effects. The selected species represent the primary fish-

Table 1.—Significance of estimated spillover effects for 13 species or groups in relation to four New England marine protected areas: Closed Area I (CA-I), Closed Area II (CA-II), Nantucket Lightship (NLS), and Western Gulf of Maine (WGOM). A log–log model of standardized catch per unit effort (CPUE; kg/h fished) in relation to distance (km) from the closed areas was estimated: $\log_e(\text{CPUE}) = a + b \log_e(x)$, where x is the distance (km) from the closed area boundary. For each model fit, the parameter estimates, significance level of the b parameter, and the number of observations (n) are given. Bold values indicate where highly significant ($P < 0.01$) negative slopes to the model were calculated.

Species or group	CA-I				CA-II				NLS				WGOM			
	a	b	P	n	a	b	P	n	a	b	P	n	a	b	P	n
All	5.1128	0.0157	0.0814	1,989	5.5768	-0.0442	0.0690	541	5.6999	-0.1210	0.0446	542	4.4873	0.0390	0.1106	992
Number of species	2.1845	0.0201	<0.0001	1,989	2.1810	-0.0069	0.3414	541	2.2976	-0.0415	0.0913	542	2.2077	0.0127	0.2179	992
Goosefish Lophius americanus[a]	2.4500	0.1658	<0.0001	1,553	1.6696	0.1797	<0.0001	316	1.7532	-0.0420	0.6763	345	2.2987	0.0742	0.0865	815
Atlantic cod Gadus morhua	2.4218	0.0173	0.2576	1,691	2.5741	0.1494	<0.0001	462	2.1384	0.1388	0.1447	447	1.9380	0.1311	0.0246	820
Winter flounder Pseudopleuronectes americanus	2.1585	0.3932	<0.0001	868	**4.3009**	**-0.2282**	**<0.0001**	**313**	2.0259	0.3518	0.0002	495	0.0080	0.4963	<0.0001	521
Witch flounder Glyptocephalus cynoglossus	1.3426	0.1684	<0.0001	1,058	0.7033	0.1659	0.0024	174	-1.0855	0.2303	0.6618	24	**2.4709**	**-0.2481**	**<0.0001**	**722**
Yellowtail flounder Limanda ferruginea	1.9447	-0.0506	0.2053	786	**3.0917**	**-0.2656**	**<0.0001**	**329**	**4.4698**	**-0.5248**	**<0.0001**	**461**	1.5495	0.1314	0.0754	576
American plaice Hippoglossoides platessoides	1.6672	0.1137	<0.0001	1,028	1.1573	0.1939	0.0005	216	-0.3291	0.2139	0.1611	15	1.9960	-0.0563	0.2433	766
Haddock	**2.8361**	**-0.3368**	**<0.0001**	**1,192**	1.6984	0.0791	0.0906	292	2.1560	0.2252	0.4162	181	1.1059	0.0113	0.8566	408
Acadian redfish Sebastes fasciatus	0.6408	-0.0072	0.8050	424	-0.2317	-0.1832	0.3750	35				0	0.9326	-0.2635	0.0129	248
Pollock Pollachius virens	1.3652	0.0501	0.0368	631	0.9089	0.0148	0.7917	120	0.7164	0.2063	0.6743	79	1.7457	-0.1430	0.0553	332
Spiny dogfish Squalus acanthias	1.7145	-0.0252	0.2572	773	1.5810	-0.1306	0.0119	134	2.0437	-0.1073	0.3203	196	0.9271	0.2211	0.0002	542
Silver hake Merluccius bilinearis	0.1453	0.0821	<0.0001	669	-0.0249	0.0528	0.1590	158	-4.0658	0.9429	0.0072	56	0.5985	-0.1400	0.0323	387

[a] Also known as monkfish.

Table 2.—Results of mixed-effects general linear models for three species caught within 50 km of two Georges Bank closed areas. The model fit relates $\log_e$(CPUE) (kg/h fished) to month, $\log_e$(distance) from the closed area, $\log_e$(depth) (m), and interaction effects between distance × month and depth × month. Data provided are the probability values for the specific effect being tested.

Source	Closed area	Atlantic cod	Haddock	Yellowtail flounder
Month	CA-I	0.0116	0.0300	<0.0001
	CA-II	<0.0001	<0.0001	0.1041
$\log_e$(distance)	CA-I	0.0001	0.1697	0.5355
	CA-II	0.1080	<0.0001	0.5408
$\log_e$(depth)	CA-I	<0.0001	0.1214	<0.0001
	CA-II	0.0001	0.8291	0.1530
$\log_e$(distance × month)	CA-I	<0.0001	0.0597	<0.0001
	CA-II	0.0956	0.1879	0.1233
$\log_e$(depth × month)	CA-I	0.0019	0.0155	<0.0001
	CA-II	<0.0001	<0.0001	0.1396

ery target species and a number of bycatch species that are widely distributed and represent a variety of life histories and habitat preferences. The selected species included: goosefish, Atlantic cod, winter flounder, witch flounder, yellowtail flounder, American plaice, haddock, Acadian redfish, pollock, spiny dogfish, and silver hake. While other species (such as some species of skate) were abundant in trawl catches, we chose not to analyze their catch rate trends due to some uncertainties in species identification. In addition to the 11 species noted above, we also considered aggregate species catch rates (all species caught; Figure 2), and the total number of species present in each trawl haul (an index of finfish species richness) as response variables.

Models of Spillover and Dispersion

We conducted a variety of empirical and model-based analyses of sea sampling and bottom trawl survey data to evaluate the potential for reserve and spillover effects. These included correlation studies for catch rates as a function of distance from each of the closed areas, evaluation of the role of environmental covariates for a subset of species, and estimation of alternative formulations of dispersion models (Beverton and Holt 1957; Quinn and Deriso 1999; Kruse et al. 2001). Reserve effects were evaluated by examining patterns of trawl survey catch rates in and near the four closed areas.

Empirical Analyses—Spillover Effects

As discussed above, one important and necessary condition indicating resource spillover from MPAs is a density gradient originating at the edge of the closed area and declining as a function of distance away from the boundary (Beverton and Holt 1957; McClanahan and Mangi 2000; Zeller et al. 2003). In situations where spillover occurs, the slope of the density–distance relationship is related to the instantaneous rate of dispersion across the boundary. The process can also be cast as one of a transfer rate (*T*) from one box (closed) to another (open), but such coefficients do not explicitly account for distance traveled as a function of time or dispersion (diffusion) rate (Beverton and Holt 1957; Quinn and Deriso 1999).

Plots of catch rates as a function of distance from closed area boundaries exhibit, in some cases, complex relationships at multiple distance scales (Figures 2–4). In the case of haddock, for example, the relationship between catch rate and minimum distance from CA-I shows two strong modes at less than 20 km and again around 100 km (Figure 3). Similarly, data for CA-II show two patterns of abundance at about the same spatial scales. For haddock, there appears to be a density–distance relationship for CA-I and CA-II but not for the NLS and WGOM MPAs. Thus, tests of the relationship between density and distance may be confounded if all the data (extending to nearly 400 km from the closed area boundaries) are considered. It is apparent that the impacts of closed areas on densities outside closed areas occur on spatial scales of tens of kilometers. While such effects may extend farther from the boundaries, they are likely confounded with effects from the other three closed areas and, thus, difficult to discern.

In order to define a usable subset of data for further analyses that is likely not confounded by multiple closed area effects, we examined several empirical relationships between catch rates and distances from closed areas, using the CA-I haddock information (Figures 3, 4). Linear regressions of catch rates versus distance (all data out to nearly 300 km from CA-I were used). Additionally, we fit lowess smoothers to the data using variable percentages of the data set in the smoothing window (Figure 4). Haddock catch rates near CA-I decline with distance from the closed areas and exhibit few high catch rate values at distances greater than about 20 km. Using a more sensitive smoother (10% of data), the decline in catch rate seems to occur at scales of about 3–10 km, while 30% smoothing results in a gradual rate of catch decline at the scale of 10–30 km (Figure 4). Based on these results, we limited the data sets used for the detection of density–distance relationships to catches of each species–closed area combination for distances within 50 km of the closed area. While these relationships are doubtlessly species-specific, this arbitrary exclusion of data was made to limit the potential confounding effects of the interaction of multiple boundaries on catch rates and is intended to be a preliminary data screening device. Additionally, because each species has different habitat requirements, and some tows within the 50-km radius may be beyond the depth ranges in which individual species are found, we used only nonzero catch tows when examining density–distance relationships for individual species–area combinations (Table 1).

A robust test of the density–distance relationship is

$$\log_e(CPUE_i) = a + b \times \log_e(X_i) + \varepsilon_i, \qquad (1)$$

where, $CPUE_i$ is the catch per hour fished (kg) for tow i, X_i is the minimum linear distance (km) between the closed area boundary and the start location of tow i, and a, b, and ε are the intercept, slope, and error, respectively, estimated from fitting the model. This model was applied to all possible combinations of data (e.g., ≤50 km from the closed area boundary using only nonzero catches for a particular species; Table 1). A total of 51 species (or species group)–closed area combinations had sufficient data for fitting the above relationship (Table 1).

A total of 19 of the 51 possible regression model fits resulted in highly significant ($P < 0.01$) estimates of the slope parameter (b) of the density–distance relationship (Table 1). Of the 19 significant relationships, 14 indicated a positive slope, and only 5 had significant negative slopes. None of the tests using aggre-

gate species catches resulted in significant relationships, and only one test for the number of species versus distance from the closed area was significant and positive (e.g., for CA-I). The species–area tests that resulted in negative density–distance relationships were: haddock for CA-I, yellowtail flounder and winter flounder for CA-II, yellowtail flounder for the NLS area, and witch flounder for the WGOM closure (Table 1). Nearly three times as many combinations resulted in positive density–distance relationships, suggesting that factors other than spillover from the reserves was influencing this relationship. One likely confounding factor responsible for many of the positive slope values is the fact that preferred depths for some species occurred primarily outside some of the closed areas. In this case, depth is likely to be a confounding variable, masking the underlying effects of dispersion from closed areas. To examine the potential confounding effects of depth preferences, we fit a series of models with covariates to simultaneously examine their effects on the strength of the density–distance relationship.

Depth preferences for many species approximate a lognormal distribution. Accordingly, a simple quadratic relationship can be used to examine the relationship between $\log_e$ catch rate and depth (Figure 5):

$$\log_e(CPUE_i) = a + b_1 Z_i - b_2 Z_i^2 + \varepsilon_i, \qquad (2)$$

where, Z_i is the depth (m) at tow i. We fit this relationship to two of the species–area combinations that resulted in significant negative density–distance relationships—haddock around CA-I (Figure 5) and yellowtail flounder near CA-II. In both cases, the quadratic regressions were significant, as were estimates of the b_1 and b_2 parameters. Thus, catch rates were significantly related to depth preferences—a well-known phenomenon for groundfishes (Swartzman et al. 1992; Perry and Smith 1994). Given the potentially confounding effects of depth preferences on results from our simple density–distance relationship (equation 1), we fit more complex models to estimate the simultaneous effects of distance and depth preference on catch rates for haddock at CA-I (Figure 6) and yellowtail flounder at CA-II, viz,

$$\log_e(CPUE_i) = a + b_1 Z_i - b_2 Z_i^2 + \\ b_3 \times \log_e(X_i) + \varepsilon_i. \qquad (3)$$

When the simultaneous effects of distance and depth were considered, parameter estimates for the

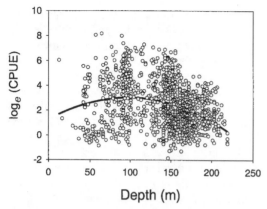

Figure 5.—Catch rate ($\log_e$[kg/h]) of haddock as a function of water depth (m). A significant parabolic relationship (as indicated by the line) was fit to the data. Data include all otter trawl tows within 50 km of Closed Area I. CPUE = catch per unit effort.

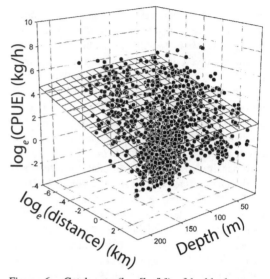

Figure 6.—Catch rate ($\log_e$[kg/h]) of haddock versus water depth (m) and $\log_e$(distance) from Closed Area I. The fitted relationships for these data indicated a significant negative effect of distance, but the parameters of the depth–density parabola were not significant.

depth effects (b_1, b_2) were not significant for either of the cases we examined (Figure 6). Thus, in these specific instances, depth preferences do not appear to be resulting in a spurious density–distance relationship. We did not examine such models for the cases where slopes of the density–distance relationships were positive, but it is likely that for some stocks with relatively deepwater preferences, this factor is important, since the four New England MPAs are primarily in relatively shallow waters.

In addition to the regression models summarized above and in Table 1 and Figures 5 and 6, we fit a series of log-linear mixed-effects models (Millar and Willis 1999) to examine factors influencing catch rates for Atlantic cod, haddock, and yellowtail flounder near CA-I and CA-II. These models examined the effects of calendar month, distance from closed areas, depth, and interaction effects between distance and month and depth and month on $\log_e$ CPUE (Table 2). These results reveal the complex interplay between variables that potentially confound the interpretation of simple density–distance relationships. In most cases, one or more of the interaction effects between depth, distance, and month were significant, which confuses the interpretation of main effects (e.g., distance, month, depth). Results of these general linear model analyses confirm the difficulty in interpreting spillover effects in situations where multiple factors are simultaneously influencing catch rates and emphasize the importance of carefully considering the basis for assuming data support a spillover hypothesis.

Estimating Dispersion Models

The above models attempt to estimate empirical relationships between variables potentially confounding the density–distance relationship. It is also useful to cast the problem in first principles related to dispersal and diffusive properties of a stock in relation to density. Consider a simple dispersion model (Beverton and Holt 1957, equation 10.2) of catch rate (C) as a function of distance from a release point in two dimensions (x,y) at any time (t):

$$\frac{\partial C}{\partial t} = \frac{1}{4}\frac{V^2}{n}\left(\frac{\partial^2 C}{\partial x^2} + \frac{\partial^2 C}{\partial y^2}\right), \qquad (4)$$

where V equals the velocity per movement event, and n is the number of movement events per unit time. Then V^2/n is defined as D, the dispersion coefficient. Collapsing the distance measures into a single vector (x, assuming dispersive movement at a constant rate regardless of vector) results in Beverton and Holt's (1957) equation 15.2:

$$C_{x,t} = \frac{C_0 \exp[-(x^2 / D \times t)]}{\sqrt{t}}. \qquad (5)$$

Assuming that the observed pattern occurs over a unit time step (1), the equation simplifies to

$$C_x = C_0(\exp - [x^2/D]), \qquad (6)$$

where C_0 is the catch rate at the boundary of the closed area, C_x is the catch rate at distance x from the boundary, and D is the dispersion coefficient. An additional factor that potentially confounds this relationship is the fact that mortality (due to natural and fishing effects) occurs across the density field. Accordingly, an additional mortality rate term can be included to describe the relationship between catch rate, distance, and time (similar to Beverton and Holt 1957, equation 15.4), viz,

$$\frac{\partial C}{\partial t} = \frac{D}{4} \frac{\partial^2 C}{\partial x^2} - ZC, \qquad (7)$$

where Z is the instantaneous rate of total mortality. If fishing effort differentially aggregates near closed areas, then Z may not be a constant function of distance, requiring a relationship between cumulative fishing mortality rate at distance x. Assuming unit time ($t = 1$) and accounting for spatial variation in F results in

$$C_x = C_0 \exp - \left\{ (x^2/D) + M + \left[\left(\sum_{x=0}^{x} pf_x \right) \times F \right] \right\}, \qquad (8)$$

where pf_x is the cumulative proportion of total standardized fishing effort at distance x from the closed area. If M is assumed constant over all distance, then the relationship simplifies to

$$C_x = C_0 \exp - \left\{ (x^2/D) + \left[\left(\sum_{x=0}^{x} pf_x \right) \times F \right] \right\}. \qquad (9)$$

We fit equation 6 to the five combinations of species and closed areas that resulted in significant negative slopes to the density–distance relationships (Tables 1, 3). In three of the cases, highly significant ($P < 0.01$) estimates of the dispersion coefficient (D) were obtained (Table 3), although the coefficient of determination of these model fits was very low (<9%), reflecting the highly variable CPUE data and the potentially confounding effects of covariates (Table 2). We also fit equation 9 for CA-I haddock using the proportion of total haddock catch as a function of distance from the closed area and assuming two fishing mortality rates ($F = 0.20$ and 1.00). In both cases, the model fits were slightly poorer than that obtained from equation 6 (ignoring the effects of the spatial pattern of fishing mortality), and no further tests of this model elaboration were attempted. Nevertheless, the development of spatially explicit models accounting for dispersive and random movements and the fishing mortality rate field represents a potentially significant development in fishery science and one that is now within our capabilities in data and models (see Beverton and Holt 1957 for a more complete theoretical treatment of dispersion rates and calculation of transfer rates between an arbitrary number of spatial zones).

Empirical Analyses—Reserve Effects

As noted in the background section, a primary prerequisite for spillover effects is the establishment of a density differential between the closed area and the surrounding open areas. We evaluated the potentials for differential densities between closed and open areas using the time series of spring and autumn research vessel bottom trawl survey data (Figures 7–9). Data collected near CA-I and CA-II are summarized herein (Figures 7, 8). Because of the low sampling intensity within any one closed area in a survey cruise, we averaged the most recent 3 years of catch per tow (kg) data to provide average ratios of recent relative biomass within versus adjacent to the closed areas (Figures 7, 8). These analyses considered most of the species analyzed from commercial data (except Acadian redfish and silver hake) but added five species for which surveys provide reliable abundance measures for some important Georges Bank fishes that may not be landed or targeted by fisheries: longhorn sculpin *Myoxocephalus octodecemspinosus,* little skate *Leucoraja erinacea,* winter skate *Leucoraja ocellata,* windowpane *Scophthalmus aquosus,* and Atlantic herring. We also applied lowess smoothers to the time series of such data to examine changes in the catch ratios inside and outside closed areas, particularly as they relate to the establishment of the reserves (Figure 9).

For CA-I, only haddock and yellowtail flounder exhibited consistently and significantly higher catch ratios inside versus outside the closure in both seasonal surveys. For other species (winter flounder, spiny dogfish, American plaice, and pollock), catch ratios were elevated within the closed area in one of the seasonal surveys but not the other. Atlantic cod catch rates were two to three times higher inside the closure as compared to outside the area, but the absolute catch rates were low and highly variable. Total numbers of fish showed no seasonal pattern or differences between inside and outside the closed areas, although total catch weight was higher in autumn, probably because of the influence of the haddock catches.

Table 3.—Parameter estimates and diagnostics for dispersion models fit for five stocks exhibiting potential spillover effects in relation to New England marine protected areas (NLS = Nantucket Lightship; WGOM = Western Gulf of Maine) (Table 1): $C_x = C_0 \exp[-(x^2/D)]$, where C_x is the catch (kg/h fished) at distance x from the closed area boundary, C_0 is the estimated density at distance $x = 0$, and D is the dispersion coefficient. Data are catches per hour fished by trawl vessels sampled with an observer onboard.

Species and closed area	Significance of model fit (P)	Coefficient of determination (r^2)	Intercept (C_0)	Dispersion coefficient (D)	Significace of D (P)
Haddock Closed Area I	<0.0001	0.0883	165.7245	26.3927	0.0036
Yellowtail flounder Closed Area II	<0.0001	0.0455	66.4203	780.6645	0.0183
Winter flounder Closed Area II	<0.0001	0.0556	235.5524	503.8814	0.0048
Yellowtail flounder NLS area[a]	1.0000	0.0000			
Witch flounder WGOM area	<0.0001	0.0338	17.3462	1,843.4141	<0.0001

[a] Model results and data indicated potential interference due to close proximity of Closed Area II.

Results for CA-II showed elevated catch rates inside the closure in both seasons for yellowtail flounder, longhorn sculpin, and winter flounder. Catch ratios were also higher inside the closure in both seasons for winter skate and windowpane, although not markedly so in spring. Atlantic cod, haddock, and pollock catch ratios were elevated in the spring but extremely low in the autumn, reflecting the seasonal use of CA-II by these species (Murawski et al. 2000). Total catch numbers and weights did not show a strong density differential between open and closed areas in either season (Figure 8).

Historically, patterns of haddock catches were similar between the area that was eventually closed in

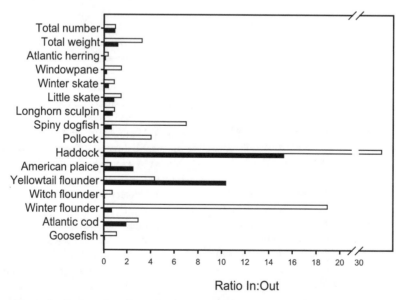

Figure 7.—Ratios of catches of various species and species groups inside and adjacent to Closed Area I during 2000–2003, based on National Marine Fisheries Service bottom trawl surveys in spring (filled bars) and autumn (open bars). Data are the ratios of mean weight per tow (kg) for intersecting survey strata, with data poststratified to inside or outside the closed area.

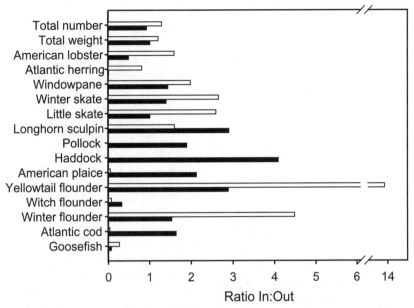

Figure 8.—Ratios of catches of various species (including American lobster *Homarus americanus*) and species groups inside and adjacent to Closed Area II during 2000–2003, based on National Marine Fisheries Service bottom trawl surveys in spring (filled bars) and autumn (open bars). Data are the ratios of mean weight per tow (kg) for intersecting survey strata, with data poststratified to inside or outside the closed area.

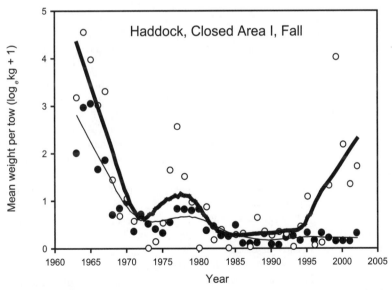

Figure 9.—Changes in the mean density of haddock (mean weight per standardized research vessel trawl tow) inside and just adjacent to Closed Area I, 1963–2002. Closed Area I was closed year-round in December 1994. Open circles and heavy solid line are for the area inside CA-I, closed circles and light line are for catches adjacent to the closed area. Note that most of the resource recovery since closure has occurred within the closed area.

1994 and the adjacent portions of intersecting survey strata that were not closed (Figure 9). Abundance indices declined sharply in the 1960s, followed by a brief resurgence in the 1970s and a long period of relatively low abundance through the early 1990s. Following the institution of the closed areas, however, the biomass trajectories inside the closed area footprint significantly diverged from those in the open area (Figure 9). Currently, the ratio biomasses during autumn, inside versus outside the closed areas, is more than 10:1. These data convincingly suggest a reserve effect of haddock associated with CA-I.

Discussion

Interpreting reserve and spillover effects for MPAs in temperate seas presents a particularly daunting set of problems related to the confounding effects of environmental covariates and behavioral adaptations to seasonally varying environments. Species may be distributed along environmental gradients that may or may not be well represented within particular closed area boundaries. Similarly, ontogenetic movements, spawning migrations, and random or environmentally induced distribution shifts belie simple interpretations of density patterns in relation to closed area boundaries.

The first requirement for density-driven spillover to occur is that a density differential exists between open and closed areas. While such gradients have been clearly established for sedentary species in temperate oceans, such as the Atlantic sea scallop on Georges Bank (Murawski et al. 2000), evidence for reserve effects on more mobile invertebrate species and finfishes is more equivocal (Figures 7–9; Murawski et al. 2000; Link et al., in press). Some stocks with limited movement potentials established density differentials within closed areas as compared to the adjacent open areas (Figures 7–9), and in most cases, these stocks were among those exhibiting significant negative density–distance relationships (Table 1). It is clear, however, that reserve effects and spillover (as evidenced by significant negative density-distance relationships) appear to be the exception rather than the rule for MPAs in these waters. The fact that generally similar results were obtained from four pseudo-replicated closed areas within this ecosystem strengthens our overall conclusion that reserve and spillover effects are not a universal consequence for temperate water MPAs. Rather, the efficacy of a particularly configured reserve is related to the specifics of the degree of random and directional movements, seasonal migration patterns, and optimal habitat preferences in relation to the reserve boundaries. The fishing intensity field in areas open to exploitation may also confound the interpretation of spillover if fishing effort is differentially attracted to the boundaries. We propose new models to account for differential fishing effects across the density field in open areas adjacent to MPAs. Likewise, statistical methods to separate the effects of species distributions along environmental gradients from reserve and spillover effects are presented and applied.

The example of haddock in and around CA-I presents a particularly compelling demonstration of reserve siting that is in concordance with the life history, habitat preferences, and movement patterns of one of the primary resource species the reserve was intended to help conserve. This is not surprising since a seasonal closed area in the vicinity of CA-I has been used for over 30 years to protect spawning haddock (Murawski et al. 2000). After the area was closed year-round, haddock biomass built up in the reserve, which now shows a substantial density differential year-round. Despite this density differential, sufficient stock travels beyond the closed area boundaries to increase haddock catch rates in the local region (<20 km) surrounding the closure. Importantly, the apparent movement of haddock to the open area surrounding CA-I is of such localized extent that abundance indices outside have not improved markedly there, as compared to inside the reserve boundaries (Figure 9). Relatively high catches just at the boundary (Figures 4, 10, 11) doubtlessly are removing production that would otherwise result in higher resource abundance in the open region around CA-I. Movement of haddock outside the boundaries of CA-I is not necessarily just dispersive in nature. Monthly patterns of haddock catch rates within 20 km of the boundary suggest that movements out of, and perhaps into, the area are related to spawning (Figure 10). Catch rates adjacent to CA-I are lowest in March and April, when spawning activity peaks. Highest observed individual catches (>3,000 kg/h) occur in the months following spawning (June–August), as water temperatures in shallow areas of CA-I warm (Figure 10).

An important consideration in using MPAs as a tool in fishery management is the extent to which effort is attracted to the closed area boundaries once the reserve is in place (Wilcox and Pomeroy 2003). Effort may concentrate at the boundaries of MPAs either due to higher CPUE, larger average fish sizes, or the *perception* of enhanced benefits to be achieved by

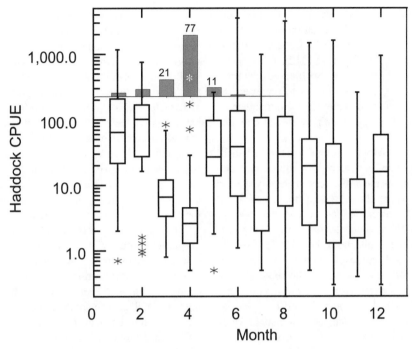

Figure 10.—Monthly catch per unit effort (CPUE; kg/h fished) for haddock caught by otter trawlers fishing within 20 km of Closed Area I, 2001–2002. The inset is the average monthly density of haddock eggs (numbers/10 m²) from ichthyoplankton surveys off the northeastern USA, 1977–1987 (from Cargnelli et al. 1999). Note that catch rates decline in the area surrounding Closed Area I when spawning is apparently most intense.

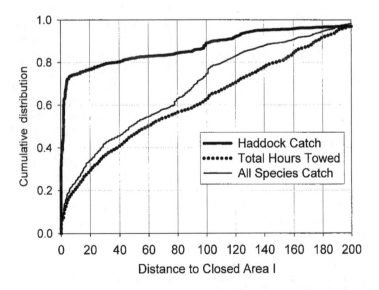

Figure 11.—Cumulative haddock catch, all species catch, and hours fished by otter trawlers as a function of distance from Closed Area I. Data are derived from observed otter trawl tows (*n* = 4,050) conducted in 2001 and 2002.

fishing near the reserve. In order to assess these considerations, we calculated concentration profiles of total effort, total catch, and haddock catch in relation to distance from CA-I (Figures 11, 12). These analyses are based on the subsample of observed otter trawl trips, which are generally thought to represent the overall patterns of effort in the fishery, although there may be some biases by home port, vessel size, and fishing area.

Proportions of total fishing effort (hours fished) and total mixed-species catch show some concentration near CA-I, but the profiles are not highly skewed (Figure 11). Thus, about 20% of hours fished and about 25% of total species catch occurs within about 10 km of CA-I. The concentration profiles are somewhat gradual and parallel until about 100 km from the boundary of CA-I (Figure 11), at which the catch increases faster than does the effort. This second area of relatively high catch rates is associated with the vicinity around CA-II. As compared with total species catches, the cumulative catch distribution of haddock is highly skewed to the closed area boundary. In fact, 39% of the total U.S. haddock catch (all areas) sampled by observers occurred within 1 km and 72% within 5 km of the boundary of CA-I (Figures 11, 12). This represents a substantial concentration of haddock catch and directed effort as compared with distributions that occurred prior to the institution of the closed areas

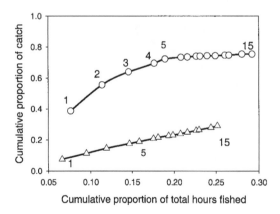

Figure 12.—Joint cumulative distribution functions depicting the relationship between haddock (circles) catch and all species (triangles) otter trawl catch and fishing effort (hours fished) as a function of distance (1–15 km) from Closed Area I. Distances are given as the numbers above and below the plotted lines. Note that for all species catch the relationship between catch and effort is approximately proportional, while haddock catches are highly skewed to the vicinity of the closed area boundary.

(National Marine Fisheries Service, Northeast Fisheries Science Center file data). Several reasons are responsible for the concentration of haddock fishing effort and catch in the vicinity of CA-I, including the institution of restrictive days at sea effort limits that tend to concentrate effort closer to port and the liberalization of haddock trip catch limits as the resource has increased. However, while directed haddock fishing has concentrated near the closed area boundaries, in response to higher catch rates of the species there, overall catch rates (Figure 2; Table 1) do not appear elevated in relation to any of the closed area boundaries, and thus, effort is not entirely clustered around the periphery of the closures (Figures 1, 12).

Interpretation of reserve and spillover effects using surveys with low spatial resolution and commercial catch and effort data, which may be problematic in terms of standardization and biases, requires a high degree of caution. Nevertheless, conclusions regarding the overall efficacy of the closed areas are generally consistent between the four unique MPAs we evaluated. There was no global buildup of biomass in the closed areas, although individual species with life history and habitat preferences that matched the habitats available within particular areas showed the most effect. It is clear that the more sedentary the species, the more likely these temperate water MPAs were to achieve a demonstrable reserve effect (a "sedentary spectrum" along which reserve effects become more or less important). Low rates of dispersive movements (e.g., for adult Atlantic sea scallops) translate into low spillover potentials. Also, given the long potential lifespans of some temperate water fishes, the full effects of MPAs in these areas may not be realized for a relatively long time after closure, especially when resources are heavily exploited and age–size distributions are truncated due to intensive fishing. An important conclusion from our work is that the placement of reserves in temperate seas requires considerable forethought regarding the species for which protection is sought, importance of fisheries enhancement versus biodiversity protection, and the benefits of MPAs and other alternative management tools relative to the costs of implementation and their likelihood of success.

Acknowledgments

We thank our colleagues at the Northeast Fisheries Science Center for their unflagging efforts in the collection of high-quality fishery-dependent and fishery-independent data. Sea sampling data were expertly extracted by Susan Wigley and Kathy Sosebee. Interpretation of the spatial geography of groundfish catches benefited from the GIS skills of Chad Keith. Finally, we appreciate the constructive reviews of the manuscript provided by anonymous referees, as well as the good-humored persistence of our editor, J. Brooke Shipley.

References

Allison, G. W., J. Lubchenco, and M. H. Carr. 1998. Marine reserves are necessary but not sufficient for marine conservation. Ecological Applications 8(Supplement 1):S79–S92.

Azarovitz, T. R. 1981. A brief historical review of the Woods Hole Laboratory trawl survey time series. Canadian Special Technical Publication of Fisheries and Aquatic Sciences 58:62–67.

Beverton, R. J. H., and S. J. Holt. 1957. On the dynamics of exploited fish populations. Chapman and Hall, London. (Facsimile reprint 1993.)

Botsford, L. W., F. Michell, and A. Hastings. 2003. Principles for the design of marine reserves. Ecological Applications 13(Supplement 1):S25–S31.

Cargnelli, L., S. Griesback, P. Berrien, W. Morse, and D. Johnson. 1999. Essential fish habitat document: haddock, *Melanogrammus aeglefinus*, life history and habitat characteristics. NOAA Technical Memorandum NMFS-NE-128.

Fisher, J. A. D., and K. T. Frank. 2002. Changes in finfish community structure associated with an offshore fishery closed area on the Scotian Shelf. Marine Ecology Progress Series 240:249–265.

Fogarty, M. J. 1999. Essential habitat, marine reserves and fishery management. Trends in Evolution and Ecology 14:133–134.

Fogarty, M. J., J. A. Bohnsack, and P. K. Dayton. 2000. Marine reserves and resource management. Pages 375–392 in C. R. C. Sheppard, editor. Seas at the millenium: an environmental evaluation. Global Issues and Processes, volume 3. Pergamon, Elsevier, New York.

Fogarty, M. P., and S. A Murawski. 1998. Large-scale disturbance and the structure of marine systems: fishery impacts on Georges Bank. Ecological Applications 8(Supplement 1):S6–S22.

Frank, K. T., N. L. Shackell, and J. E. Simon. 2000. An evaluation of the Emerald/Western Bank juvenile haddock closed area. ICES Journal of Marine Science 57:1023–1034.

Halpern, B. S. 2003. The impact of marine reserves: do reserves work and does reserve size matter? Ecological Applications 13(Supplement 1):S117–S137.

Halpern, B. S., and R. R. Warner. 2003. Matching marine reserve design to reserve objectives. Proceedings of the Royal Society of London 270:1871–1878.

Horwood, J. W., J. H. Nichols, and S. Milligan. 1998. Evaluation of closed areas for fish stock conservation. Journal of Applied Ecology 35:893–903.

Kruse, G. H., N. Bez, A. Booth, M. W. Dorn, S. Hills, R. N. Lipcius, D. Pelletier, C. Roy, S. J. Smith, and D. Witherell, editors. 2001. Spatial processes and management of marine populations. 17th Lowell Wakefield Fisheries Symposium. University of Alaska Sea Grant College Program Report AK-SG-01-02.

Link, J., F. Almeida, P. Valentine, P. Auster, R. Reid, and J. Vitaliano. In press. The effect of area closures on Georges Bank. In P. W. Barnes and J. P. Thomas, editors. Benthic habitats and the effects of fishing. American Fisheries Society, Symposium 41, Bethesda, Maryland.

McClanahan, T. R., and S. Mangi. 2000. Spillover of exploitable fishes from a marine park and its effects on the adjacent fishery. Ecological Applications 10(6):1792–1805.

Millar, R. B., and T. J. Willis. 1999. Estimating the relative density of snapper in and around a marine reserve using a log-linear mixed-effects model. Australian and New Zealand Journal of Statistics 41(4):383–394.

Murawski, S. A. 1996. Factors influencing bycatch and discard rates: analyses from multispecies/multifishery sea sampling. Journal of Northwest Atlantic Fishery Science 19:31–39.

Murawski, S. A., R. Brown, H.-L. Lai, P. J. Rago, and L. Hendrickson. 2000. Large-scale closed areas as a fishery management tool in temperate marine systems: the Georges Bank experience. Bulletin of Marine Science 66:775–798.

NRC (National Research Council). 2001. Marine protected areas: tools for sustaining ocean ecosystems. National Academy Press, Washington, D.C.

Overholtz, W. J. 1985. Seasonal and age-specific distribution of the 1975 and 1978 year-classes of haddock on Georges Bank. NAFO (Northwest Atlantic Fisheries Organization) Scientific Council Studies 8:77–82.

Perry, R. I., and S. J. Smith. 1994. Identifying habitat associations of marine fishes using survey data: an application to the Northwest Atlantic. Canadian Journal of Fisheries and Aquatic Sciences 51:589–602.

Piet, G. J., and A. D. Rijnsdorp. 1998. Changes in the demersal fish assemblage in the south-eastern North Sea following the establishment of a protected area ("plaice box"). ICES Journal of Marine Science 55:420–429.

Quinn, T. J., II, and R. Deriso. 1999. Quantitative fish dynamics. Oxford University Press, New York.

Roberts, C. M., J. A. Bohnsack, F. Gell, J. P. Hawkins, and R. Goodridge. 2001. Effects of marine reserves on adjacent fisheries. Science 294:1920–1923.

Russ, G. R., A. C. Alcala, and A. P. Maypa. 2003. Spillover from marine reserves: the case of *Naso vlamingii* at Apo Island, the Philippines. Marine Ecology Progress Series 264:15–20.

Shipley, J. B., editor. 2004. Aquatic protected areas as fisheries management tools. American Fisheries Society, Symposium 42, Bethesda, Maryland.

Swartzman, G., C. Huang, and S. Kaluzny. 1992. Spatial analysis of Bering Sea groundfish survey data using general additive models. Canadian Journal of Fisheries and Aquatic Sciences 52:369–380.

Ward, T. J., D. Heinemann, and N. Evans. 2001. The role of marine reserves as fisheries management tools: a review of concepts, evidence and international experience. Bureau of Rural Science, Canberra, Australia.

Wilcox, C., and C. Pomeroy. 2003. Do commercial fishers aggregate around marine reserves? Evidence from Big Creek Marine Ecological Reserve, central California. North American Journal of Fisheries Management 23:241–250.

Willis, T. J., R. B. Millar, R. C. Babcock, and N. Tolimeiri. 2003. Burdens of evidence and the benefits of marine reserves: putting Decartes before des horse? Environmental Conservation 30(2):97–103.

Zeller, D., S. L. Stoute, and G. R. Russ. 2003. Movements of reef fishes across marine reserve boundaries: effects of manipulating a density gradient. Marine Ecology Progress Series 254: 269–280.

American Fisheries Society Symposium 42:185–193, 2004

Why Have No-Take Marine Protected Areas?

JAMES A. BOHNSACK[1]

Southeast Fisheries Science Center, NOAA Fisheries,
75 Virginia Beach Drive, Miami, Florida 33149, USA

JERALD S. AULT

University of Miami, Rosenstiel School of Marine and Atmospheric Science,
Division of Marine Biology and Fisheries, 4600 Rickenbacker Causeway, Miami, Florida 33149, USA

BILLY CAUSEY

Florida Keys National Marine Sanctuary,
Post Office Box 500368, Marathon, Florida 33050, USA

Abstract.—Although the title of this symposium implied a focus on fully protected marine areas, most presentations actually dealt with a range of traditional "marine protected areas" or "marine managed areas" that offer less than "full" resource protection. Some presentations noted a backlash against establishing no-take reserves. Here we provide 17 reasons why there is a strong scientific, management, and public interest in using no-take marine reserves to build sustainable fisheries and protect marine ecosystems. We also discuss some underlying technical and philosophical issues involved in the opposition to their usage.

Introduction

Marine protected areas are used increasingly to manage marine resources, but they often mean different things to different people, based primarily on the level of protection they provide. The World Conservation Union defined marine protected areas (MPAs) as "any area of the intertidal or subtidal terrain, together with its overlying water and associated flora, fauna, historical and cultural features, which has been reserved by law or other effective means to protect part or all of the enclosed environment" (IUCN 1994; Kelleher 1999). In the USA, Presidential Executive Order 13158 provided a similar definition: "any area of the marine environment that has been reserved by Federal, State, territorial, tribal or local laws or regulations to provide lasting protection for part or all of the natural and cultural resources therein." Under these broad definitions, a wide variety of sites could be considered as MPAs.

We focus on "marine reserves," here defined as marine protected areas permanently closed to all fishing and other extractive uses with limited exceptions for research and education by permit (Ballantine 1997). Because of the many different terms that have been used to describe marine reserves, the terminology is often confusing to both scientists and the public. Common terms used to describe marine reserves include no-take areas, nonconsumptive areas, fishery reserves (PDT 1990), marine ecological reserves, sanctuary preservation areas (USDOC 1996), research natural areas (Brock and Culhane 2004, this volume), fully protected areas (Roberts and Hawkins 2000), and sanctuary, outside the USA.

Closing areas to fishing has long been widely practiced in fishery management in historical and modern times to protect critical habitat, restore depleted species, and protect vulnerable stocks at spawning aggregation sites (e.g., Beverton and Holt 1957). Most closures, however, have been either seasonal, applied only to specific species, or have been limited to restrict certain destructive or wasteful fishing methods. Rarely have areas been permanently closed to all types of fishing. Modern fisheries interest in marine reserves began in the 1980s as a way to both protect marine ecosystem biodiversity and build sustainable fisheries (PDT 1990; Bohnsack 1996; Bohnsack and Ault

---

[1] Corresponding author: jim.bohnsack@noaa.gov

1996). This interest has accelerated after failures of traditional fishery effort and size control measures to support sustainable fisheries and prevent collapses of fisheries and coastal ecosystems (Ludwig et al. 1993; Russ 1996; Botsford et al. 1997; Jackson 1997; Guénette et al. 1998; Pauly et al. 1998, 2002; Jackson et al. 2001; Christensen et al. 2003; Myers and Worm 2003; Rosenberg 2003).

Marine reserve implementation remains a rare and controversial measure despite support from numerous theoretical and empirical studies (Johnson et al. 1999; Murray et al. 1999; Fogarty et al. 2000; Roberts et al. 2001; Halpern and Warner 2002; Halpern 2003) and reviews that call for their expanded application in resource management (PDT 1990; NRC 1999, 2001; Roberts and Hawkins 2000; Ward et al. 2001; Pew Oceans Commission 2003; Pauly 2004, this volume). In response to the rare use of marine reserves, 161 academic scientists took the unusual step of issuing a signed consensus statement supporting the specific use of no-take marine reserves at the 2001 annual meeting of the American Association of Science (NCEAS 2001). Widespread concerns over marine resource protection in the USA resulted in Presidential Executive Order 13158, which seeks to inventory and assess existing MPAs (U.S. Office of the Federal Register 2000), and the adoption of a goal to protect 20% of U.S. coral reefs with marine reserves by 2010 by the U.S. Coral Reef Task Force (USCRTF 2000). The two largest U.S. marine reserve networks were established only recently in Florida and California. Two ecological reserves covering 280 km^2 (151 nautical mi^2) in the Tortugas region of the Florida Keys National Marine Sanctuary were established in 2001 (USDOC 2000). A contiguous 87-km^2 (47-nautical-mi^2) no-take research natural area was also approved for Dry Tortugas National Park but has not yet been implemented (Brock and Culhane 2004, this volume). Most recently, 10 reserves covering 244.5 km^2 (132 nautical mi^2) in the Channel Islands, California, were established in 2002 (McArdle et al. 2003).

Application of marine reserves has been controversial and has generated a backlash at times by those who favor continued use of other traditional fishery management actions (Shipp 2003) or multiple-use MPAs with only limited restrictions (Agardy et al. 2003; Clark 2003). Some concerns are that marine reserves may not be effective for biological (Carr and Reed 1993) or other reasons (Jameson et al. 2002); could be counter productive to conservation for social reasons (Agardy et al. 2003); and could threaten

fishing "rights" of recreational anglers as expressed in the proposed Freedom to Fish Act (Lydecker 2004, this volume).

Here, we present reasons why there is a high degree of scientific, management, and public interest in using permanent no-take protection compared to using "multiple-use" zoning or other traditional fishery management measures. Our intent is to clarify the issues in the continuing debate on appropriate use of marine reserves and spatial management in marine fishery and conservation management.

Results

Permanent, no-take marine reserves have certain unique qualities with potential benefits that are not necessarily provided by other types of marine protected or managed areas. Below we describe 17 unique attributes of marine reserves roughly organized into categories under fundamental, scientific, and management considerations.

Fundamental Considerations

(1) High Level of Ecosystem Protection

Fishing is a known major threat to marine populations and ecosystems (Dayton et al. 1995; Pauly et al. 1998, 2002, 2003). By removing fishing, no-take reserves potentially provide a high level of resource protection by eliminating threats from directed take of targeted organisms, bycatch mortality of nontarget organisms, and habitat damage from fishing activities. In an endless gradation between totally open and completely closed, marine reserves provide a high level of protection but not total protection. They do not, for example, directly protect against regional pollution, climate change, natural disturbance, or human disasters (Jameson et al. 2002). Other provisions can be added that provide higher levels of resource protection, such as prohibiting touching, diving, research, or even human entry, but with potential social and economic costs in terms of reduced benefits from nonextractive activities.

(2) Potential Ecological Integrity

Because no-take marine reserves protect all species, habitats and populations impacted by previous fishing can eventually recover and restore ecological integrity to reflect "natural" ecosystem structure and function. Permanent protection allows ecological integrity to ultimately persist in reserves.

(3) Precautionary Approach
The precautionary approach can be stated simply: when in doubt, be cautious. In practice, if you don't have a complete understanding about the functioning and dynamics of natural systems or their management, then some resources should be withheld from exploitation until a complete understanding is obtained (Bohnsack 1999a). Lauck et al. (1998) demonstrated how marine reserves can mitigate the effects of uncertainty associated with fishery exploitation.

(4) Shifted Burden of Proof
Compared to other types of managed areas, marine reserves shift the burden of proof from proving that fishing causes an adverse impact to proving that it does not (Dayton 1998). The result is that, in reserves, management focus shifts from a risk-prone approach, in which actions are taken only after resource impacts are demonstrated, to a more risk-averse approach, in which resources are protected until it can be demonstrated that an activity is not harmful.

(5) Existence and Future Value
Marine reserves help protect existence value for people who do not directly use resources and for future generations. Aldo Leopold (1949) noted that we cannot prevent the alteration, management, and use of resources, but we need to affirm their right to continued existence, and in some places, their continued existence in a natural state. His biotic ethic requires human obligation, responsibility, and self-sacrifice to preserve ecosystems for present and future generations. This mantra needs to be adopted for effective management of marine ecosystems.

(6) Increased Public Understanding and Appreciation
Marine reserves provide opportunities for quality formal education at the primary, secondary, and graduate levels. With public access, they also provide better public understanding and appreciation of marine ecosystems and marine reserves and the importance of effective resource management. Pauly (1995) described the shifting baseline problem in which each generation develops lower expectations about natural resources based on its own direct experience with depleted resources. Marine reserves with public access offer an opportunity to reverse this trend by restoring areas with more natural and healthy ecosystems. They also provide citizens an opportunity to directly observe the effectiveness of resource management and understand its importance by comparing reserves to surrounding areas.

(7) Enhanced Nonextractive Human Uses
By separating incompatible activities and protecting some areas from fishing and depletion, no-take reserves can support nonextractive uses that have ecological, social, genetic, economic, educational, scientific, recreational, aesthetic, spiritual, and wilderness importance (Bohnsack 1998). They can diversify the economy by providing new social and economic opportunities. This is especially important for activities that require high resource quality. Otherwise, only those activities that depend on depleted or low quality resources can persist.

(8) Better Resource Protection
Unlike many other measures, there are no legal ways to avoid or circumvent the no-take provision which offers the possibility of better overall resource protection than do other measures. Trip limits and bag limits for a recreational fishery, for example, are popular conservation measures, but their effectiveness can be circumvented by making more fishing trips. Similarly, the effectiveness of gear restrictions and minimum size limits can be negated by increased fishing effort. Marine reserves also offer better resource protection because they buffer against changes in total effort or fishing practices in surrounding areas.

Scientific Considerations

(9) Objective Criterion
The no-extraction criterion prohibiting any activity that intentionally removes organisms or habitat is objective and easy to determine as compared to many other criteria that are subjective or difficult to define. Allowing "limited extraction" in a multiple-use MPA, for example, is problematic because there is no clear definition of what "limited" means. Accurately determining a level of extraction that is "not harmful" to a population or an ecosystem is difficult and mostly unknown. Also, monitoring or controlling the amount of take is not practical in most cases.

(10) Simplicity
Compared to other criteria, it is easy to determine whether an activity is extractive or not and fundamentally simpler to explain than why some users are allowed to remove resources and not others. Note, nonextractive, is not the same as, nor should it be confused with, nonconsumptive. Nonextractive recreational diving, for example, could be considered consumptive as the result of repeated contact and damage to the benthos. Allowing diving and other

nonextractive uses within marine reserves assumes that their impacts are either controllable or have much less significant impact than fishing. If not, additional protective measures may be necessary to confine, reduce, eliminate, or mitigate those nonextractive impacts. In the Florida Keys, for example, divers are also prohibited from touching coral as an added protection. One suggestion is to call these "kapu zones," after the Hawaiian word "kapu" (meaning "do not touch" or "forbidden"; Bohnsack 2000a). Kapu was historically used in Hawaii to protect marine areas.

(11) Control Sites

One of the most important tools in science is the experimental control, in which the influence of a variable is either controlled for or eliminated. By eliminating fishing, marine reserves provide control sites to objectively evaluate the effects of extractive impacts on marine ecosystems. They also provide a comparative basis for assessing the effectiveness of various fishery management measures in surrounding areas. Without control sites, it is almost impossible to scientifically address larger questions about how much resource can be removed from a marine ecosystem and still maintain the biological productivity, persistence, and ecological integrity.

(12) Distinguish between Natural and Anthropogenic Disturbance

Scientists and managers often need to distinguish between changes caused by natural versus anthropogenic events. Without marine reserves, environmental signals can become hopelessly confounded with fishing impacts. Observed higher abundance of exploited species in no-take reserves compared to similar habitats in surrounding areas, for example, indicates that fishing is the primary factor influencing the observed differences and has more impact on those species than other anthropogenic forcing factors such as regional pollution. In contrast, data showing no differences between reserves and surrounding areas may indicate that regional factors (either natural or anthropogenic) are more important influences on populations.

(13) Increased Scientific Knowledge and Understanding

Marine reserves can facilitate the elucidation of natural processes and enhance scientific knowledge and understanding of marine ecosystems by providing comparative areas with minimal human disturbances. Certain scientific experiments and observations involving biodiversity, behavior, and ecosystem processes can only be conducted in reserves.

Management Considerations

(14) Public Acceptance

Although large land areas in the United States have been protected from hunting and other extraction for well over a century, few aquatic areas have received similar protection. This fact that protected areas are widely used and accepted on land suggests that similar protections could be applied and accepted in the sea. The fact that they have not yet been widely applied in the ocean can be attributed in part to a historical lack of understanding and awareness of marine ecosystems, mistaken beliefs that marine resources are unlimited and impervious to human impacts, and what some consider inalienable rights to fish anywhere.

When high levels of protection are necessary, marine reserves may cause less social and economic disruption and receive better public acceptance than other measures that provide a similar level of resource protection (unless the closed area happens to be a predominantly favorite fishing area). Marine reserves, for example, become an attractive alternative when compared to closing down a fishery entirely or severely reducing bag limits, increasing minimum size limits, and restricting the number of participants. Potentially, reserves could allow more people to participate in a fishery than would otherwise be possible because total fishing mortality is less if some areas are highly protected (Bohnsack 2000b).

(15) Simplified Enforcement

As a management tool, reserves can potentially simplify enforcement by making violations easier to detect. Since the act of fishing is a violation, it is not necessary to obtain, identify, or measure catch. Violations can be detected by surface, aerial, or satellite surveillance, using a variety of technology and vessel monitoring systems. Because permanent no-take provisions apply to all species, there may be less public confusion and better compliance than if different closed areas were established for individual species in multiple-species fisheries. Establishing different seasons or closed areas with overlapping or conflicting boundaries for each species could be much more confusing and impractical.

The legal authority to close significant areas to fishing and technological means to monitor compliance and ensure enforcement have advanced in recent decades. The legal authority changed with the widespread expansion of national exclusive economic zones in 1977 (Bohnsack 1996). Technological advances in navigation, surveillance, and vessel tracking, as well as a new emphasis on homeland security, make monitoring and enforcement of marine reserves more practical.

(16) Direct Fishery Benefits
Marine reserves potentially can provide many direct fishery benefits (Bohnsack 1998). The five most important benefits follow. Reserves can reduce the chances of overfishing by providing refuges from population exploitation. Compared to having all areas exploited under one set of regulations, reserves potentially can provide greater fishery yields in the long-term by having a larger and more dependable supply of eggs and larvae dispersed to fishing grounds. Reserves can also potentially increase yield from spillover, where animal emigration exports biomass from reserves through to surrounding fishing grounds (PDT 1990; Roberts et al. 2001). Reserves also can provide insurance to sustainable stocks by potentially accelerating stock recovery following natural disturbance, human accidents, management errors, or years of poor stock–recruitment (PDT 1990). Finally, they may be the only measure that can effectively preserve stock genetic structure from detrimental effects of selective fishing practices (Conover and Munch 2002).

(17) Indirect Fishery Benefits
Fishery stock assessment and management models depend on obtaining accurate estimates of critical population parameters of growth, natural mortality, and fecundity. If all areas are subjected to fishing, measuring these parameters and gaining an essential understanding of trophic and habitat relationships, recruitment variations, behavior, and population response to environmental variability are difficult, if not impossible, to obtain. Marine reserves can potentially benefit fisheries indirectly by allowing some critical population dynamic and fishery parameters to be estimated independent of fishery influences with a rigorous sampling design (Ault et al. 2002).

Discussion and Conclusions

The main priority of permanent no-take marine reserves is to protect biodiversity: ecological structure and function at the genetic, species, community, seascape, and ecosystems levels (NRC 2001). Their use has generated considerable scientific, management, and public interest because the no-extraction provision is simple and objective and offers a high level of resource protection that can potentially restore and maintain ecological integrity in areas with minimum human disturbance. Many scientific questions can best or only be examined using marine reserves. From a management perspective, marine reserves are attractive because they potentially provide a win–win conservation alternative that offers a high level of ecosystem protection while providing fishery benefits and enhancing and diversifying nonextractive human uses.

Much, however, remains to be learned because the science of marine reserves is new and most existing reserves are rare, small, recently established, limited to few habitats, or cover only very small portions of the total managed area (Pauly 2004, this volume). Because they are rare, more need to be implemented if they are to provide anything more than a token role in protecting marine biodiversity. Because marine reserves are rare and recently established, few scientific studies exist (Halpern and Warner 2002; Halpern 2003), leaving many questions and uncertainty concerning their application to biodiversity and fishery protection. More research is needed to address questions concerning individual reserve size, total number, location, total area, and habitats that need to be included to be truly effective. In addition, more replicated research is needed, especially at larger and more ecologically relevant spatial and temporal scales, to address questions of costs and benefits, effectiveness, and necessary design features for reserve networks. Many questions remain unresolved concerning social and ecological impacts of fishing displacement, applications to highly migratory species, and social acceptance, compliance, and enforcement. Thus, considerable scientific interest exists in establishing reserves in different regions and habitats and under different biological, oceanographic, and physical environments as well as in different social and economic environments.

Even though they prohibit fishing, marine reserves do not conflict with "multiple-use MPAs" because they create or enhance many kinds of activities within and outside their boundaries that conflict with fishing. When embedded in larger MPAs such as the Florida Keys National Marine Sanctuary, for example, they also support multiple human uses by separating incompatible activities and increasing total resource protection. A belief that fishing and other human activities can be practiced simultaneously in all areas without conflict is becoming far less realistic considering growing human population demands and the intensity of resource usage. Likewise, allowing all areas to be exploited with "limited restrictions" demands a high level of knowledge and human control that at present is essentially nonexistent.

Despite offering many potential benefits, marine reserves have generated considerable opposition (Norse et al. 2003; Shipp 2003). Most opposition has focused on technical issues about the applicability of reserves to different species and habitats, proof of fishery benefits, and the quantifying of design features (number, size, location, spacing, boundary configurations, and total area covered) for individual reserves and networks (Carr and Reed 1993; Botsford et al. 2001). Other issues involve enforcement, impacts of displacing fishing on people and resources outside of reserves (Bohnsack 2000b), and how to incorporate reserves into comprehensive management programs (Jameson et al. 2002). Some opposition simply reflects resistence to changing the status quo because it creates winners and losers. Fishers, who effectively have had historical access to the entire ocean, can be expected to aggressively oppose any changes that restrict that access (Lydecker 2004, this volume). Although this is not a scientific issue, such shifts are common and routinely handled by political and government institutions.

Philosophical opposition has received less attention but ultimately may be more important than the technical issues. While much attention has focused on economic costs and benefits, for example, relatively little attention has been paid to conflicts caused by wide differences in conservation ethics (Callicott 1992; Bohnsack 2003). As Leopold (1949) recognized, economics is not an ethic, and basing management decisions solely on economic self-interest is unwise. Inevitably, it leads to failure because elements without economic value eventually will be eliminated to the detriment of the economic parts. Leopold's biotic ethic led to a shift in management emphasis from "sustained production of resources or commodities, to a recognition that true sustained yield requires preservation of the health of the entire system" (Leopold 1949). Much of the current controversy over marine reserves appears to be a result of philosophical failures to recognize that people are part of marine ecosystems, that limits to human usage exist, and that human well-being is dependent on maintaining ecosystem health. Protecting marine biodiversity and maintaining sustainable fisheries are not mutually exclusive problems.

A key philosophical issue involves human dominance. Can marine ecosystems be manipulated and controlled at will and, if so, should all areas be exploited? Marine reserve application is based, in part, on a simple premise that if protected from human interference, nature has evolved to take care of itself.

This premise conflicts with the top-down "command and control" engineering approaches that attempt to control complex human and ecological systems (Holling and Meffe 1996). This human control view is reflected in concerns that some resources may be underutilized in terms of total yield and, therefore, wasted by using marine reserves. An extreme example of this thinking is the position that marine reserves are not "management" tools because they do not involve active human manipulation.

Another issue is the philosophical dichotomy between fisheries and ecosystem management perspectives. In fisheries, marine reserves are usually considered a "tool" to be used independently of other fishery management options (Norse et al. 2003; Shipp 2003) and not as part of an integrated management system (Norse et al. 2003). The assumption is that fisheries are independent of biodiversity and ecosystem management. In an ecosystem perspective, fishery productivity is directly derived from ecosystem biodiversity, and the two must be managed together. Thus, much of the conflict between ecosystem and fishery management is an artifact of separating these two functions. We give three examples that elucidate this philosophical conflict: the amount of area needed for marine reserve networks, the displacement of fishing effort by marine reserves, and the current efforts to shift fisheries from single-species to ecosystem-based management.

First, considerable angst has been generated over questions concerning how much area should be included in marine reserve networks. Proponents of marine reserves usage argue that substantial portions of marine environments need reserve protection (Bohnsack et al. 2002; Pauly et al. 2003; Pew Oceans Commission 2003), but they generate considerable criticism when attempting to apply principals as guidelines using area percentages (Agardy et al. 2003; Norse et al. 2003; Shipp 2003). The critics are correct in that no one percentage will apply to all ecosystems or areas. However, the same critics ignore the fact that there is a need for a minimum percentage and that no biological, social, or economic theory exists showing that all areas should be exploited. Thus, while there should be agreement that fixed percentages of reserve area will not apply to all marine ecosystems, there should also be agreement that there is no support for zero as a percentage either. Ideally, adaptive management should be used to fine-tune protection to specific habitats and areas (Walters 1986; Murray et al. 1999).

Second, marine reserves are often criticized for not directly addressing human and environmental impacts of fishing effort displacement to areas outside reserve boundaries (Norse et al. 2003). This problem is not a failure of marine reserves per se but a failure to include marine reserves as part of comprehensive resource management strategies. Despite claims by some opponents, we know of no statements that marine reserves alone will solve all fishery problems. If overfishing is a problem, effort controls and other traditional fishery measures are also needed, including size limits, bag limits, quotas, limited entry, closed seasons, gear restrictions, and closed areas for specific fisheries (Bohnsack 2000b). If these other fishery measures are not effective, larger proportions of habitats may need to be closed. Relying solely on no-take protection, however, may reduce options and flexibility for optimizing social and economic benefits (Murray et al. 1999).

Third, use of marine reserves represents a philosophical shift from single-species and reactive fishery management to a more precautionary approach using proactive spatial and ecosystem-based management (Bohnsack 1999b). Although many practical details still need to be worked out to make this shift operational, at the theoretical level it requires integrating fishery and ecosystem considerations.

In conclusion, no-take marine reserves are primarily intended to protect ecosystem biodiversity. They offer qualitative and quantitative qualities that are more than simply sequestering populations in no-take areas (Norse et al. 2003) or providing just another fishery management tool (Norse et al. 2003). Fundamentally, marine reserves use a simple, ecosystem-based, and precautionary approach to offer a high level of resource protection that benefits present human activities and future generations. Marine reserves increase human knowledge, understanding, and appreciation of marine ecosystems and their management by offering a high and objective level of protection and a scientific basis for assessing human impacts and management effectiveness. Reserves potentially can simplify enforcement, benefit fisheries, and eventually achieve wide public acceptance. We suggest that advancing the science of resource management requires considering people a fundamental part of marine ecosystems, shifting the focus of fishery management from resources as mere commodities to sustaining functional ecosystems, and incorporating marine reserve concepts and networks into comprehensive marine resource management.

Acknowledgments

We thank W. J. Ballantine, W. J. Richards, R. L. Shipp, and an anonymous reviewer for providing critical comments. This research was partially supported by the National Oceanic and Atmospheric Administration (NOAA) Coastal Ocean Program South Florida Program Grant NA17RJ1226, the NOAA Caribbean Reef Ecosystem Study Grant NA17OP2919, the National Park Service Cooperative Ecosystem Studies Unit Grant H500000B494, the National Undersea Reserach Center, and the NOAA Coral Reef Program.

References

Agardy, T., P. Bridgewater, M. P. Crosby, J. Day, P. K. Dayton, R. Kenchington, D. Laffoley, P. McConney, P. A. Murray, J. E. Parks, and L. Peau. 2003. Dangerous targets? Unresolved issues and ideological clashes around marine protected areas. Aquatic Conservation: Marine and Freshwater Ecosystems 13:353–367.

Ault, J. S., S. G. Smith, G. A. Meester, J. Luo, J. A. Bohnsack, and S. L. Miller. 2002. Baseline multispecies coral reef fish stock assessment for Dry Tortugas. NOAA Technical Memorandum NMFS-SEFSC-487.

Ballantine, W. J. 1997. "No-take" marine reserve networks support fisheries. Pages 702–706 in D. A. Hancock, D. C. Smith, A. Grant, and J. P. Beumer, editors. Developing and sustaining world fisheries resources: the state and management, 2nd world fisheries congress proceedings. CSIRO Publishing, Collingwood, Australia.

Beverton, R. J. H., and S. J. Holt. 1957. On the dynamics of exploited fish populations. Ministry of Agriculture, Fisheries and Food, Fishery Investigations Series II, volume XIX, London.

Bohnsack, J. 2000a. Kapu zones. MPA News 1(5):5–6.

Bohnsack, J. A. 1996. Maintenance and recovery of fishery productivity. Pages 283–313 in N. V. C. Polunin and C. M. Roberts, editors. Tropical reef fisheries. Chapman and Hall, Fish and Fisheries Series 20, London.

Bohnsack, J. A. 1998. Application of marine reserves to reef fisheries management. Australian Journal of Science 23:298–304.

Bohnsack, J. A. 1999a. Incorporating no-take marine reserves into precautionary management and stock assessment. Pages 8–16 in V. R. Restrepo, editor. Providing scientific advice to implement the precautionary approach under the Magnuson-Stevens Fishery Conservation and Management Act. NOAA Technical Memorandum NMFS-F/SPO-40.

Bohnsack, J. A. 1999b. Ecosystem management, marine reserves, and the art of airplane maintenance. Proceedings of the Gulf and Caribbean Fisheries Institute 50:304–311.

Bohnsack, J. A. 2000b. A comparison of the short term impacts of no-take marine reserves and minimum size limits. Bulletin of Marine Science 66:615–650.

Bohnsack, J. A. 2003. Shifting baselines, marine reserves, and Leopold's biotic ethic. Gulf and Caribbean Science 14(2):1–7.

Bohnsack, J. A., and J. S. Ault. 1996. Management strategies to conserve marine biodiversity. Oceanography 9:73–82.

Bohnsack, J. A., B. Causey, M. P. Crosby, R. G. Griffis, M. A. Hixon, T. F. Hourigan, K. H. Koltes, J. E. Maragos, A. Simons, and J. T. Tilmant. 2002. A rationale for minimum 20–30% no-take protection. Pages 615–619 in Proceedings of the 9th International Coral Reef Symposium. Indonesian Institute of Sciences, Society for Reef Studies, Indonesia.

Botsford, L. W., J. C. Castilla, and C. H. Petersen. 1997. The management of fisheries and marine ecosystems. Science 177:509–515.

Botsford, L. W., A. Hastings, and S. D. Gaines. 2001. Dependence of sustainability on the configuration of marine reserves and larval dispersal distance. Ecology Letters 4:144–150.

Brock, R., and B. Culhane. 2004. The no-take research natural area of Dry Tortugas National Park (Florida): wishful thinking or responsible planning? Pages 67–74 in J. B. Shipley, editor. Aquatic protected areas as fisheries management tools. American Fisheries Society, Symposium 42, Bethesda, Maryland.

Callicott, J. B. 1992. Principal traditions in American environmental ethics: a survey of moral values for framing an American ocean policy. Ocean and Coastal Management 17:299–325.

Carr, M. H., and D. C. Reed. 1993. Conceptual issues relevant to marine harvest refuges: Examples from temperate reef fishes. Canadian Journal of Fisheries and Aquatic Sciences 50:2019–2028.

Christensen, V., S. Guénette, J. J. Heymans, C. J. Walters, R. Watson, D. Zeller, and D. Pauly. 2003. Hundred-year decline of North Atlantic predatory fishes. Fish and Fisheries 4:1–14.

Clark, J. R. 2003. Letter to the editor. MPA News 5(1):6.

Conover, D. O., and S. B. Munch. 2002. Sustaining fisheries yields over evolutionary time scales. Science 297:94–96.

Dayton, P. K. 1998. Reversal of the burden of proof in fisheries management. Science 279:821–822.

Dayton, P. K., S. F. Thrush, M. T. Agardy, and R. J. Hofman. 1995. Environmental effects of marine fishing. Aquatic Conservation: Marine and Freshwater Ecosystems 5:205–232.

Fogarty, M. J., J. A. Bohnsack, and P. K. Dayton. 2000. Marine reserves and resource management. Pages 283–300 in C. Sheppard, editor. Seas at the millennium: an environmental evaluation. Pergamon, Elsevier, New York.

Guénette, S., T. Lauck, and C. Clark. 1998. Marine reserves: from Beverton and Holt to the present. Reviews in Fish Biology and Fisheries 8:1–21.

Halpern, B. 2003. The impact of marine reserves: do reserves work and does reserve size matter? Ecological Applications 13:S117–137.

Halpern, B. S., and R. R. Warner. 2002. Marine reserves have rapid and lasting effects Ecology Letters 5:361–366.

Holling, C. S., and G. K. Meffe. 1996. Command and control and the pathology of natural resource management. Conservation Biology 10:328–337.

IUCN (International Union for the Conservation of Nature and Natural Resources). 1994. Guidelines for protected area management categories. IUCN, Gland, Switzerland and Cambridge, UK.

Jackson, J. B. C. 1997. Reefs since Columbus. Coral Reefs 16:S23–S32.

Jackson, J. B. C., M. X. Kirby, W. H. Berger, K. A. Bjorndal, L. W. Botsford, B. J. Bourque, R. H. Bradbury, R. Cooke, J. Erlandson, J. A. Estes, T. P. Hughes, S. Kidwell, C. B. Lang, H. S. Lenihan, J. M. Pandolfi, C. H. Peterson, R. S. Steneck, M. J. Tegner, and R. R. Warner. 2001. Historical overfishing and the recent collapse of coastal ecosystems. Science 293:629–638.

Jameson, S. C., M. H. Tupper, and J. M. Ridley. 2002. The three screen doors: can marine "protected" areas be effective? Marine Pollution Bulletin 44:1177–1183.

Johnson, D. R., N. A. Funicelli, and J. A. Bohnsack. 1999. The effectiveness of an existing estuarine no-take fish sanctuary within the Kennedy Space Center, Florida. North American Journal of Fisheries Management 19:436–453.

Kelleher, G. 1999. Guidelines for marine protected areas. International Union for the Conservation of Nature and Natural Resources, Gland, Switzerland, and Cambridge, UK.

Lauck, T., C. W. Clark, M. Mangel, G. R. Munro. 1998. Implementing the precautionary principle in fisheries management through marine reserves. Ecological Applications 8(Supplement 1):S72–S78.

Leopold, A. 1949. A Sand County almanac. Oxford University Press, London.

Ludwig, D., R. Hilborn, and C. Walters. 1993. Uncertainty, resource exploitation, and conservation: lessons from history. Science 260:17–18.

Lydecker, R. 2004. How the organized recreational fishing community views aquatic protected areas. Pages 15–19 in J. B. Shipley, editor. Aquatic protected areas as fisheries management tools. American Fisheries Society, Symposium 42, Bethesda, Maryland.

McArdle, D., S. Hastings, and J. Ugoretz. 2003. California marine protected area update. California Sea Grant College Program, University of California Publication T-051, La Jolla.

Murray, S. N., R. F. Ambrose, J. A. Bohnsack, L. W. Botsford, M. H. Carr, G. E. Davis, P. K. Dayton, D. Gotshall, D. R. Gunderson, M. A. Hixon, J. Lubchenco, M. Mangel, A. MacCall, D. A. McArdle, J. C. Ogden, J. Roughgarden, R. M. Starr, M. J. Tegner, and M. M. Yoklavich. 1999. No-take reserve networks: protection for fishery populations and marine ecosystems. Fisheries 24(11):11–25.

Myers, R. A., and B. Worm. 2003. Rapid worldwide depletion of predatory fish communities. Nature (London) 423:280–283.

NCEAS (National Center for Ecological Analysis and Synthesis). 2001. Scientific consensus statement on marine reserves and marine protected areas. American Association for the Advancement of Science. Available: www.nceas.ucsb.edu/consensus (March 2004).

Norse, E. A., C. B. Grimes, S. R. Ralston, R. Hilborn, J. C. Castilla, S. R. Palumbi, D. Fraser, and P. Karieva. 2003. Marine reserves: the best option for our oceans? Frontiers in Ecology and the Environment 1(9):495–502.

NRC (National Research Council). 1999. Sustaining marine fisheries. National Academy Press, Washington, D.C.

NRC (National Research Council). 2001. Marine protected areas: tools for sustaining ocean ecosystems. National Academy Press, Washington, D.C.

Pauly, D. 1995. Anecdotes and the shifting baseline syndrome of fisheries. Trends in Ecology and Evolution 10:430.

Pauly, D. 2004. On the need for a global network of large marine reserves. Abstract only. Page 63 in J. B. Shipley, editor. Aquatic protected areas as fisheries management tools. American Fisheries Society, Symposium 42, Bethesda, Maryland.

Pauly, D., J. Alder, E. Bennett, V. Christensen, P. Tyedmers, and R. Watson. 2003. The future for fisheries. Science 302:1359–1361.

Pauly, D., V. Christensen, J. Dalsgaard, R. Froese, and F. Torres. 1998. Fishing down marine food webs. Science 279:860–863.

Pauly, D., V. Christensen, S. Guenette, T. J. Pitcher, U. R. Sumaila, C. J. Walters, R. Watson, and D. Zeller. 2002. Towards sustainability in world fisheries. Nature (London) 418:689–695.

PDT (Plan Development Team). 1990. The potential of marine fishery reserves for reef fish manage-ment in the U.S. southern Atlantic. Snapper–grouper plan development team report for the South Atlantic Fishery Management Council. NOAA Technical Memorandum NMFS-SEFC-261.

Pew Oceans Commission. 2003. America's living oceans: chartering a course for sea change. Pew Oceans Commission, Arlington, Virginia.

Roberts, C. M., J. A. Bohnsack, F. Gell, J. P. Hawkins, and R. Goodridge. 2001. Effects of marine reserves on adjacent fisheries. Science 294:1920–1923.

Roberts, C. M., and J. P. Hawkins. 2000. Fully protected marine reserves: a guide. Endangered Seas Campaign, World Wildlife Fund, Washington, D.C., and University of York, UK.

Rosenberg, A. A. 2003. Managing to the margins: the overexploitation of fisheries. Frontiers in Ecology and the Environment 1(2):102–106.

Russ, G. R. 1996. Fisheries management: what chance on coral reefs? NAGA, the ICLARM Quarterly July 1996:5–9.

Shipp, R. L. 2003. A perspective on marine reserves as a fishery management tool. Fisheries 28(12):10–21.

USCRTF (U.S. Coral Reef Task Force). 2000. The national action plan to conserve coral reefs. USCRTF, Washington, D.C.

USDOC. (U.S. Department of Commerce). 2000. Strategy for stewardship: Tortugas ecological reserve final supplemental environmental impact statement/final supplemental management plan. National Oceanic and Atmospheric Administration, National Ocean Service, Office of National Marine Sanctuaries, Silver Spring, Maryland. Available: www.fknms.nos.noaa.gov/regs/FinalFSEIS.pdf (March 2004).

U.S. Office of the Federal Register. 2000. Executive Order 13158 of May 26, 2000 on marine protected areas. Federal Register 65(105):34909–34911.

Walters, C. J. 1986. Adaptive management of renewable resources. MacMillan Publishing Company, New York.

Ward, T. J., D. Heinemann, and N. Evans. 2001. The role of marine reserves as fisheries management tools: a review of concepts, evidence and international experience. Bureau of Rural Sciences, Canberra, Australia.

American Fisheries Society Symposium 42:195–210, 2004

A Reexamination of Monitoring Projects of Southern Florida Adult Spiny Lobster *Panulirus argus*, 1973–2002: The Response of Local Spiny Lobster Populations, in Size Structure, Abundance, and Fecundity, to Different-Sized Sanctuaries

RODNEY D. BERTELSEN,[1] CARROLLYN COX, RICHARD BEAVER, AND JOHN H. HUNT

Florida Fish and Wildlife Conservation Commission, Florida Marine Research Institute, South Florida Regional Laboratory, 2796 Overseas Highway, Suite 119, Marathon, Florida 33050, USA

Abstract.—Lobster populations in southern Florida fall into three size-classes: less than 10 km², 10–100 km², and more than 100 km². Databases spanning the past 30 years are being reexamined to investigate the relationship of size of sanctuary on size structure, density, and fecundity of populations of spiny lobster *Panulirus argus* (also known as Caribbean spiny lobster). The density of the lobster population in small sanctuaries (established in 1997) has not changed; however, a few larger males may be protected. In the medium-sized sanctuary, Western Sambo Ecological Reserve (WSER), the density of male lobsters has roughly doubled and that of females has quadrupled. When the large sanctuary, Dry Tortugas National Park (DTNP), was established in 1974, large females (>120 mm carapace length) did not bear eggs, perhaps because there were no large males there. Today, all large females bear eggs and contribute 31% of the total fecundity. Fecundity estimates for area and season are difficult to calculate because they are sensitive to size distribution, fecundity-to-size relationships, and other factors. Our preliminary estimate is that 28 million eggs are produced per season per ha in the fore reef of DTNP, 21 million in WSER, 18 million in small sanctuary preservation areas, and 14 million in the Florida Keys fishery.

Introduction

The waters surrounding southern Florida, including the Florida Keys and the Dry Tortugas, support a US$20 million fishery for spiny lobster *Panulirus argus* (also known as Caribbean spiny lobster; Florida Marine Fisheries Information System, unpublished data maintained by the State of Florida). As part of the management strategy of this fishery, and sometimes to meet other conservation goals, lobster sanctuaries have been established throughout the region.

In southern Florida, there are currently 27 lobster sanctuaries covering an area of 2,816 km² (Table 1; Figure 1). The oldest sanctuaries are the Dry Tortugas National Park and Everglades National Park (established in 1974). The newest are the Tortugas Banks

North and Tortugas Banks South, known collectively as the Tortugas Ecological Reserve (established in 2001). Although Everglades National Park and the Tortugas Ecological Reserve are the largest sanctuaries, lobsters are found only in relatively small portions of these sanctuaries. Juvenile lobsters are confined mainly to the southern rim of Florida Bay in the Everglades National Park, and adult lobsters are confined to the shallower portions of the Tortugas Banks North portion of the Tortugas Ecological Reserve.

Most spiny lobster sanctuaries are in areas containing either nursery or adult habitat. Nursery habitats include shallow, hard-bottom areas with red algae *Laurencia* spp. and sponges (phylum Porifera), although other complex structures can provide shelter (Marx and Herrnkind 1985; Herrnkind and Lipcius 1989; Childress and Herrnkind 1996; Acosta and Butler 1997). Adult sheltering habitats include hard-

---

[1] Corresponding author: Rod.Bertelsen@fwc.state.fl.us

Table 1.—The creation date, size, administration (FKNMS = Florida Keys National Marine Sanctuary, NPS = National Park Service, FDEP = Florida Department of Environmental Protection), years of lobster surveys, and life stages of lobsters present for all current lobster sanctuaries in southern Florida.

Name	Established	Size (km²)	Currently administered by	FMRI[a] lobster surveys	Habitat[b]	Notes[c]
Everglades National Park (excluding land and freshwater)	1974	1,543	NPS	1991–present	NJ	
Biscayne Bay/ Card Sound	1984	447	Florida Fish and Wildlife Commission	1993; 2000– present	NJ	
Carysfort	1997	5.2	FKNMS	1997–present	S	SPA
The Elbow	1997	0.9	FKNMS	1995–1997	S	SPA
Dry Rocks	1997	0.2	FKNMS		S	SPA
Grecian Rocks	1997	1.1	FKNMS	1997–2002	S	SPA
French Reef	1997	0.4	FKNMS	1995–1997	S	SPA
Pennekamp exclusion zones[d]	1992		FDEP	1992	J	All patch reefs protected
Conch Reef	1997	0.2	FKNMS		S	SPA
Conch Reef research	1997	0.1	FKNMS	1995–2002	S	Special use area
Davis Reef	1997	0.6	FKNMS		S	SPA
Hens and Chickens	1997	0.6	FKNMS	1997–1999	J	SPA
Cheeca Rocks	1997	0.2	FKNMS		J	SPA
Alligator Reef	1997	0.6	FKNMS	1995–2002	S	SPA
Tennessee Reef	1997	0.5	FKNMS	1995–2002	S	Special use area
Coffins Patch	1997	1.5	FKNMS	1995–2002	S	SPA
Sombrero Reef	1997	0.7	FKMNS	1995–2002	S	SPA
Newfound Harbor	1997	0.4	FKNMS	1997–1999	J	SPA
Looe Key	1981	1.2	FKNMS	1986–1989; 1995–2002	S	SPA enlarged from 0.5 km² in 1997
Looe Key research only	1997	0.3	FKNMS	1997–2002	JS	Special use area
Eastern Sambo research only	1997	0.3	FKMNS	1995–1997	S	Special use area
Western Sambo Ecological Reserve	1997	30	FKNMS	1995–present	JS	ER
Eastern Dry Rocks	1997	0.3	FKNMS	1995–1997	S	SPA
Rock Key	1997	0.3	FKNMS	1995–1997	S	SPA
Sand Key	1997	1.5	FKNMS	1995–2002	S	SPA
Dry Tortugas National Park	1974	261	NPS	1995–2000; 2002	NJS	
Tortugas Ecological Reserve	2001	518	FKNMS	1999–2000; 2002	S	ER

[a] Some of these surveys were conducted with other research partners, especially Florida State University (William Herrnkind) and Old Dominion University (Mark Butler IV).
[b] Newly settled and juvenile lobsters can be found in all areas not marked with an N or J (including the Tortugas Banks); however, those life stages are rarely observed. Legal-sized (>76 mm carapace length) lobsters can be found in all areas; however, reproductively active adults are not found in a given sanctuary above, unless marked with an S.
[c] SPA = Sanctuary Preservation Area (non-consumptive areas or public visitation); Special use area = permitted research only, no public visitation; ER = ecological reserve (same restrictions as SPAs, larger area).
[d] Lobsters may not be captured from any patch reef or coral head inside Pennekamp Park at any time. During the open season, lobsters may be captured in other benthic habitats such as grass beds and sand.

bottom reef or rocky substrates with cavities in both coral and rock that serve as daytime sheltering dens (Herrnkind 1980; Herrnkind and Lipcius 1989). At night, lobsters leave their dens to forage for food

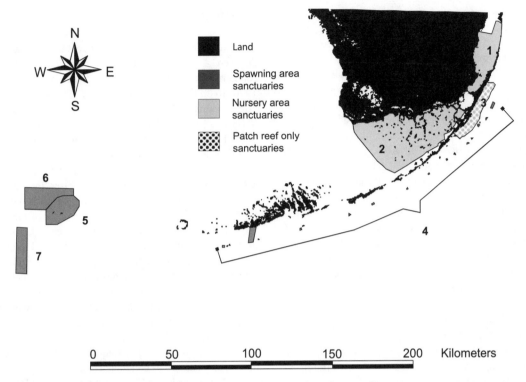

Figure 1.—Locations of all the southern Florida lobster sanctuaries: (1) Biscayne Bay and Card Sound; (2) Everglades; (3) Pennekamp; (4) Florida Keys National Marine Sanctuary; (5) Dry Tortugas; (6) Tortugas Banks North; and (7) Tortugas Banks South and Riley's Hump.

(Herrnkind and Redig 1975; Cox et al. 1997) in rubble-covered areas and elsewhere. We have observed foraging behavior in deeper (25 m) hard-bottom, rubble, and sandy areas (authors' personal observations). Adults are found in depths ranging from intertidal to more than 40 m in some areas, with water temperature likely being a limiting factor (see Herrnkind 1980). In southern Florida, nursery and adult habitats are typically separated by many kilometers, with nursery habitat found in shallow waters near the islands and Florida Bay and adult habitat found along the reef tract and in deeper waters. However, the Dry Tortugas is somewhat unique: on several occasions we have observed small juveniles in the same areas as adults during diver-based surveys.

Beginning in 1974, following a study and recommendations by Gary Davis (U.S. National Park Service), the Dry Tortugas National Park (DTNP) became a 261-km^2 lobster sanctuary. It is located approximately 120 km from the nearest inhabited land, Key West. Everglades National Park (ENP) also became a lobster sanctuary at that time. Although ENP does not harbor any large breeding lobsters, portions of the park contain lobster nursery habitat. In 1981, a 0.5-km^2 portion of the Looe Key fore reef was set aside as a sanctuary. It is located within the Florida Keys fishery. More recently, in 1997, the Florida Keys National Marine Sanctuary (FKNMS) established 20 small sanctuaries (each <10 km^2) between Carysfort Reef and Sand Key and the medium-sized (30 km^2) Western Sambo Ecological Reserve (WSER) within the Florida Keys fishery. Most of these sanctuaries encompass adult habitat along the reef tract. The rationale for creating the small sanctuaries was to protect the shallow living corals in places that receive many visitors, not to protect the lobsters. But because lobsters are often found in dens within the corals, prohibiting lobster-fishing protects the coral. No changes were expected in the size structure and abundance of these lobsters because typical lobster movements would likely expose most lobsters inside these small sanctuaries to the fishery within a relatively short period of time. The rationale for creating the medium-sized sanctuary included both protecting shallow corals and other habitats and allowing natural spawning of the resident fauna. Here, changes in the local lobster population were expected. In addition, FKNMS enlarged the Looe Key sanctuary to 1.5 km² following the recommendations of a study conducted by John Hunt (Hunt et al. 1991).

The objective of this paper is to collectively examine the large-scale adult lobster monitoring projects that have been conducted since the 1970s. Both the size structure and abundance of lobsters have been monitored as one means of evaluating the performance of these sanctuaries. Extensive lobster-monitoring projects have been conducted in southern Florida during every decade since the 1970s. Large-scale monitoring of lobsters in DTNP was conducted during the mid 1970s and again during the late 1990s. Similar monitoring of the Looe Key sanctuary was conducted during the late 1980s and the late 1990s. The entire Florida Keys reef tract was monitored from 1995 to 2002 by two different projects. Our goal is to reassemble these data in a manner that permits population size structure, abundance, and fecundity in sanctuaries of different sizes to be compared. The common denominator for all these projects was that lobsters were observed on adult fore reef habitat during the breeding season (approximately March through July; see Bertelsen and Cox 2001), and our comparisons will be based solely on observations of lobsters during this time period and in this habitat. Consequently, the sanctuaries protecting juveniles and subadults discussed in the overview of sanctuaries are not considered further.

Methods

Estimations of size structure, abundance, and fecundity were generated from data assembled from all major southern Florida diver-based lobster surveys from the 1970s to 2002. These data include a 1973–1975 Dry Tortugas project by Gary Davis of the National Park Service; a 1987–1989 Looe Key project by John Hunt of the Florida Marine Research Institute; and two recent Florida Fish and Wildlife Conservation Commission data sets: (1) a 1995–1999 Keys-wide and Dry Tortugas project for the Marine Fisheries Initiative (MARFIN), and (2) a 1996–2002 Keys-wide project for FKNMS. Table 2 provides a summary of these projects and terminology that references each data set in this paper. Other data sets that were used to intercalibrate these main data sets are listed below. Because the fore reef habitat is the only habitat common to all these sanctuaries and because fecundity estimates were made, data were filtered to include only fore reef surveys during the breeding season (March through July; see Bertelsen and Cox 2001).

An examination of size frequency distribution of harvested lobsters was conducted using data collected by the National Marine Fisheries Service's Trip Interview Program. These data include size measurements of commercially harvested and landed lobsters. These data were filtered to include only Keys ocean-side fishing trips. This filter best compares with size frequency data from the fore reef diver surveys.

Table 2.—Characteristics of the data collected for projects used in this review.

Location	Years	Frequency of surveys, seasonality	Agency[a]	Project as identified in this article
Dry Tortugas	1973–1975	Monthly, year-round	NOAA, NPS	1970s Dry Tortugas
Looe Key	1987–1989	Bi-weekly, quad-weekly, year-round	FWC-FMRI	1980s Looe Key
Keys reef-tract/Dry Tortugas/Tortugas Banks	1995–2000	6-week cycle, March–September	FWC–FMRI	Marine Fisheries Initiative (MARFIN)
Keys reef-tract	1996–2001[b]	One set of surveys in July and September	FWC–FMRI	Marine Reserves
Keys reef-tract/Dry Tortugas/Tortugas Banks[c]	2002	July	FWC–FMRI/ RSMAS/NURC	Spree

[a] NOAA = National Oceanic and Atmospheric Administration; NPS = National Park Service; FWC = Florida Fish and Wildlife Commission; FMRI = Florida Marine Research Institute; RSMAS = Rosenstiel School of Marine and Atmospheric Science; NURC = National Undersea Research Center.
[b] Plus one trip in 2000 to Dry Tortugas and Tortugas Banks; also, Western Sambo Ecological Reserve monitoring continues to present.
[c] These data and portions of other data sets as described in Methods were used to calibrate timed surveys with area surveys.

Size–frequency tables for male and female lobsters were constructed for each project directly from each project's original data. Because the 1970s Dry Tortugas lobster data were based on 10-mm-carapace-length (CL) intervals, size–frequency histograms for all projects were depicted on this scale.

Because different field methodologies were used by the projects, estimates of lobster abundance for each project were also calculated by using different methods. All estimates were scaled to lobsters/ha. Abundance could not be estimated for the 1970s Dry Tortugas data because the field methods did not include either an area or search time.

The 1980s Looe Key project contained two data sets, used herein to estimate abundance. One portion of the fore reef was sampled in a way that highly accurate counts could be obtained without disturbing the lobsters. Because surveys were made every 2 weeks, the minimal-impact survey method should have minimized diver impact on subsequent surveys. In this portion, lobsters were counted but not captured, so an accurate count was made, but gender or egg-bearing status information could not be determined. The area searched was measured. The other portion of the fore reef was surveyed by capturing lobsters. Size, gender, and egg-bearing status of these lobsters were determined. The area of the search was measured here as well. For this paper, overall abundance of all lobsters was calculated directly from the minimal-impact portion, then the sex ratio as calculated from the second portion was applied to the overall abundances to estimate abundance of males and females.

The 1990s MARFIN and Marine Reserves projects based their surveys on 1-h timed searches. A conversion factor of 1 h to 2,500 m^2 was applied to estimate the area covered (see below). All lobsters found in these surveys were captured, but time spent capturing lobsters was not included in search time. The MARFIN sites were surveyed on a 6-week cycle from March to September. Marine Reserve sites were sampled in July and September. These data were merged for this paper.

Estimating fecundity of lobsters for each project required the following elements: (1) a size–frequency table for female lobsters, (2) an estimate of the abundance of female lobsters, (3) the expected percentage of egg-bearing female lobsters by size-class, and (4) a fecundity–size relationship for female lobsters. Issues specific to the various projects regarding size–frequency and abundance are given above.

For spiny lobsters in southern Florida during the breeding season, the percentage of egg-bearing females within a given size-class increases from smaller to larger size-classes. The most detailed information on female size and percentage of egg-bearing females comes from the MARFIN project, which was designed to estimate a fecundity ratio for females in the Keys and Dry Tortugas. The equations that predict the percentage of egg-bearers by size in the Keys (% egg bearers = $0.02 + 0.10 \times CL_{mm}$) and the Dry Tortugas (% egg bearers = $-0.60 + 0.18 \times CL_{mm}$) (see Bertelsen and Matthews 2001) were applied to female lobsters in all projects using the appropriate equation with respect to the origin of the data. The minimum observed size of egg bearers was 57 mm CL in the Keys and 70 mm CL in the Dry Tortugas. One important modification was made in our analysis of the 1970s Dry Tortugas project. Although Gary Davis found a minimum size of egg-bearing female lobsters in the 1970s that is nearly identical to that found in the Dry Tortugas today, he found a dramatic decline in the percentage of egg bearers larger than 120 mm CL. That drop in the percentage of large egg-bearing female lobsters was incorporated into our fecundity estimates for the 1970s Dry Tortugas lobster population.

The fecundity-to-size relationship in southern Florida spiny lobster was described during the MARFIN project (Cox and Bertelsen, abstract from the Fifth International Conference and Workshop on Lobster Biology and Management, 1997). This single relationship between fecundity and size (egg count = $-231,212 + 91.88\ CL_{mm}^{2}$) is valid for both the Keys and Dry Tortugas and was used in all fecundity estimates.

The fecundity of a local population was estimated using the following calculations. First, the estimated abundance of females within a given size-class is equal to the total female population abundance estimate times the percentage of females in a given size-class. Second, the expected percentage of egg bearers in a given size-class was estimated using the appropriate Dry Tortugas or Keys equation for percentage of egg bearers given above. Third, the expected number of eggs per egg bearer in a given size-class was calculated from the fecundity-to-size relationship. Next, the estimated number of eggs for a given size-class was calculated by multiplying the results of the first three steps. Finally, the estimated total fecundity of a local population became the sum of fecundity estimates for each size-class.

To estimate abundance and fecundity on an area basis for projects that used timed searches, we converted time of search to area searched using a conversion factor based on data collected during three separate projects. In the Spree data set (see Table 2), lobster count, time of search, and area searched were recorded during primarily deep (20–30 m), hard-bottom habitat dives. In the 1980s Looe Key data set, some surveys included additional measurements that provided the position of each lobster den on a measured grid. During some Marine Reserves surveys, the minute within an hour-long search was recorded when a den was discovered. These various estimates of density, distance between dens, and minutes between dens, were used to develop a computer simulation to create a virtual lobster population based on the timed surveys of the MARFIN and Marine Reserves projects. The time-to-area conversion was iteratively adjusted until a simulated sampling program reproduced an expected abundance estimate. A conversion of 1 h = 2,500 m^2 produced the best abundance estimates across a variety of expected densities.

As stated above, there are no data from the 1970s Dry Tortugas surveys that would permit an estimate of abundance. Lacking this, we nevertheless provide two estimates of fecundity using the following assumptions about abundance. For the first case, we used recent density estimates from surveys conducted in the Tortugas Banks. The rationale for using Tortugas Banks density estimates are (1) it is a remote area with a lightly fished lobster population that may currently be similar to that found in the Dry Tortugas during the 1970s (i.e., currently, two or three charter boats take recreational spear-fishing and lobster-catching trips to the Tortugas Banks), and (2) length–frequency distributions in the Dry Tortugas during the 1970s were more similar to recent length–frequencies on the Tortugas Banks than to recent length–frequencies in the Dry Tortugas. This suggests that the fishing mortality was similar for both the 1970s Dry Tortugas and today's Tortugas Banks populations. Thus, we estimated fecundity of the 1970s Dry Tortugas lobsters using the recent abundance estimates of Tortugas Banks.

For the second estimate, we assumed that, without fishing, the 1970s Dry Tortugas lobster population would be roughly equivalent to the current Dry Tortugas population. When a small area closed to fishing was temporarily opened, lobster abundance declined by 60% (Davis 1974). To estimate lobster abundance in the 1970s, we applied a 60% reduction of abundance of all legal-sized lobsters in the current Dry Tortugas population to the size–frequency distribution of the 1970s population.

The time-to-area conversion was performed to provide a crude estimate of abundance and fecundity across various projects that used different field methods. These conversions regrettably invalidate statistical tests for abundance and fecundity; however, we will present estimates of abundance and fecundity as a guideline.

In order to organize our characterization of local lobster populations in sanctuaries and the fishery, we have grouped sanctuaries into three size-classes: (1) small (<10 km^2), (2) medium (10–100 km^2), and (3) large (>100 km^2; Table 1). All 18 small spawning-area sanctuaries and the single medium sanctuary are found in the Keys. The only large spawning sanctuary for which we have sufficient data for our analyses is the DTNP. The nearby Tortugas Ecological Reserve is only 2 years old so not enough time has passed to present any meaningful trends or changes. However, we will present the Tortugas Banks (part of which became the Tortugas Ecological Reserve) surveys from the late 1990s as presanctuary baseline information.

Results

Size Differentials

One useful performance measure for a lobster sanctuary is the average difference in size between males and females: adult males tend to grow faster than do adult females (Hunt and Lyons 1986; Herrnkind and Lipcius 1989). Fishing mortality tends to reduce size differences between males and females. Today, the largest size differential (21.7 mm CL) is found in the largest sanctuary, DTNP (Table 3). At the creation of the DTNP lobster sanctuary in 1974, the size differential was 6.7 mm CL. The smallest size differential (4.2 mm CL) is found in the fishery. An intermediate size differential (5.9 mm CL) was found in both the small and medium sanctuaries. Looe Key, which qualifies as a small sanctuary but was created in 1981, has a slightly higher male–female differential, 8.0 mm CL, than do the other small sanctuaries (established in 1997). This differential in Looe Key did not change between the 1987–1989 and the 1996–2002 surveys.

Size–Frequency Histograms

Size–frequency histograms are also useful for evaluating the performance of lobster sanctuaries. Histograms for male and female lobsters in the DTNP have a distinctive shape (Figure 2), with the mode occurring well above the minimum legal harvestable size. For males, the frequency of the very large lobsters (>140 mm CL) is higher than that for females, and the

Table 3.—Comparison of densities, fecundity, and the male–female size differential for various regions and years on the fore reef during the breeding season. Note that for a sanctuary such as Dry Tortugas National Park (DTNP), fecundity estimates do not include deeper waters (>10 m). Also note that the Tortugas Banks region contains deep reef and hard-bottom habitat only. na = can not be calculated; r = marine reserve (lobster sanctuary); WSER = Western Sambo Ecological Reserve; SPAs = Sanctuary Preservation Areas.

Location	Time period	Density/ha Overall	Males[a]	Females[a]	Millions of eggs/ha per year	Male–female size differential (mm)
DTNP	1973–1975	na	na	na	11.8[b]/13.4[c]	6.7
DTNP (r)	1996–2000	72	39	33	33.8	21.7[d]
WSER (r)	1996–2002	145	46	99	44.0	5.9[d, e]
Looe Key (r)	1987–1989	69	14	55	27.6	8.0
Looe Key (r)	1996–2002	87	17	70	35.4	8.0[d]
SPAs (r)	1997–2002	68	26	42	16.6	5.9[e]
Tortugas Banks	1999–2002	18	7	11	11.1	9.5[d]
Fishery	1996–2002	69	21	48	15.4	4.2[d]

[a] Adjusted proportionally to account for observed lobsters whose gender was not determined.
[b] Assumed a Tortugas Banks density and percentage of egg bearers reported by Gary Davis (U.S. National Park Service) in the 1970s.
[c] Assumed current Dry Tortugas density with a 60% reduction as reported by Gary Davis (see text for details).
[d] Differential includes data from both MARFIN and Marine Reserve databases.
[e] Average differential for all 5 years of data starting at the time the sanctuaries were less than a year old.

frequency remains high through 180 mm CL. In contrast, histograms for both sexes in the Keys fishery show a rapid decline in frequency just above the legal harvestable size. This basic pattern of rapid decline above the minimum size is mirrored in both the small and medium sanctuaries. Histograms for the Tortugas Banks (surveys were performed before the area was designated as a sanctuary) show a typical fishery characteristic in that frequency of large lobsters declines rapidly above the mode in contrast to the size frequency of large lobsters in DTNP. However, the decline occurs at roughly the same size-class as the mode found in DTNP rather than just above the legal size limit as in the Keys fishery.

Mean Size and 90th Percentile Size

Overall, the mean size of lobsters measured in our monitoring surveys in the Florida Keys has increased virtually everywhere, in both sanctuary and fishery, from 1997 to 2002 (Figure 3A). Of all these increasing trends in mean size, only the mean size of males in the WSER shows an ordinary least squares significant positive slope (Table 4), but this result is somewhat misleading because nearly half of the overall increase occurred between the first and second year after the creation of the sanctuary.

The 90th percentile for sizes of male and female lobsters in the fishery has remained remarkably constant, but the 90th percentile for sizes of males in both the small sanctuaries and WSER has risen dramatically (Figure 3B). These two positive trends are statistically significant ($P < 0.05$; Table 4), and in both the small sanctuaries and WSER, the typical large (i.e., 90th percentile) male lobster has increased at a rate of more than 3 mm CL per year. The 90th percentile size of females in both small and medium sanctuaries has also risen but not as much as that of male lobsters. Although a significant positive trend for female lobsters was found in WSER (+1.44 mm CL/year; $P = 0.04$), the positive trend was not significant in the small sanctuaries.

In the fishery, the mean size of lobsters in our surveys increased, but the 90th percentile size did not change. This suggests that the frequency of smaller lobsters decreased. In a fishery, this would indicate a reduction of recruitment to the fishery. One way to examine this is to determine the ratio of new recruits to legal-sized lobsters. Our diver-based surveys do not contain enough observations to test this hypothesis however, a large number of size observation for lobsters collected at commercial fish houses over the same time period. We found that the proportion of new recruits (76–80 mm CL) in the catch declined signifi-

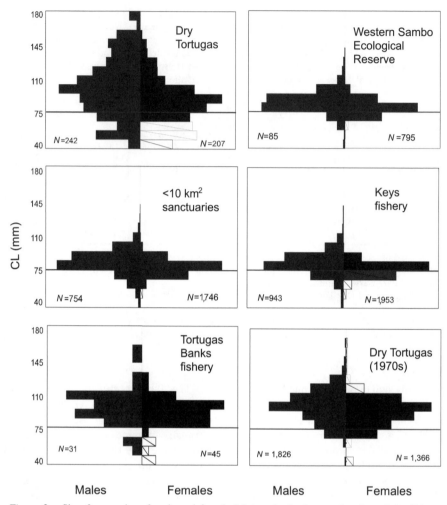

Figure 2.—Size–frequencies of male and female lobsters in the fore reef region of the fishery; small and medium-sized sanctuaries; the Dry Tortugas National Park Sanctuary, both recently and in the 1970s; and the Tortugas Banks, recently, but prior to becoming a sanctuary. The Tortugas Banks does not contain shallow fore reef habitat, and the 1970s Dry Tortugas data were collected throughout the park and cannot be sorted by habitat. White bars with a black diagonal in the female histograms indicate size-classes where the percentage of egg bearers is very low (<30%) or zero. Horizontal line indicates legal harvestable size (76 mm CL).

cantly for both female lobsters (–4.09 per season; r^2 = 0.81; P = 0.05) and male lobsters (–2.26 per season; r^2 = 0.56; P = 0.04) (Figure 4).

Abundance

Abundance of lobsters within a sanctuary should exceed that of lobsters in the surrounding fishery if the spatial extent of the sanctuary is sufficient to encompass the typical daily home ranges of some of the local lobsters. The greater the percentage of home ranges that are protected, the greater the enhancement of abundance should be. In all regions, the estimates of abundance range between 69 lobsters/ha and 87 lobsters/ha, with two exceptions (Table 3). One exception is the Tortugas Banks which has only 18 lobsters/ha. Factors that may explain this lower abundance figure include the relatively deep water (although all our surveys were conducted in areas less than 30 m deep), the relatively large distance between Tortugas Banks and juvenile habitats in the DTNP (the Tortugas Banks' population likely depends on adult spillover from DTNP for a majority of its recruitment) and fishing occurring on the banks during the times of our surveys.

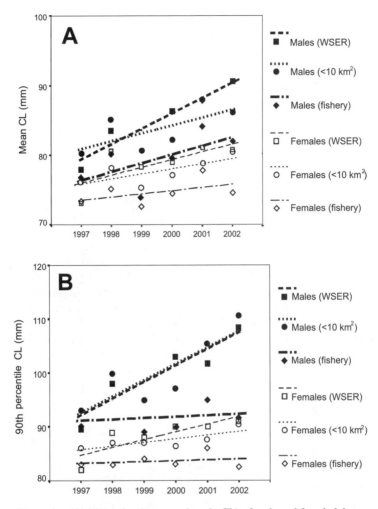

Figure 3.—(A) Mean size (carapace length, CL) of male and female lobsters in the Florida Keys fore reef habitat during the breeding seasons in the fishery, in small (<10 km²) sanctuaries, and in the medium-sized (10–100 km²) sanctuary from 1997 (the inception of the sanctuaries) to 2002. Lines represent ordinary least squares regression. (B) The 90th percentile sizes of male and female lobsters along the Florida Keys fore reef habitat during the breeding seasons in the fishery, in small (<10 km²) sanctuaries, and in the medium-sized (10–100 km²) sanctuary from 1997 (the inception of the sanctuaries) to 2002. Lines represent ordinary least squares regressions. WSER = Western Sambo Ecological Reserve.

The other exception is WSER, which has 145 lobsters/ha. The enhanced abundance within WSER has been remarkable and is not replicated in any other sanctuary along the Florida Keys, with the exception of the nearby (<3 km) Eastern Sambo sanctuary.

On the fore reef during the breeding season, the abundance of female lobsters exceeds the abundance of male lobsters everywhere except in Dry Tortugas National Park, where the densities of male and fe-

male lobsters is similar (Table 3). It has been suggested that female lobsters in the Dry Tortugas move to deeper water to lay eggs then return in the fall (Davis 1974). This hypothesis came from comparing the sex ratio in the summer and fall: the relatively low number of females in the summer returns to a near 1:1 ratio in the fall. A similar temporal sex-ratio pattern has been reported for lobsters in the Florida Keys in the 1920s (Crawford and DeSmidt 1922). Although this pattern

Table 4.—Analysis of trends (ordinary least squares linear regression) of male and female mean size and the 90th percentile size in the fishery, in small sanctuaries (<10 km²), and in the Western Sambo Ecological Reserve (WSER; 10–100 km²). Carapace length (CL) in mm.

Location	Sex	Mean size				90th percentile size			
		Trend (CL/year)	R^2	F	P	Trend (CL/year)	R^2	F	P
Fishery	Female	+0.46	0.227	1.177	0.339	+0.16	0.055	0.232	0.655
	Male	+1.25	0.403	2.705	0.175	+0.24	0.027	0.122	0.744
Small	Female	+0.72	0.526	4.437	0.103	+0.67	0.592	5.796	0.074
sanctuaries (<10 km²)	Male	+1.13	0.460	3.403	0.139	+3.03	0.718	161.121	0.033
WSER	Female	+1.13	0.512	4.320	0.106	+1.44	0.679	36.144	0.044
(10–100 km²)	Male	+2.21	0.882	22.515	0.018	+3.14	0.875	172.198	0.006

of sex-ratio change was observed in the Dry Tortugas in 1995 (Bertelsen and Hunt 1996), we did not observe it in subsequent years nor did we find a clear pattern in the Marine Reserves project. Both the abundance of lobsters and the sex ratio have remained consistent at Looe Key through the 1980s and 1990s.

Fecundity

The population fecundity estimates are derived from estimates of the abundance of female lobsters, the size

distribution of female lobsters, the size–fecundity relationship equation, and the likelihood that a particular-sized lobster will produce eggs. The greatest population fecundity estimates were found in the WSER (44 million eggs/ha each year) and the two oldest sanctuaries, Looe Key (35 million eggs/ha each year) and DTNP (34 million eggs/ha each year) (Table 3). Fecundity estimates are roughly halved in the other small (<10 km²) sanctuaries and in the fishery. The Tortugas Banks, with a relatively low abundance of females, produces only 11 million eggs/ha each year. The fe-

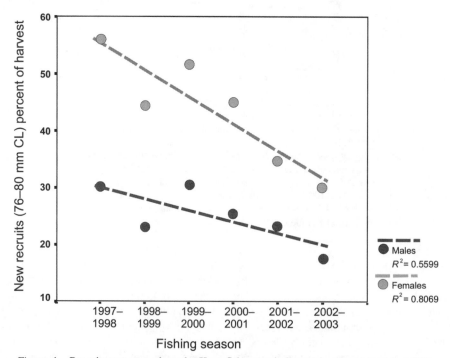

Figure 4.—Recruitment strength to the Keys fishery as indicated by the percentage of 76–80-mm-carapace length (CL) male and female lobsters in the total harvest (Oceanside Keys Trip Interview Program data, *www.sefsc.noaa.gov/tip.jsp*).

cundity estimates for Looe Key showed an increase from 28 to 35 million eggs/ha from the 1980s and 1990s. This may be due to a small increase in mean size and estimated abundance of female lobsters at Looe Key. However, this increase may also be an artifact of the time-to-area conversion we employed for the 1990s data.

We have calculated a crude estimate of the population fecundity at DTNP during the 1970s. Although we have good information regarding the size–frequency distribution for female lobsters and an estimate of the percentage of egg bearers by size-class, we do not have the data necessary to estimate density. As detailed in the methods, we estimated density based on two different sets of assumptions: (1) we assume that the current Tortugas Banks density could also apply to DTNP during the 1970s because both have similar size–frequency distributions, suggesting both were fished similarly; and (2) we imposed a 60% reduction in legal sized-lobsters using the current Dry Tortugas National Park density. We also imposed the decline in percentage of egg bearers in larger females as described by Davis (1974). Under the first set of assumptions, the DTNP lobsters produced 13 million eggs/ha per year, and under the second assumption, 12 million eggs/ha per year were produced. If these assumptions are realistic, then the Dry Tortugas fore reef population has tripled egg production since the inception of the sanctuary in 1974.

In the early 1970s Gary Davis observed a dramatic reduction in fecundity starting with female lobsters about 120 mm CL. By the 140-mm-CL size-class, female lobsters did not bear eggs. He coined the term "reproductive senility" to describe the phenomenon, suggesting that this was part of the female lobster life history. In the 1990s, we returned to the sanctuary and found no evidence of loss of fecundity in large females (Bertelsen and Matthews 2001). Every female larger than 120 mm CL carried eggs. We also observed the return of large males as described above. We suggested that the absence of these males prevented the large females from mating. We have modified our fecundity estimates to assess the overall loss of fecundity in the 1970s population and to estimate what the loss of fecundity might be if large males suddenly became very rare again. Without the fecundity loss of large female lobsters, we estimate that the 1970s Dry Tortugas lobsters should have produced 12.5 million eggs per ha (i.e., there was a 10% loss). A similar loss of fecundity applied to today's lobster population would reduce fecundity to only 24.7 million eggs per ha (a 27% loss). The greater percentage

loss that would occur in today's Dry Tortugas population is due to the greater proportion of large females found today than were found in the 1970s Dry Tortugas population.

Discussion

We have used the term lobster sanctuary to refer to places where lobsters cannot be fished; however, lobster protection was not always the reason for creating these sanctuaries. When the DTNP lobster sanctuary was established, it was established specifically as a lobster sanctuary. The national park already had other use restrictions, but in the mid-1970s, lobsters were added as a no-take species. The Looe Key sanctuary was established to protect the entire reef community, so no special attention was given to the design of the sanctuary with regard to lobsters. However, when the sanctuary was expanded from 0.5 km² to 1.2 km², part of the rationale for the expansion was to include lobster-foraging habitat (Hunt, abstract submitted to the American Fisheries Society's International Symposium on Marine Harvest Refugia, 1991). The small sanctuaries administered by FKNMS were established in 1997 to protect small portions of the reef community in high-use zones. No special attention was given to lobster, although lobsters within the sanctuaries were protected. Most of these small areas are called Sanctuary Preservation Areas (SPAs), and public visitation is permitted. Some fishing activities (e.g., catch and release) are permitted in a few of these SPAs. There are four Special Use Areas where only permitted researchers are allowed (see Table 1 for a listing). There are two Ecological Reserves (Western Sambo and Tortugas Banks), and although they are regulated similarly to an SPA, their design and size are intended to promote a natural ecosystem and maintain diversity. The WSER is one order of magnitude larger than the SPAs, whereas the Tortugas Ecological Reserve is two orders of magnitude larger than the SPAs (Table 1).

Tagging studies suggest that lobsters are capable of movements that exceed the dimensions of any sanctuary in southern Florida (Davis 1977; Gregory and Labisky 1986; Hunt et al. 1991); therefore, lobster sanctuaries do not absolutely protect a local population of lobsters from harvest, but the probability of harvest is lower for those lobsters that encounter a sanctuary. The resulting population in a sanctuary over time reflects the accumulation of those probabilities applied to lobsters of different sizes and movement behaviors associated with gender, growth, reproductive status, foraging habitats, and all other things that help shape a lobster's home range. The ecological landscape, the boundaries of the sanctuary itself, and proximity of each sanctuary to other sanctuaries also play a role in determining the abundance and size-structure of a local lobster population. For example, a lobster sanctuary established in the middle of a sand flat would provide little protection because lobsters spend little time in such habitats. A lobster sanctuary placed in a hard-bottom region with ample daytime shelters but no nighttime foraging areas would offer, in a simple sense, 12 h of protection for the time a lobster sheltered in the sanctuary and 12 h of exposure to a fishery during foraging activities (Cox et al. 1997). So, what can we infer from size–frequency information, abundance estimates, and fecundity estimates of the lobster population in the fishery and in small, medium, and large sanctuaries?

The size–frequency distribution of male and female lobsters in DTNP is probably similar to that found in a natural, unfished population in an offshore spawning environment. Because the sanctuary is now 30 years old, this size–frequency distribution is probably "dynamically stable" – in other words, all of the factors that changed the shape and density of this particular lobster population after it became a sanctuary have equilibrated (Bertelsen and Matthews 2001). The skewing of the male size–frequency distribution toward larger individuals is the result of different growth rates. Male lobsters, from subadults through adults, grow at a faster rate than do female lobsters (Hunt and Lyons 1986). Consequently, all other things being equal (such as mortality), in a mature adult lobster population, the average-sized male will be larger than the average-sized female. A fishery with a minimum size limit imposes an additional mortality risk for all lobsters over the size limit. Fishery mortality thus tends to equalize the size distribution for both males and females above the size limit. The greater the fishery mortality, the closer the male and female size distributions will become. The size–frequency distributions of the small and medium-sized sanctuaries still appear to be heavily influenced by fishing pressures, and we suspect that the "dynamic stability" of these sanctuaries is still several years away.

Tortugas Banks size–frequency distributions from our surveys prior to the creation of the Tortugas Ecological Reserve in 2003 have the appearance of a fished population because the relative frequency of both male and female lobsters decline dramatically above a specific size-class. The decline, however, occurs at the same size-class as does the mode in the frequency dis-

tributions of DTNP. This suggests that the Tortugas Banks may be a spillover destination for lobsters from the DTNP lobster sanctuary. The size–frequency distributions at DTNP during the early 1970s, when it was fished, support this hypothesis. As in the Tortugas Banks, the size frequencies at the Dry Tortugas of the 1970s showed a mode well above the legal minimum size and a rapid decline in frequencies in larger size-classes. If the Tortugas Banks is truly a spillover destination of the Dry Tortugas population, then the size–frequency distribution of the Tortugas Banks lobster population should, over time, take on more characteristics of the size–frequency distribution of the Dry Tortugas lobsters and abundance of lobsters on Tortugas Banks should increase.

The size–frequency distributions for the fishery, small sanctuaries, and the medium-sized WSER all show a distinct drop in frequencies just above the legal harvestable size limit. This suggests that either the sanctuaries have not been separated from the fishery long enough to develop their own distinct size–frequency distributions or that they are not sufficiently separated from the fishery in space (i.e., the sanctuary is not large enough or shares too much border with the fishery). We believe it unlikely that the size structure of lobsters in the small sanctuaries will ever differ much from the fishery. The Looe Key sanctuary was established in 1981 (albeit at a smaller size than today), but the resident lobster population's size structure there is very much like that of the fishery. The size structure of the WSER lobster population, however, may still be changing. Recent personal observations suggest that more large male and female lobsters may be present at WSER than in previous years.

The distinct difference in the size structures of the male and female lobsters in DTNP, with the high frequencies of very large (>140 mm CL) males, is found nowhere else in the southern Florida fore reef. The absence of these distinctively larger males in the Dry Tortugas may account for the lack of fecundity in the largest (>120 mm CL) females during the 1970s. A similar account of a lack of fecundity in larger females was reported for female lobsters larger than 130 mm CL in Antigua and Barbuda during the early 1970s (Peacock 1974). From the size–frequency distributions reported from the fishery, there appeared to be an adequate supply of large males present in the trap fishery; however, a dramatic decline in the harvest (15% per year) was occurring in that fishery at that time, and perhaps the large males were quickly fished out of these islands.

The size differential between male and female lobsters may be a key performance measure to use in the evaluation of lobster spawning sanctuaries. This differential increased markedly in the DTNP (Table 3) between 1974 and the present and is much larger than that in any of the other sanctuaries or in the Keys fishery. We do not know how long it may take for the size differentials to stabilize in these new sanctuaries. The Dry Tortugas studies were done 25 years apart, but we assume the size differential had already changed prior to our studies. Further, although the areal extent of a sanctuary undoubtedly plays a role in shaping the resident lobster population, so too should factors such as the arrangement and quantity of sheltering and foraging habitat, the amount of recruitment that normally occurs, and the fishing intensity surrounding each sanctuary. Hence, the stable size differential for all the small sanctuaries is not likely to be 8.0 mm CL as in the case of the older small Looe Key sanctuary. Looe Key contains ample high-quality shelter and foraging habitat within the protected area (Cox et al. 1997). Consequently, one should not conclude that any 1-km² parcel of fore reef will result in a size differential equal to that found at Looe Key.

No single size parameter can be used to effectively monitor all aspects of a local lobster population through time. For example, the mean size of a lobster population is influenced by the frequency of both large and small lobsters. Hypothetically, we could witness, in three different areas, an increase in the mean size of lobsters due to any of the following: (1) a reduction in the frequency of small lobsters and no change in the frequency of large ones, (2) an increase in the frequency of large lobsters and no change in the frequency of small ones, and (3) a shift in the entire size–frequency distribution. In all three cases, an identical rise in mean size could be recorded, but factors influencing the change in each area would be fundamentally different.

To monitor changes in the larger lobsters in a local population, we propose using the 90th percentile as a way to characterize the size of the typical "large" lobster. The maximum size is not a satisfactory indicator because it is too easily influenced by chance encounters. Our decision to use the 90th percentile was somewhat arbitrary, but we believe it is far removed enough from the maximum to not be influenced by chance. It is also less influenced by fluctuations in abundance of smaller lobsters.

Remarkably, the 90th percentile male size increased identically in both the small sanctuaries and the medium-sized WSER, whereas the 90th-percentile-sized male lobster in the fishery remained constant. This finding suggests that, in the fore reef, sanctuaries as small as 0.5 km² are of sufficient size to encompass the home range of some males. Earlier lobster tagging studies in southern Florida foretold this possibility. Davis (1977) found that larger lobsters in the Dry Tortugas moved generally shorter distances than did smaller lobsters. Hunt et al. (1991) found the same 110-mm-CL male on the same fore reef site through most of 1987. In addition, both Davis (1977) and Little (1972) reported that larger lobsters tended to move less frequently.

Although it appears that movement patterns (distance and frequency) of female Caribbean spiny lobsters *Panulirus argus* are such that they do not receive much protection from small sanctuaries, this was not the case with a spiny lobster species in New Zealand, *Jasus edwardsii*, on a 15-ha patch within the approximately 6-km² Leigh Marine Reserve (Kelly and MacDiarmid 2003). Here, the larger the female, the more associated she was to a site; no real site association was seen in males until they attained a threshold size of 130 mm CL, whereupon male site association increased.

Throughout these first 5 years since the creation of the small sanctuaries, we have not seen any significant increase in lobster abundance in the small sanctuaries. The overall average density of lobsters within these small sanctuaries appears to have remained roughly the same as in the surrounding fishery, but within a single small sanctuary, large twofold to fourfold fluctuations regularly occur in a single year, and tenfold changes in density have been recorded. The strong aggregating tendencies and the overall mobility of spiny lobster probably account for these fluctuations, and because of this, single-year abundance assessments of spiny lobsters in small sanctuaries should not be used to make management decisions.

The Western Sambo Ecological Reserve is the only medium-sized (10–100 km²) sanctuary, so the benefits of survey replication cannot be explored here, and we cannot know if what we have observed in the Western Sambos would be typical for any hypothetical 10-km² to 100-km² sanctuary in southern Florida. Nonetheless, since its creation, this medium-sized sanctuary has shown increases in the 90th percentile size of large males identical to those found in small sanctuaries. The greatest change, however, has occurred in the density of male and female lobsters and in female fecundity. The density of male and female lobsters has risen to twice that of the fishery and fecundity to nearly three times that of the fishery. Fecundity and lobster density estimates are greater in the WSER than in any other location in southern Florida. There is no way to know if these high levels can be maintained over the next few years, but experience elsewhere suggests a decline may be inevitable. MacDiarmid and Breen (1993) observed a fourfold increase in lobsters in a then newly created Cape Rodney to Okakari Point Marine Reserve in New Zealand. However, after 4 years, the population levels declined when increased "predation" (in this case, fishers, discovered that traps placed along the sanctuary boundary harvested much of the spillover) lowered abundance. Lobster traps have been placed along the border of the WSER since its inception, as has been the case in DTNP. Also, within the WSER, though we have not quantified it, the number and average size of grouper (family Serranidae; a lobster predator) appear to have recently risen (authors' personal observations).

As stated above, other factors that undoubtedly play a role in determining the resulting size structure and abundance of a local population within a sanctuary are the landscape of a sanctuary and the distribution of sheltering and foraging habitats in the area. With the exception of the 1980s Looe Key project, where information about den size and substrate (for example, what species of coral covered a given den) was collected, none of these monitoring projects was designed to incorporate properties of habitat quality and landscape of the sanctuary. However, the small sanctuaries may simply be too small for landscape factors to influence size distribution or abundance in any predictable way other than what we have witnessed regarding the significant increase in the 90th percentile male size. While the average abundance of all the small sanctuaries indicates that abundance has not increased above that of the fishery, hidden beneath those reported means are large variations in abundance estimates (sometimes over tenfold) from year to year and sanctuary to sanctuary (Cox and Hunt, unpublished). We attempted to evaluate changes in abundance with an a priori view that large, year-to-year swings in abundance might indicate a small sanctuary that contains abundant high-quality habitat and is capable of temporarily supporting a large aggregation of lobsters, whereas areas with consistently low abundance estimates and small fluctuations might indicate comparatively poor habitat. However, the results of this exercise did not produce results consistent with our perceptions of habitat quality (i.e., perceptions of complexity and rugosity) in the various small sanctuaries. For example,

the greatest variance in abundance estimates was found in a small sanctuary with low relief. Also, two of the three lowest-ranked sanctuaries in year-to-year abundance estimates contain some of the highest relief.

Incorporating and understanding the role of landscape and habitat in the sanctuaries will likely require more spatially explicit studies involving tracking lobsters, determining home ranges, and developing accurate habitat maps of a study area. Studies incorporating quality of habitat have proven valuable in defining the interplay between subadult and adult lobsters and survival (Lipcius et al. 1997, 1998) and social behavior (Childress and Herrnkind 1994, 2001). Future objectives regarding evaluation of sanctuaries should incorporate both detailed habitat maps and lobster-movement patterns with respect to gender, size, and reproductive status.

Summary and Final Notes

The density of lobsters during the breeding season on the fore reef does not appear to be enhanced in small-sized (<10 km^2) sanctuaries but has risen to twice that of the fishery within the single medium-sized (10–100 km^2) sanctuary, WSER. The density of lobsters during the breeding season on the fore reef in the single large-sized (>100 km^2) sanctuary, DTNP, is approximately equal to that of the fishery.

Because adult male lobsters grow faster than do adult female lobsters, sexual dimorphism will be present in an unfished population. The difference in average size between male and female lobsters is greatest in the large sanctuary (DTNP). The difference in average size in the medium-sized (WSER) and small sanctuaries is only slightly greater than that in the fishery. The 90th percentile male lobster size has risen significantly in both the small and medium-sized sanctuaries. This factor could lead to greater sexual dimorphism in some sanctuaries.

Fecundity is enhanced in the medium-sized sanctuary (WSER) to a level two and a half times greater than that in the surrounding fishery. Fecundity is not enhanced in the small sanctuaries. Although the level of fecundity reported from the DTNP fore reef zone is twice that reported from the fishery and is only slightly lower than from WSER; the estimate of fecundity presented herein does not include an important deep zone that is part of the DTNP and is not part of the WSER or the other small sanctuaries (see below).

Because all the small sanctuaries are located along shallow fore reefs (with the exception of three inshore sanctuaries), we have limited our comparisons regarding other sizes of sanctuaries to that portion that also occupies fore reef habitat (with the single exception of Tortugas Banks, which contains only deep reef habitat). A simple examination of fecundity estimates along the fore reef omits an even larger portion of the overall egg-production story in the Dry Tortugas region. Since the 1970s, it has been known that the number of female lobsters in the shallow reef environment declines during the egg-bearing season and increases in the fall after egg laying has concluded. The leading hypothesis is that females mate and lay eggs in shallow water, then move to deeper waters to release them. During the MARFIN study, we had the opportunity to survey deeper regions around the Dry Tortugas and found that the number of eggs found per unit effort (search time) was nearly double the number of eggs found per unit effort in the shallow areas.

Over the course of the first 5–6 years of these newly created sanctuaries, the constancy of the 90th percentile size and the rise in mean size of male and female lobsters in the fishery, coupled with a reduction of new recruits to the fishery as shown in the trap fishery data, suggest that all the changes we have detected within these sanctuaries have occurred in spite of an overall Keys-wide "recruitment failure." One of the many functions of marine protected areas is to protect against recruitment failure (Baker 2000). Perhaps the Western Sambo Ecological Reserve has already helped in that function.

Acknowledgments

Dr. Gary Davis was the first pioneer regarding spiny lobster research in the Dry Tortugas. His unselfishness in providing historical data to these authors requires more than an acknowledgment, but we are limited to this. The 1980s Looe Key research was supported by the National Oceanic and Atmospheric Administration (NOAA) grant number 50-DGNC-6-00093. The 1990s Keys-wide and Dry Tortugas research was supported by a MARFIN grant from NOAA (NA57FF0300). The 1990s Keys-wide and Western Sambo Ecological Reserve research was supported by NOAA grants 7-WC-A-90005, 43-WC-NC-900917, 40-WC-NA-006223, and 40-WC-NA-1A0341. The 1970s Dry Tortugas research was supported by a grant from Florida Sea Grant (SUSF-SG-74-201). We also gratefully acknowledge support from the National Park Service, the Florida Keys National Marine Sanctuary, and the National Undersea Research Center. We also thank the anonymous reviewers.

References

Acosta, C. A., and M. J. Butler, IV. 1997. Role of mangrove habitat as a nursery for juvenile spiny lobsters, *Panulirus argus*, in Belize. Marine and Freshwater Research 48:721–728.

Baker, J. L. 2000. Guide to marine protected areas. Department for Environment and Heritage, Adelaide, South Australia, Australia.

Bertelsen, R. D., and C. Cox. 2001. Sanctuary roles in populations and reproductive dynamics of spiny lobster. Pages 591–605 *in* G. H. Kruse, N. Bez, A. Booth, M. W. Dorn, S. Hills, R. N. Lipcius, D. Pelletier, C. Roy, S. J. Smith, and D. Witherell, editors. Spatial processes and management of marine populations. University of Alaska Sea Grant, Alaska Sea Grant AK-SE-01-02, Fairbanks.

Bertelsen, R. D., and J. H. Hunt. 1996. Spiny lobster spawning potential and population assessment: a monitoring program for the South Florida fishing region. National Oceanic and Atmospheric Administration/National Marine Fisheries Service, Final Report, MARFIN Grant 0518, Miami.

Bertelsen, R. D., and T. R. Matthews. 2001. Fecundity dynamics of female spiny lobster (*Panulirus argus*) in a south Florida fishery and Dry Tortugas National Park lobster sanctuary. Marine and Freshwater Research 52:1559–1565.

Childress, M. J., and W. F. Herrnkind. 1994. The behavior of juvenile Caribbean spiny lobsters in Florida Bay: seasonality, ontogeny and sociality. Bulletin of Marine Science 54:819–827.

Childress, M. J., and W. F. Herrnkind. 1996. The ontogeny of social behaviour among juvenile Caribbean spiny lobsters. Animal Behaviour 51:675–687.

Childress, M. J., and W. F. Herrnkind. 2001. The influence of conspecifics on the ontogenetic habitat shift in juvenile Caribbean spiny lobsters. Animal Behaviour 62:465–472.

Cox, C., J. H. Hunt, W. G. Lyons, and G. E. Davis. 1997. Nocturnal foraging of the Caribbean spiny lobster (*Panulirus argus*) on offshore reefs of Florida, USA. Marine and Freshwater Research 48:671–680.

Crawford, D. R., and W. J. J. DeSmidt. 1922. The spiny lobster, *Panulirus argus*, of southern Florida: it's natural history and utilization. U.S. Bureau of Fisheries Bulletin 38:284–310.

Davis, G. E. 1974. Notes on the status of spiny lobsters *Panulirus argus* at Dry Tortugas, Florida. Pages 22–34 *in* W. Seaman, Jr., and D. Y. Aska, editors. Research and information needs of the

Florida spiny lobster fishery. State University System, Florida Sea Grant Program SUSF-SG-74-201.

Davis, G. E. 1977. Effects of recreational harvest on a spiny lobster, *Panulirus argus*, population. Bulletin of Marine Science 27:223–236.

Gregory, D. R., Jr., and R. F. Labisky. 1986. Movements of the spiny lobster, *Panulirus argus*, in south Florida. Canadian Journal of Fisheries and Aquatic Sciences 43:2228–2234.

Herrnkind, W. F. 1980. Spiny lobsters: patterns of movement. Pages 349–407 in J. S. Cobb and B. F. Phillips, editors. The biology and management of lobsters, volume 1, physiology and behavior. Academic Press, New York.

Herrnkind, W. F., and R. N. Lipcius. 1989. Habitat use and population biology of Bahamian spiny lobster. Proceedings of the Gulf and Caribbean Fisheries Institute 39:265–278.

Herrnkind, W. F., and M. X. Redig. 1975. Preliminary study of establishment of den residency by spiny lobster, *Panulirus argus*, at Grand Bahama Island. Hydrolab Journal 3:96–101.

Hunt, J. H., and W. Lyons. 1986. Factors affecting growth and maturation of spiny lobsters, *Panulirus argus*, in the Florida Keys Canadian Journal of Fisheries and Aquatic Sciences 43:2243–2247.

Hunt, J. H., T. R. Matthews, D. Forcucci, B. S. Hedin, and R. D. Bertelsen. 1991. Management implications of trends in the population dynamics of the Caribbean spiny lobster, *Panulirus argus*, at Looe Key National Marine Sanctuary. National Ocean Service, Final Report 50-DGNC-6-00093, Rockville, Maryland.

Kelly, S., and A. B. MacDiarmid. 2003. Movement patterns of mature spiny lobsters, *Jasus edwardsii*, from a marine reserve New Zealand Journal of Marine and Freshwater Research 37:149–158.

Lipcius, R. N., D. B. Eggleston, D. L. Miller, and T. C. Luhrs. 1998. The habitat-survival function for Caribbean spiny lobster: an inverted size effect and non-linearity in mixed algal seagrass habitats. Marine and Freshwater Research 49:807–816.

Lipcius, R. N. W. T. Stockhausen, D. B. Eggleston, L. S. Marshall, Jr., and B. Hickey. 1997. Hydrodynamic decoupling of recruitment, habitat quality and adult abundance in the Caribbean spiny lobster: source-sink dynamics? Marine and Freshwater Research 48:807–815.

Little, E. J., Jr. 1972. Tagging of spiny lobsters (*Panulirus argus*) in the Florida Keys, 1967–1969. Florida Department of Natural Resources Marine Research Laboratory Special Scientific Report 31.

MacDiarmid, A. B., and P. A. Breen. 1993. Spiny lobster population change in a marine reserve. Pages 47–56 in C. N. Battershill, D. R. Schiel, G. P. Jones, R. G. Creese, and A. MacDiarmid, editors. Proceedings of the Second International Temperate Reef Symposium. NIWA Marine, Wellington, New Zealand.

Marx, J. M., and W. F. Herrnkind. 1985. Macroalgae (Rhodophyta: *Laurencia* spp.) as habitat for young juvenile spiny lobsters, *Panulirus argus*. Bulletin of Marine Science 36:423–431.

Peacock, N. A. 1974. A study of the spiny lobster fishery of Antigua and Barbuda. Proceedings of the Gulf and Caribbean Fisheries Institute 26(1973):117–130.

American Fisheries Society Symposium 42:211–223, 2004

The Emerald and Western Banks
Closed Area Community:
Changing Patterns and Their Comparison

JONATHAN A. D. FISHER[1]

*Department of Biology, University of Pennsylvania,
Philadelphia, Pennsylvania 19104-6018, USA*

Abstract.—A 13,700-km² area including both Emerald and Western banks (Northwest Atlantic Ocean) was closed to groundfish trawling in 1987 to conserve haddock *Melanogrammus aeglefinus*. This closure provided the opportunity to examine changes in the fish community using multispecies abundance data that have been collected annually since 1970. Indices of community similarity among years were calculated for the closed area and potential fished reference areas on the eastern Scotian Shelf. Based on results from cluster analysis and nonmetric multidimensional scaling of annual community similarities, there are clear differences between the closed area and candidate reference areas. These differences are explained by variations in local oceanographic conditions that have led to the development and maintenance of different communities at nearby locations. In contrast, a previous investigation reported increased community similarity between the closed area and a fished reference area on the western Scotian Shelf. This finding is discussed with regard to the long-distance dispersal of fish from the closed area, which may homogenize communities across space. These two findings suggest that multiple, antagonistic factors influence community dynamics through space and time. Without prior knowledge of closed and reference area communities and the dispersal potential from the closed area, one may draw incomplete or misleading conclusions on the effects of closures. Species-specific associations between local abundance and geographic distribution are also discussed as important variables in the use of closed areas for multispecies protection. Lastly, changes to the benthic community may be taking place within the closed area. Research questions directed at the benthic community are outlined and advocated.

Introduction

A large area known for high catches of adult haddock *Melanogrammus aeglefinus*, including Emerald and Western banks and some surrounding deeper water on the central Scotian Shelf (Northwest Atlantic Ocean), was closed year-round to trawling in 1987 (Figure 1). This measure was the result of a fishing-industry-led attempt to reduce the catching and discarding of juvenile haddock. Frank et al. (2000) reviewed temporary groundfish closures on the Scotian Shelf and the history of the 13,700-km² Emerald and Western banks closed area. Since its designation, the closed area has been scrutinized in relation to its ef-

fects on the haddock stock within the Northwest Atlantic Fisheries Organization statistical divisions 4Vn, 4Vs, and 4W (Figure 1; Frank and Simon 1998; Frank et al. 2000, 2001). Recently, this Division 4VW (including Divisions 4V and 4W) haddock stock has demonstrated strong recruitment and an increase in spawning stock biomass, though the fishery remains closed due to poor individual growth and high natural mortality (Frank et al. 2001).

Community and ecosystem-level data sources have not been used to guide management decisions involving Scotian Shelf fisheries (Zwanenburg et el. 2002). Unlike many tropical and temperate reef closed areas (Halpern 2003), temperate continental shelf fishery closed areas have mainly been used to address single species research questions in accordance with their management objectives (but see Piet and

[1] E-mail: jfisher2@sas.upenn.edu

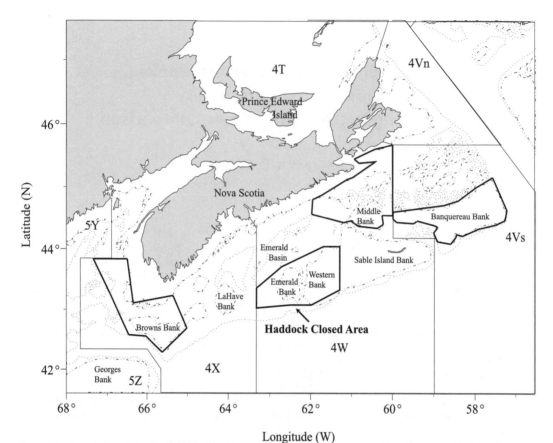

Figure 1.—Geography of the Scotian Shelf, with Northwest Atlantic Fisheries Organization fishery statistical divisions also shown. The locations of the Emerald and Western banks closed area, the Browns Bank reference area, the Middle Bank area, and the Banquereau Bank area are denoted by heavy solid lines. Dashed and dotted lines represent the 100-m and 200-m depth contours, respectively.

Rijnsdorp 1998; Murawski et al. 2000). Such emphasis on single species neglects possible community changes that may result from the removal of fishing disturbance and bycatch mortality that affect multiple species. Community-level investigations have revealed fishery impacts in a variety of geographical areas (Hall 1999; Kaiser and de Groot 2000; Zwanenburg 2000). Community-level approaches may similarly provide a means to understand the effects of cessation of fishing on target and nontarget species.

Data preceding and following the removal of trawling on the central Scotian Shelf provide information on multispecies abundance and distribution. The examination of these data through time and space provides additional information on fish assemblages not evident from a series of single species examinations (Mahon and Smith 1989; Clarke 1993; Fisher and Frank 2002). The Emerald and Western banks area supports high larval fish diversity (Shackell and

Frank 2000) and is a spawning source for many species (Scott 1983; Reiss et al. 2000). These characteristics suggest that the closed area might have a positive effect on other species in addition to haddock.

A long-term Scotian Shelf and Bay of Fundy groundfish trawl survey conducted by the Canadian Department of Fisheries and Oceans has provided opportunities to investigate species and community responses to the area closure at multiple spatial scales. Using these data from the Emerald and Western banks haddock closed area, the effects of the closure have recently been evaluated at the community level, based on changing abundance, distribution, and size frequencies of the component species (Fisher and Frank 2002).

This chapter builds on the previous investigation of community change by highlighting changes in the finfish community since the removal of groundfish trawling. These observations are compared to changes occurring across the remainder of the eastern Scotian Shelf, with emphasis on the importance of reference area choice and the implications for comparing closed and fished assemblages. Secondly, examples and discussion of some species-specific associations between local abundance and geographic distribution are included, with emphasis on species' responses to area closures. Finally, the need for research on benthic assemblages is considered. Understanding benthic community change may provide evidence of habitat change that affects fish community change within closed areas.

The Emerald and Western Banks Fish Community in the Context of the Eastern Scotian Shelf

In an earlier evaluation using multivariate techniques, the Emerald and Western banks closed area was compared to a Browns Bank reference area on the western Scotian Shelf in which trawling was never prohibited year-round (Fisher and Frank 2002). The closed area community underwent a significant shift in species dominance structure, driven primarily by increased abundance of several species, including Atlantic herring *Clupea harengus*, winter flounder *Pseudopleuronectes americanus*, redfish *Sebastes* spp., silver hake *Merluccius bilinearis*, red hake *Urophycis chuss*, Atlantic mackerel *Scomber scombrus*, and haddock. The Browns Bank fished reference area underwent a similar shift. This was surprising given the contrasting fishery regimes between the two areas. Although these two areas lie in different management areas for many species (Figure 1), the Browns Bank reference area was likely influenced by the population dynamics of the closed area via the distribution dynamics of some species. This influence was identified by the positive relationship between local abundance and regional distribution in and around the closed area by haddock, winter flounder, pollock *Pollachius virens*, Atlantic herring, and silver hake (Fisher and Frank 2002). As these species became more abundant locally, they were also caught in a greater number of areas around the closed area. The time spans until species in the reference area increased abundance were discussed in terms of the possible interaction between these two areas. Increases in abundance generally occurred first within the closed area. From these observations, it was inferred that the closed area likely supplies some individuals to the fished

Browns Bank reference area, thereby leading to the observed increase in similarity between the closed and fished areas.

Throughout the current examination, year-to-year variation of the finfish community is portrayed using nonmetric multidimensional scaling (Clarke 1993). First, the Bray-Curtis measure of community similarity was calculated for all years based on species abundance from all species collected within the closed area during the annual survey. Next, only the ranks of the Bray-Curtis similarities were used to determine the relative positioning of the annual community similarities in multidimensional space, which are presented as plots in two dimensions. These techniques produce nonmetric multidimensional scaling plots, where points closest together represent those years or locations with the greatest relative community similarity and those farthest apart indicate samples with the least community similarity (Kruskal and Wish 1978; Clarke 1993). For example, the samples from the years 1983 and 2002 have communities that are ranked as least similar within the closed area, and, consequently, these samples are positioned very far apart (see Figures 2, 3). Also, based on the measures of Bray-Curtis community similarity among years, clustering was used to determine the natural groupings of multiple years based on community similarities. Clustering was performed in order to determine whether the groupings of years fall into distinct before closure and after closure groups or whether the designation of the closed area did not follow a similar timeline to observed changes to the community. Additionally, examination of the clusters and multidimensional scaling plots formed using data from candidate reference areas that have never been closed year-round to all groundfish trawling was undertaken in order to determine whether the cluster groups formed by the closed area were unique to that region of the eastern Scotian Shelf (Division 4VW, Figure 1).

The changes in rank similarity between annual points are due to changes in the component species' abundance, and these individual species' contributions can be quantified using the similarity percentages procedure outlined by Clarke (1993). A species' contribution to the overall similarity in a cluster group is based both on its abundance and its consistency in the group of annual samples. This is referred to as "percent similarity" in Tables 1 and 2. Additionally, to identify species responsible for the temporal changes occurring within the closed area and elsewhere on the Scotian Shelf, "discriminating species" (those species that consistently occur within one cluster group but

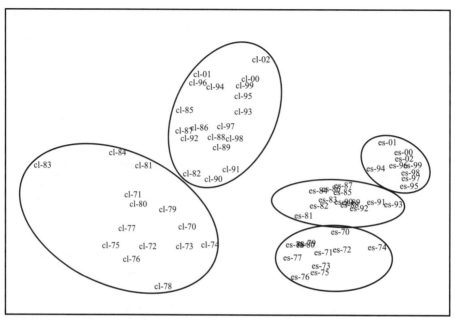

Figure 2.—Nonmetric multidimensional scaling plot of the Emerald and Western banks closed area and the remainder of the eastern Scotian Shelf (Division 4VW) sampled from 1970 to 2002. Distances between samples on the plot indicate the underlying relative community similarity, with points closest together most similar to each other. Closed area relative similarities are shown with the prefix "cl" accompanying the two-digit sample year. The eastern Scotian Shelf prefix is "es." The two closed area cluster groups are indicated by ovals. The eastern Scotian Shelf clusters into three temporal groups, also identified by their cluster groups. Cluster group relatedness is described in the text. Kruskal's stress formula one = 0.10 indicates that the two-dimensional plot produces an accurate portrayal of the multidimensional data (Kruskal and Wish 1978).

not another) are also listed together with their relative ability to contribute to the dissimilarity between two cluster groups (Clarke 1993). As all species ultimately contribute to within-cluster-group similarity and discrimination between cluster groups, only the top ranked species are tabulated and discussed.

Currently, the closed area community continues on a trajectory away from the preclosure state (Figure 2). The current state is due to continued high abundance of the above mentioned species relative to the preclosure years (Table 1). In addition, the continued decline in abundance of some large species that characterized the assemblage during the preclosure years continues to occur (Atlantic cod, thorny skate, and cusk). Atlantic cod were reported at their lowest levels within the closed area during the 2001 survey, and no thorny skate were captured within the closed area for the first time in 1999 and again in 2002. These observations demonstrate that some community members continue to display no apparent recovery and continue to decline, despite the closed area's large size and long duration.

The historically similar communities in the closed area and Browns Bank reference area became increasingly similar following closure (Fisher and Frank 2002). In contrast to this pattern, reference areas of similar size on the eastern Scotian Shelf, including the Banquereau Bank and Middle Bank areas (Figure 1), do not display such community similarities with the closed area at any time during the survey's history (Figure 3). There is little relative overlap in their communities with that of the closed area, or even between the Banquereau Bank and Middle Bank areas, compared to the overlap between the closed area and Browns Bank assemblages (Figure 3). Similarly, Mahon and Smith (1989) reported finfish communities on Emerald and Western banks were more similar to those from Browns Bank and from other banks on the western Scotian Shelf than to those on the eastern Scotian Shelf. These observations were linked primarily to differences in water temperature, with relatively low water temperatures persisting in the eastern areas of Division 4VW (Mahon and Smith 1989). If the closed area had been compared to these similar-sized

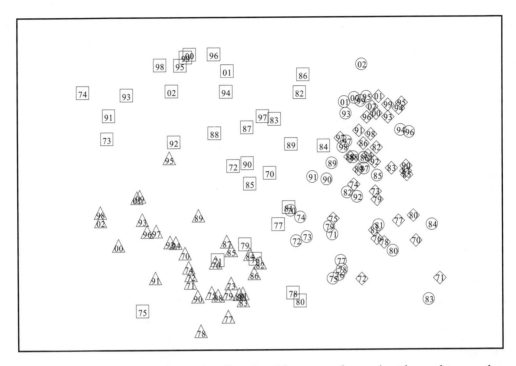

Figure 3.—Nonmetric multidimensional scaling plot of four areas of approximately equal area on the Scotian Shelf. Distances between samples on the plot indicate the underlying relative community similarity, with points closest together most similar to each other. Circles = Emerald and Western banks closed area; diamonds = Browns Bank area; squares = Middle Bank area; and triangles = Banquereau Bank area. Two-digit sample labels indicate years, ranging from 1970 to 2002. Kruskal's stress formula one = 0.17 and indicates that the two-dimensional plot produces a suitable portrayal of the multidimensional data.

areas from the eastern Scotian Shelf instead (either Banquereau Bank area or Middle Bank area), the direction of relative community change would have been reversed. Rather than increased similarity, as between the closed and Browns Bank areas, communities in these areas decreased in similarity relative to the closed area community. This occurred despite the fact that both the Banquereau Bank and Middle Bank areas are geographically closer to the closed area than the Browns Bank area is to the closed area (Figure 1).

In contrast to the divergence from the closed area observed in the Middle and Banquereau banks areas, the eastern Scotian Shelf (Division 4VW) exclusive of the closed area shows temporal change in its community from the earliest years in the survey, as does the closed area (Figure 2). This trend occurs despite the overall community differences between these areas due to differences in fishing activity, an increased range of habitats across the larger area encompassing the rest of the eastern Scotian Shelf, and sample size differences.

The temporal trends in the communities suggest that the closed area and the remaining eastern Scotian

Shelf communities may be responding to large-scale oceanographic or fishery-induced changes. However, the eastern Scotian Shelf cluster groups occur during distinctly different periods relative to the closed area changes. The dominant pattern in the closed area is two cluster groups formed at the 67% similarity level, closely matching the implementation of the closed area (Figure 2). The dominant cluster pattern of community change within the rest of the eastern Scotian Shelf is three distinct periods, with groups defined between 1970 and 1980, 1981–1993, and 1994–2002 (Figure 2; Table 2). The first two temporal groups are more closely aligned to each other (at 77% similarity) than they are to the most recent group (at 72% similarity). Within these groups on the eastern Scotian Shelf, some species remained characteristic through the first two periods. Redfish contributed most to the within groups similarity through periods 1 and 2, followed by American plaice, silver hake, Atlantic cod, and yellowtail flounder (Table 2).

During the most recent cluster group from the eastern Scotian Shelf, some small pelagic species, forage

Table 1.—Closed area similarity percentages summaries from preclosure (1970–1986) and postclosure (1987–2002) survey data. Only species contributing 3% or greater average similarity to the within-periods structure during the preclosure and postclosure periods or 3% or greater to the dissimilarity (between periods difference) are quantified. An asterisk indicated the listed species contributed less than 3% to the average similarity or dissimilarity.

| Species | Preclosure | | Postclosure | | Preclosure and postclosure difference |
	Average number per tow	Percent similarity	Average number per tow	Percent similarity	Percent dissimilarity
Haddock	81.5	13.1	133.9	11.0	3.4
Silver hake	29.5	8.5	55.6	8.4	4.1
Atlantic cod *Gadus morhua*	11.5	8.3	8.6	5.0	*
Yellowtail flounder *Limanda ferruginea*	16.3	7.8	16.0	6.4	*
American plaice *Hippoglossoides platessoides*	6.6	6.8	9.8	5.9	*
Longhorn sculpin *Myoxocephalus octodecemspinosus*	4.3	5.3	5.8	4.9	*
Pollock	2.9	4.1	7.1	5.1	3.2
White hake *Urophycis tenuis*	1.3	3.9	1.0	*	*
Thorny skate *Amblyraja radiata*	3.4	5.6	0.3	*	3.3
Witch flounder *Glyptocephalus cynoglossus*	1.2	4.3	0.9	*	*
Goosefish *Lophius americanus*[a]	0.8	3.9	0.6	*	*
Atlantic halibut *Hippoglossus hippoglossus*	0.5	3.8	0.5	*	*
Cusk *Brosme brosme*	0.6	3.6	0.1	*	*
Atlantic herring	2.4	*	64.3	7.0	9.8
Redfish	0.4	*	14.1	4.4	4.7
Red hake	1.4	*	4.5	4.3	4.0
Winter flounder	0.1	*	3.1	3.7	5.0
Atlantic mackerel	1.2	*	4.0	3.3	3.8
Atlantic argentine *Argentina silus*	6.4	*	5.7	*	4.3
Moustache sculpin *Triglops murrayi*[b]	0.9	*	1.2	*	3.1
Gulf Stream flounder *Citharichthys arctifrons*	0.2	*	1.1	*	3.3
Blackbelly rosefish *Helicolenus dactylopterus*	0.1	*	1.2	*	3.1

[a] Also known as monkfish.
[b] Also known as mailed sculpin.

fish, and coldwater species increased and remained in high abundance to capture numerical dominance from some of the former large species. This change in species composition followed a coldwater intrusion onto the northeastern Scotian Shelf (Frank et al. 1996). Northern sand lance, silver hake, Atlantic herring, redfish, American plaice, and capelin characterize this final group. On the eastern Scotian Shelf, exclusive of the closed area, the first of these two transitions between cluster groups can be described as a result of the introduction of coldwater and "forage" species and to an increase in abundance of some large predators (Table 2). There followed the proliferation of coldwater species and the reduction in abundance of some large predators during the final transition to the most recent cluster group. Increases in capelin and Atlantic herring, a decrease in redfish, an increase in daubed shanny, a decrease in Atlantic cod, and increases in snakeblenny and Greenland halibut distinguished this group from the second grouping (Table 2). If the eastern Scotian Shelf is forced into the "closed" and "open" groupings characteristic of the closed area, the results are similar, with increases in capelin, northern sand lance, Atlantic herring, and silver hake, unchanged abundance of redfish, and increased haddock distinguishing between these two time periods.

Based on the changes on the eastern Scotian Shelf, it is apparent that the closed area response shares few parallels with the response in the larger area of the remaining eastern Scotian Shelf. Apart from increased

Table 2.—Eastern Scotian Shelf area (exclusive of the closed area) similarity percentages summaries during the three periods listed below, showing only species contributing 2.5% or greater average similarity or dissimilarity. An asterisk indicates the listed species contributed less than 2.5% to the average similarity or dissimilarity.

Species	1970–1980		1981–1993		First and second period differences	1994–2002		Second and third period differences
	Average number per tow	Percent similarity	Average number per tow	Percent similarity	Percent dissimilarity	Average number per tow	Percent similarity	Percent dissimilarity
Redfish	115.0	8.1	165.1	6.9	3.1	49.1	4.5	3.1
American plaice	56.1	7.7	45.2	5.7	*	38.5	4.5	*
Silver hake	18.1	5.0	85.0	6.2	4.4	64.6	4.9	*
Yellowtail flounder	37.0	6.7	27.3	5.0	*	24.2	3.9	*
Atlantic cod	31.3	6.3	53.6	5.5	*	10.7	3.2	2.8
Thorny skate	9.0	4.8	8.0	3.6	*	3.6	*	*
Haddock	9.1	4.0	43.0	5.4	3.9	47.6	4.3	*
Witch flounder	6.3	4.0	4.1	3.1	*	8.5	3.0	*
White hake	3.3	3.7	9.4	3.6	*	5.6	2.7	*
Red hake	0.4	2.5	2.5	2.5	2.9	1.8	*	*
Longhorn sculpin	5.2	3.9	4.2	*	*	4.8	*	*
Smooth skate Malacoraja senta	0.9	2.7	0.4	*	*	0.4	*	*
Longfin hake Urophycis chesteri	2.1	3.0	3.8	*	*	1.8	*	*
Checker eelpout Lycodes vahlii[a]	1.1	*	1.3	*	2.9	3.7	*	*
Northern sand lance Ammodytes dubius	21.5	*	7.0	*	3.8	217.9	5.0	7.4
Atlantic herring	2.7	*	10.4	*	3.2	73.1	4.8	4.8
Capelin Mallotus villosus	0.2	*	15.8	*	4.4	163.1	4.5	7.0
Daubed shanny Leptoclinus maculatus	0.4	*	0.1	*	*	2.6	*	2.9
Snakeblenny Lumpenus lampretaeformis	0.2	*	0.1	*	*	3.4	*	2.6
Greenland halibut Reinhardtius hippoglossoides	0.3	*	0.2	*	*	3.5	*	2.5

[a] Also known as shorttailed eelpout or Vahl's eelpout.

abundance of Atlantic herring and haddock on the eastern Scotian Shelf, the closed area is characterized roughly between a closed state and the historic fished state. The remainder of the Scotian Shelf has undergone temporal changes in the relative abundance of species and changes in community composition that do not correspond with changes in the closed area. These differences underscore the fact that, although a baseline of change is occurring on the eastern Scotian Shelf (Zwanenburg et al. 2002), the closed area community is on a distinct trajectory from the remaining areas on the eastern Scotian Shelf. The potential causes and implications of these differences are better understood in the context of dispersal and differences in local oceanographic conditions.

One major concern in the evaluation of closed area effects is separating local population responses within the closed area from regional increases in one or more species, given the capacity for dispersal as larvae and expected large home ranges for groundfish and pelagic species on continental shelves. Fisher and Frank (2002) attributed the observed increase in Atlantic herring abundance within the closed area to the removal of trawling based on examinations of annual Atlantic herring distributions on the Scotian Shelf. Evidence for the closed area as a source for Atlantic herring production was also provided by Reiss et al. (2000), who recorded the highest concentrations of larval Atlantic herring and the smallest sized Atlantic herring larvae on Western Bank when they sampled the central Scotian Shelf. These observations led Reiss et al. (2000) to conclude that Western Bank is an offshore fall spawning source for this species (as well as for Atlantic cod and silver hake). Other investigations of community change in the Northwest Atlantic (Fogarty and Murawski 1998) and specific examination of Atlantic herring spatial and temporal trends (Overholtz 2002) have indicated that this species was undergoing increases in abundance at many locations, coincident with establishment of the Emerald and Western banks closed area. These observations suggest that increases in Atlantic herring abundance may have been due, in part, to changing fishing practices or trophic structure at the regional scale. Such observations highlight that multiple factors can influence both closed areas and reference areas and lead to challenges in separating local closed area effects from regional effects.

The problem of partitioning effects of the closed area from a background of regional dynamics in continental shelf systems is due to an interaction between large home ranges of some species as well as to varia-tion in oceanographic features, which influence the historic and present community structures within candidate reference locations. These additional influences on the communities in and around closed areas lead to two caveats: (1) Closed and fished areas may be geographically distant enough that the local oceanographic features are quite dissimilar. In this situation, differences in community composition derived from different oceanographic or other environmental conditions confound comparisons between closed and fished areas. (2) Based on the large home ranges and dispersal potential of some species, reference locations may be too close to the closed area to exclude the influence of species movement from the closed area. The former concern is likely the situation driving the dissimilarity between the closed area and geographically close areas on the eastern Scotian Shelf (Figure 3), as these areas differ in water temperature (Mahon and Smith 1989). The use of Browns Bank on the western Scotian Shelf as a reference area did have the advantage of similar fauna and similar oceanographic features to the closed area that were quite different in areas to the east of the closed area (Fisher and Frank 2002). In the examination of the Browns Bank reference area, spillover of some species from the closed area to the Browns Bank reference area was inferred due to the positive correlation between increasing local abundance and distribution. Given these patterns, even the Browns Bank area that was deemed a suitable reference area due to similar species composition and relative abundance with the closed area was likely not independent of the closed area's dynamics. Both of these concerns can be expected to be multiplied in other situations, given that replicated reference sites are often prescribed for comparison with an impacted area (Underwood 1992, 1994).

If demographic connections between closed areas and reference areas result from population dispersal from the closed area, and if these connections are not investigated, then studies of continental shelf communities may dismiss closed area effects as entirely due to regional influences. In an examination of changes in demersal assemblages in the North Sea, Piet and Rijnsdorp (1998) divided a large protected area "plaice box" prior to analyses due to different local conditions. These authors found that differences in species richness between fished and closed periods were influenced by the influx of southern species coincident with the closure and that these increases were obvious in both closed and fished areas. Difficulties will arise in determining closed area community changes, if dispersal connections with potential reference areas and local or regional oceanographic changes are not recognized.

While the Emerald and Western banks closed area is affording protection to some species, it is not enough to protect the full suite of species that were formerly relatively abundant within its bounds. The lack of response of some species to the removal of fishing is likely influenced by the contraction of species' ranges when fishing reduces their local abundance.

Linking Local Abundance, Geographic Distribution, and Closed Areas

Abundance is only one measure in the examination of species responses to area closures. A species' regional geographic distribution is also important and is often positively correlated with its local abundance when a time series of these variables is examined. This positive association indicates that increases in distribution are often accompanied by a greater increase in population size than is predicted from increased distribution alone—the distribution increases and the average local abundance increases at sites within the species' range (Gaston et al. 2000). This association between local abundance and distribution also occurs for many species undergoing distribution declines, so not only do some species occupy fewer sites through time, but their local abundance at the few remaining sites is also reduced. Due to both the positive and negative implications of abundance–distribution relationships (Lawton 1993), examinations of the association between these variables should be included in any assessments of population status.

At the spatial scale of Division 4W, which includes the closed area, several species demonstrated the propensity for increasing distribution concurrent with increases in local abundance (including haddock, winter flounder, pollock, Atlantic herring, and silver hake). Fisher and Frank (2002) discussed these findings as likely contributors to the increased similarity between the Emerald and Western banks closed area and the Browns Bank reference area. Some of the species that demonstrated increased local abundance within the closed area may have spread out to occupy additional areas as far away as Browns Bank. This positive abundance–distribution relationship following increased abundance within a closed area provides a "double benefit" for species undergoing abundance increases (Gaston 1999), resulting in higher local population densities over a larger area. These rela-

tionships, when examined for increasing populations associated with closed areas, have implications for transfer rates from closed areas to fished areas, as a species without a positive relationship should increase in abundance but remain only within the closed area boundaries. This type of population increase without an increase in geographic distribution must be distinguished from positive correlations between local abundance and distribution, as these scenarios result in different expectations as to population increases beyond the closed area's boundaries.

Core areas that maintain high local abundance when a population is in decline may also be very important for population growth (see Hutchings 1996 for an Atlantic cod example). If core areas of high local abundance exist for multiple species, then these areas would be ideal locations for seasonal or permanent spatial protection. If core areas for species can be identified, then depending on their locations, the spatial scale for multispecies protection can be estimated. If multispecies core areas overlap, then the spatial scale for multispecies protection would be less than if species-specific core areas are widely dispersed. In addition to the positive associations between local abundance and distribution in Division 4W around the closed area, at the stock-specific spatial scales of fisheries management, almost half of the stocks from the 24 most prevalent species on the Scotian Shelf demonstrated significant positive correlations ($P < 0.05$) between local abundance and stock distribution (Fisher 2002).

The prevalence of these positive relationships on the Scotian Shelf demonstrates that numerous stocks likely have the capacity to benefit both from increases in local abundance and increased distribution following the removal of fishing mortality. Positive correlations between local abundance and distribution have been used previously to examine the interaction between abundance and catchability in fisheries and are applicable to the designation and evaluation of area closures. As many stocks exhibit this type of relationship, mapping their distribution during the years of low local abundance may provide an understanding of where most species currently overlap and where closed areas would provide for multispecies spatial protection.

As demonstrated by both Atlantic cod and thorny skate stocks from Division 4VsW, using data from 1970 to 2001, some stocks at the management scale demonstrate a positive correlation between local abundance and distribution due to decreasing values of both of these measures (Figure 4). One notable feature of the Atlantic cod stock is that the decrease in local abundance and distribution continued after a fishery moratorium was established in 1994 (Figure 4A). Also, this plot demonstrates that Atlantic cod underwent a rapid decrease in local abundance and especially decreased in distribution within the stock area in a very short time between 1992 and 1994, which eventually led to the lowest levels of these measures. These examples have implications for community-level responses to the closed area, as species with declining range may or may not inhabit the closed area. Therefore, multispecies recovery within the closed area will be dependent on where species reside during periods of low population abundance. The hypothesized effects of density-de-

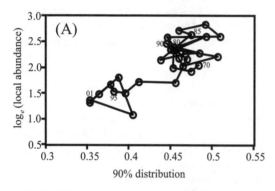

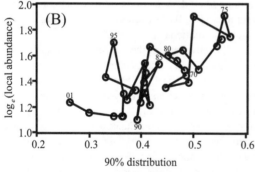

Figure 4.—Plots of the positive, significant correlation between local abundance and regional distribution for two stocks on the eastern Scotian Shelf from 1970 to 2001. Geographical distributions are quantified as the minimum proportion of 10-min geographic squares that contain greater than 90% of species abundance in each year within Division 4VsW. Note that through time, both of these stocks have undergone declines in both distribution and local abundance at the limited number of sites where they remain. Panel (A) shows Division 4VsW Atlantic cod ($r = 0.78$, $P < 0.01$). Panel (B) shows Division 4VsW thorny skate ($r = 0.71$, $P < 0.01$). The last two digits of the sample years are indicated near seven data points on each plot.

pendent habitat use and the siting and species-specific recovery rates of fishery closed areas are explored further by Jennings (2000).

The use of space is an essential parameter for the success of protected areas, which influences the timing of increases and the transport of individuals from protected areas. Clearly, interactions between local abundance and regional distribution are very important for current analyses of the Emerald and Western banks closed area's effect on the fish community and for any future application of closed areas on the Scotian Shelf for multispecies conservation.

Opportunities for Additional Investigations of Community Change

The data explored so far have focused entirely on the fish community, without reference to additional groups of species that are not adequately sampled during the groundfish research survey. In order to fully appreciate or eventually dismiss additional effects of the Emerald and Western banks closed area on the community, examination of benthic communities should be undertaken. These organisms have also been released from fisheries disturbance, and since many benthic species are site specific, they may provide the strongest evidence of community change. Therefore, comparisons of the closed area to fished regions of similar depth and surficial geology should be undertaken. As demonstrated relatively soon after the establishment of the Georges Bank closed area system, the most dramatic increase in abundance and biomass was shown for the sea scallop *Placopecten magellanicus*, a species that fortuitously benefited from the groundfish closed area (Murawski et al. 2000). In addition to sessile species of economic concern, impacts on the removal of trawling effort may be realized by nontarget benthic species. Collie et al. (1997) examined the effects of fishing on benthic megafauna by sampling areas marked by trawling disturbance and those without disturbance on Georges Bank. A similar sampling regime of untrawled areas within the Emerald and Western banks closed area and trawled sites of similar depth and sediment characteristics outside the closed area may provide evidence of the effect of the removal of trawling disturbance.

Browns Bank, which served as a fished reference area for the examination of fish species responses to the closed area, has recently been thoroughly explored through the use of multibeam sonar technology, benthic surface sampling of biotic and abiotic features, and analysis of photographic samples (Kostylev et al. 2001). The detailed examination of Browns Bank was undertaken in order to compose benthic habitat maps and to evaluate defining features of the habitats. Exploration of the Emerald and Western banks closed area using similar techniques would complement the examination of the fished area. A comparison of relative abundance and species richness of the benthic taxa in the closed area to the biota described by Kostylev et al. (2001) for Browns Bank may provide evidence of decadal-scale effects of the removal of large-scale trawling disturbance. If positive effects are apparent within the closed area, then it may serve as a reference area for documenting relative impacts in other fished areas.

The effects of biogenic habitat features may contribute locally to the success of juvenile life stages of species that demonstrate recent increased abundance within the closed area. The combination of sonar technologies, mapping overlying benthic fauna, and examining historic and present distributions of fish from trawl surveys should provide more information on fish species–habitat associations than is available (Scott 1982). This type of local-scale evaluation of fish–habitat associations has been advocated to improve fisheries management in the Northwest Atlantic (Langton et al. 1995).

A recent examination of haddock recruitment and the timing of phytoplankton blooms in an area that included Emerald and Western banks concluded that early spring blooms may be a necessary condition for increased haddock recruitment, though they may not be sufficient to explain differences among years (Platt et al. 2003). It is possible that an increase in positive biogenic habitat features within the closed area contributed to increased survival of juvenile haddock in recent years (1998 and 1999). This hypothesis is consistent with recent spatial models of habitat-mediated predation on juvenile fish (Lindholm et al. 2001). The hypotheses that biogenic habitat features are both more abundant and have positive effects on juvenile fish survival have yet to be evaluated in the Emerald and Western banks closed area. Given the removal of groundfish trawling effort over such a large area since 1987, the closed area provides an ideal location for future benthic habitat research.

Conclusions

The Emerald and Western banks closed area remains a unique section of the Scotian Shelf in which natural demographic processes and rates currently govern the population dynamics of the assemblage, due to the absence of groundfish trawling. Since the closed area's designation in 1987, its community has undergone a change in relative species abundance that is unique in comparison to other areas on the eastern Scotian Shelf. Based on local oceanographic differences and the potential for multispecies dispersal from the closed area, it is apparent that multiple influences shape both the community patterns and the interpretation of changes within and outside closed area boundaries. Therefore, it is recommended that choice of reference areas should incorporate several factors. Simple comparisons of closed and fished areas, without a clear understanding of historic differences or demographic linkages, may obscure true community-level changes following the removal of fishing disturbance.

Examination of local abundance and geographical patterns provides additional information as to the efficacy of spatial closures for the preservation and improvement of multiple populations. Depending on the strength of the correlation between local abundance and geographic distribution, different species should be expected to display a range of local and regional responses to the removal of fishing. While detailed information on abundance and distribution is available for many members of the fish community, benthic community change has not yet been quantified. However, benthic community change may also be occurring within the Emerald and Western banks closed area and may be influencing the abundance and composition of the present fish community. Elsewhere on the Scotian Shelf, methods for the evaluation of benthic biota have been demonstrated. The application of those methods within the Emerald and Western banks closed area should be undertaken to test hypotheses related to habitat change and changes in fish survival.

Acknowledgments

I thank Dr. Ken Frank of the Bedford Institute of Oceanography for offering enthusiastic discussion on these issues and for supporting my Masters degree research. Azolyn Fisher, Stan Kemp, J. Brooke Shipley, and anonymous reviewers offered many comments which improved the clarity of the manuscript.

References

Clarke, K. R. 1993. Non-parametric multivariate analyses of changes in community structure. Australian Journal of Ecology 18:117–143.

Collie, J. S., G. A. Escanero, and P. C. Valentine. 1997. Effects of bottom fishing on the benthic megafauna of Georges Bank. Marine Ecology Progress Series 155:159–172.

Fisher, J. A. D. 2002. Finfish community change in the Northwest Atlantic: responses to a long-term area closure and testing the generality of the abundance–distribution relationship. Master's thesis. Dalhousie University, Halifax, Nova Scotia.

Fisher, J. A. D., and K. T. Frank. 2002. Changes in finfish community structure associated with a large offshore fishery closed area on the Scotian Shelf. Marine Ecology Progress Series 240:249–265.

Fogarty, M. J., and S. A. Murawski. 1998. Large-scale disturbance and the structure of marine systems: fishery impacts on Georges Bank. Ecological Applications 8:S6–S22.

Frank, K. T., J. E. Carscadden, and J. E. Simon. 1996. Recent excursions of capelin (*Mallotus villosus*) to the Scotian Shelf and Flemish Cap during anomalous hydrographic conditions. Canadian Journal of Fisheries and Aquatic Sciences 53:1473–1486.

Frank, K. T., R. K. Mohn, and J. E. Simon. 2001. Assessment of the status of Div. 4TVW haddock. Department of Fisheries and Oceans, Canadian Science Advisory Secretariat Research Document 2001/100, Ottawa.

Frank, K. T., N. L. Shackell, and J. E. Simon. 2000. An evaluation of the Emerald/Western Bank juvenile haddock closed area. ICES Journal of Marine Science 57:1023–1034.

Frank, K. T., and J. E. Simon. 1998. Evaluation of the Emerald/Western Bank juvenile haddock closed area. Department of Fisheries and Oceans, Canadian Stock Assessment Secretariat Research Document 98/53, Ottawa.

Gaston, K. J. 1999. Implications of interspecific and intraspecific abundance–occupancy relationships. Oikos 86:195–207.

Gaston, K. J., T. M. Blackburn, J. J. D. Greenwood, R. D. Gregory, R. M. Quinn, and J. H. Lawton. 2000. Abundance–occupancy relationships. Journal of Applied Ecology 37:39–59.

Hall, S. J. 1999. The effects of fishing on marine ecosystems and communities. Blackwell Scientific Publications, Oxford, UK.

Halpern, B. 2003. The impact of marine reserves: do reserves work and does size matter? Ecological Applications 13:S117–S137.

Hutchings, J. A. 1996. Spatial and temporal variation in density of northern cod and a review of hypotheses for the stock's collapse. Canadian Journal of Fisheries and Aquatic Sciences 53:943–962.

Jennings, S. 2000. Patterns and prediction of population recovery in marine reserves. Reviews in Fish Biology and Fisheries 10:201–231.

Kaiser, M. J., and S. J. de Groot. 2000. Effects of fishing on non-target species and habitats: biological conservation and socio-economic issues. Blackwell Scientific Publications, Oxford, UK.

Kostylev, V. E., B. J. Todd, G. B. J. Fader, R. C. Courtney, G. D. M. Cameron, and R. A. Pickrill. 2001. Benthic habitat mapping on the Scotian Shelf based on multibeam bathymetry, surficial geology and sea floor photographs. Marine Ecology Progress Series 219:121–137.

Kruskal, J. B., and M. Wish. 1978. Multidimensional scaling. Sage Publications, Beverly Hills, California.

Langton, R. W., P. J. Auster, and D. C. Schneider. 1995. A spatial and temporal perspective on research and management of groundfish in the Northwest Atlantic. Reviews in Fisheries Science 3:201–219.

Lawton, J. H. 1993. Range, population abundance and conservation. Trends in Ecology and Evolution 8:409–413.

Lindholm, J. B., P. J. Auster, M. Ruth, and L. Kaufman. 2001. Modeling the effects of fishing and implications for the design of marine protected areas: juvenile fish responses to variations in seafloor habitat. Conservation Biology 15:424–437.

Mahon, R., and R. W. Smith. 1989. Demersal fish assemblages on the Scotian Shelf, Northwest Atlantic: spatial distribution and persistence. Canadian Journal of Fisheries and Aquatic Sciences 46:134–152.

Murawski, S. A., R. Brown, H. L. Lai, P. J. Rago, and L. Hendrickson. 2000. Large-scale closed areas as a fishery-management tool in temperate marine systems: the Georges Bank experience. Bulletin of Marine Science 66:775–798.

Overholtz, W. J. 2002. The Gulf of Maine–Georges Bank Atlantic herring (*Clupea harengus*): spatial pattern analysis of the collapse and recovery of a large marine fish complex. Fisheries Research 57:237–254.

Piet, G. J., and A. D. Rijnsdorp. 1998. Changes in the demersal fish assemblage in the south-eastern North Sea following the establishment of a protected area ("plaice box"). ICES Journal of Marine Science 55:420–429.

Platt, T., C. Fuentes-Yaco, and K. T. Frank. 2003. Spring algal bloom and larval fish survival. Nature (London) 423:398–399.

Reiss, C. S., G. Panteleev, C. T. Taggart, J. Sheng, and B. deYoung. 2000. Observations on larval fish transport and retention on the Scotian Shelf in relation to geostrophic circulation. Fisheries Oceanography 9:195–213.

Scott, J. S. 1982. Selection of bottom type by groundfishes of the Scotian Shelf. Canadian Journal of Fisheries and Aquatic Sciences 39:943–947.

Scott, J. S. 1983. Inferred spawning areas and seasons of groundfishes on the Scotian Shelf. Canadian Technical Report of Fisheries and Aquatic Sciences 1219.

Shackell, N. L., and K. T. Frank. 2000. Larval fish diversity on the Scotian Shelf. Canadian Journal of Fisheries and Aquatic Sciences 55:1747–1760.

Underwood, A. J. 1992. Beyond BACI: the detection of environmental impacts in the real, but variable, world. Journal of Experimental Marine Biology and Ecology 16:215–224.

Underwood, A. J. 1994. On beyond BACI: sampling designs that might reliably detect environmental disturbances. Ecological Applications 4:3–15.

Zwanenburg, K. C. T. 2000. The effect of fishing on demersal fish communities of the Scotian Shelf. ICES Journal of Marine Science 57:503–509.

Zwanenburg, K. C. T., D. A. Bowen, A. Bundy, K. Drinkwater, K. Frank, R. O'Boyle, D. Sameoto, and M. Sinclair. 2002. Decadal changes in the Scotian Shelf large marine ecosystem. Pages 105–150 in K. Sherman and H. R. Skjoldal, editors. Large marine ecosystems of the North Atlantic: changing states and sustainability. Elsevier, Amsterdam.

American Fisheries Society Symposium 42:225–236, 2004
© 2004 by the American Fisheries Society

A Key Role for Marine Protected Areas in Sustaining a Regional Fishery for Barramundi *Lates calcarifer* in Mangrove-Dominated Estuaries? Evidence from Northern Australia

JANET A. LEY[1]

Australian Maritime College, Post Office Box 21, Beauty Point, Tasmania 7270, Australia

IAN A. HALLIDAY

Queensland Department of Primary Industries, Post Office Box 76, Deception Bay, Queensland 4508, Australia

Abstract.—Monitoring the effectiveness of marine protected areas requires sensitivity to both resource conservation and public awareness. Nondestructive sampling methods are preferred, but potential insights into fishery management benefits may be limited. The toolbox of fisheries science contains established length-based methods that can potentially be used to extend assessment analyses. In northeastern Australia, state regulations allow commercial gill nets to be set within most regional estuaries, primarily targeting barramundi *Lates calcarifer* (Centropomidae; also known as barramundi perch). The state closed several riverine estuaries throughout this region to commercial fishing in the early 1980s. To monitor the effect of these closures on barramundi (and other) populations, our research teams deployed gill nets (mesh sizes 19–152 mm) bimonthly over 2 years within three pairs of systems (one closed and one neighboring estuary open to commercial net fishing per pair). On an hourly basis during each net set (1500–2100 hours), all fish were removed and measured. Of 1,657 barramundi netted, 96% were returned to the water alive. Catch averaged 2.8 times greater, and biomass 3.5 times greater, in closed systems compared to open ($P < 0.02$). Barramundi catch in closed systems exceeded that in open systems for each 30-mm length-class, including all classes below the legal limit of 580 mm. Barramundi are well-studied, protandrous hermaphrodites occurring in spatially discrete stocks in tropical mangrove estuaries. Applying published parameter values in a length-based analysis revealed that estimated egg-production was 21 times greater in the closed systems ($P < 0.03$). In fact, egg production in commercially fished estuaries was estimated to be near zero. The single most important factor in sustaining the commercial barramundi fishery in this region may be the function of interspersed estuarine reserves in providing sources of spillover and recruitment to neighboring open systems. Given current regulations, a network of protected estuaries may be especially beneficial, if not essential, for maintaining a sustainable fishery along this coast.

Introduction

Monitoring marine protected area (MPA) effectiveness in conserving stocks of exploited species requires an approach to data collection that is sensitive to both resource conservation and public relations. Destructive sampling methods may seriously bias monitoring

results by modifying populations under study. In addition, local supporters of MPA designations may strongly object to the use of sampling methods that require sacrifice of specimens. Within the toolbox of fisheries population biology are well-established, direct, and length-based methods that can be used with minimal impact on MPA biota (see reviews in Rochet and Trenkel 2003; Trenkel and Rochet 2003). Some of these derived measures (e.g., sex ratio, percent maturity) may be useful indicators for evaluating the

---

[1] E-mail: J.Ley@fme.amc.edu.au

effectiveness of marine reserves (Russ et al. 1998; Russ 2002). Indicators useful for MPA monitoring should quantify the impact of fishing closures, be straightforward in application, and utilize readily attainable data. Such data usually consists of a short time series (<4 years) of abundance indices by species and length-class (ages not available) obtained by a scientific survey using catch and release or visually based sampling methods (Rochet and Trenkel 2003). This study represents a test case using both direct and length-based indices derived from nondestructive sampling to evaluate the effectiveness of fishing closures in tropical estuarine systems.

Barramundi *Lates calcarifer* (Centropomidae; snooks; also known as barramundi perch), a protandrous hermaphrodite and estuarine-dependent species, is the leading target species in both recreational and commercial fisheries of northern Australian estuaries. Sustaining barramundi populations depends on both managing the fishery and protecting estuarine habitat. Adults spawn around river mouths, and resultant larval and juvenile barramundi occupy nearby marine wetlands. Juveniles then move into freshwater habitats to grow for 1 or 2 years (Grey 1987; Williams 1997). Barramundi initially breed as males, beginning at approximately 530 mm total length (TL; 2 years and older), and subsequently as females, beginning at approximately 700 mm (3 years and older; Davis 1986). By 920 mm (6 years and older), theoretically, all individuals have become mature females. Species whose life cycle includes a change of sex represent a particular challenge to fisheries managers because of the need to protect both young and old fish (i.e., both sexes) from overfishing (Bohnsack 1993; Milton et al. 1998). Barramundi populations can suffer if fishing removes numerous large female fish, which are highly fecund and vital in sustaining egg production. Thus, fishing could have direct effects on populations when removing large females and when removing males before they have had a chance to breed.

For populations comprised of many isolated stocks, overfishing can lead to localized species extirpations (Jennings et al. 2001). Barramundi populations are spatially structured into genetically discrete stocks existing in different groups of river systems (i.e., within a common catchment or embayment; Salini and Shaklee 1987). Thus, barramundi populations from geographically close estuaries (<100 km apart) are probably of a common stock (Kailola et al. 1993). Tagging studies carried out in northeastern Queensland have revealed that migration of adults may occur between neighboring estuaries if suitable coastal habitat exists along the intervening foreshores (Russell and Garrett 1988). The degree of larval dispersal between estuaries has not been directly quantified to date.

Commercial fishers use gill nets to target barramundi within estuaries and along foreshores throughout northeastern Queensland. Regulations include minimum length (580 mm) and maximum length (1,200 mm) sizes, spawning seasonal closures, and gear restrictions. However, this fishery has been characterized as a resource under pressure based on declining catch rates along the Queensland eastern coast (Williams 1997). In the late 1980s, recreational linefishers convinced State of Queensland officials to close several estuaries to commercial fishing to reduce use conflict. Here we describe fishery-independent surveys using catch-and-release sampling methods in six riverine estuaries located in tropical, northeastern Queensland, three that were closed to commercial net fisheries and three that were open. The objectives of the current study were:

(1) to compare the status of barramundi stocks in open and closed systems as measured by direct population indicators (i.e., total catch; biomass; length-frequency distributions; and average length of catch, L_{bar});

(2) to compare fishing and biological processes influencing barramundi stocks in open and closed systems as measured by population indicators derived from length-based methods and applicable published parameters, (i.e., fishing mortality and egg production); and

(3) to discuss the application of these population indicators to evaluating MPA effectiveness.

Methods

Study Area

The study area extended 400 km along the northeastern Australian coast between Cairns and Bowen in Queensland (Figure 1). Average annual rainfall is >4,000 mm in the northern part of the study area but <1,000 mm in the south. In the north, coastal rivers flow through mountainous rainforest catchments; southern rivers flow through broad, level floodplains dominated by eucalypts *Eucalyptus* spp. (as illustrated in Figure 2). Estuaries have continuous mangrove shorelines ranging from a narrow fringing band to broad-basin forests. Near their oceanic outfalls, these

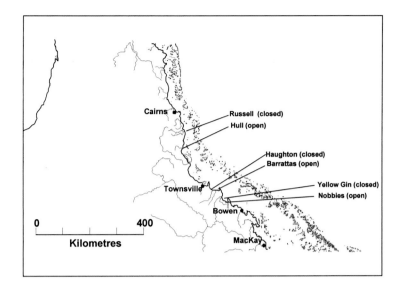

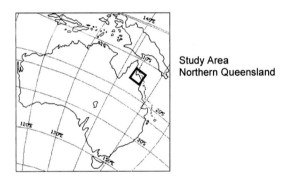

Figure 1.—Study area showing the location of the estuaries surveyed and subregions.

systems can vary in shape, size, and dynamics. Although physicochemical factors such as average salinity, mangrove area, and channel configuration tend to vary, trends were similar within pairs of systems (see further details in Ley and Halliday, in press).

Queensland's Fisheries Act of 1994 permits recreational line fishing in riverine estuaries throughout the area. Licensed commercial fishers can set gill nets in riverine estuaries if stretched-mesh sizes are between 150 and 215 mm, with most using nets near 150 mm. Nine estuaries in the region had been closed to all forms of commercial fishing for at least 7 years before our study, primarily to reduce user conflict. We focused on three of these closed systems, one in the northern region of the study area, one in the middle region, and a third in the southern region (Figure 1). For each estuary in which commercial fishing had been

banned ("closed"), a nearby (within 80 km) estuary open to commercial net fishing ("open") was selected for comparison. Sites in the three pairs were sampled upstream (2–7 km from the mouth) and downstream (within 1 km of the mouth).

Sampling Procedures

Bimonthly sampling of the six estuaries was conducted over the 7-d period of neap tides (0.5–1.8-m range). At each of the upstream and downstream sampling sites, replicate monofilament gill nets were deployed; these consisted of two nets each of the following mesh sizes: 152 mm, 102 mm, 51 mm, and multi-panel (19, 25, and 32 mm; Table 1). Nets were set from 1500 to 2100 hours and checked on an hourly basis. Ninety-six percent of all barramundi (and most other larger specimens)

Figure 2.—Aerial photograph of the Haughton River, Australia, estuarine system located in the middle region and closed to net fishing. An area approximately 10 km × 8 km is depicted.

caught were returned to the water alive and in good condition, as noted by independent observers from the local recreational fishing organization, Sunfish. Throughout the net-soak period, aquatic conditions were measured at 5-min intervals on one or two Hydrolab Datasonds (Loveland, Ohio). Salinity, temperature, turbidity, dissolved oxygen, water level, and pH were recorded near the net-deployment sites. Further details about the study sites and sampling procedures are available in Halliday et al. (2001) and Ley et al. (2002).

Analysis

Following natural log transformation, catch of barramundi from all nets combined was compared between open and closed systems using a paired two-sample *t*-test for means. For each specimen, weight was estimated by applying the length-weight equation

$$W = 0.00005 \times L^{2.8602}.$$

Parameters for this equation were derived from dead individuals supplemented with data from other specimens collected in northeastern Queensland in previous studies (Ley, unpublished). Following square-root transformation, mean biomass for open and closed systems was compared using a paired *t*-test.

Length frequencies were analyzed by 30-mm length-classes. Frequencies by length categories were compared for each pair of open and closed systems using a paired *t*-test. The average length of the catch is denoted as L_{bar} (King 1995; Trenkel and Rochet 2003). To derive L_{bar}, the following equation was applied to the length-frequency data for each system:

$$L_{bar} = \frac{1}{C} \times \sum_{l=1}^{L} C_l \times l,$$

where C is the total catch in the population and C_l is the catch in length-class l.

Length-based population indicators were calculated for catch in the largest mesh nets used in the research surveys (152-mm mesh). This net is equivalent to the most common mesh used by commercial fishers directly within the estuarine systems open to gillnetting. Calculation of these indicators required the standardization of the catch data to eliminate net selectivity bias.

Selectivity

The design specifications of the nets are detailed in Table 1. The selectivity calculation procedure assumed there were no differences among individual barramundi relative to probability of encounter with the gill nets due to variables other than size (e.g., variation in swimming speed was not accounted for). Only the probability of being caught given an encounter with the nets was considered. Fish may be caught in a gill net when (1) their heads slip through a mesh and, upon attempting to back out, the mesh catches the fish

Table 1.—Specifications of the gill nets used in the fishery-independent study. Note the 19-mm-mesh, 25-mm-mesh, and 32-mm-mesh panels were hung end to end to make one continuous net 38.1 m long.

Mesh size		Ply	Number of meshes deep	Number of meshes long	Hanging ratio	Fishing depth (m)	Fishing length (m)
mm	in						
152	6	40	33	650	0.4	4.6	38.5
102	4	24	50	500	0.4	4.7	21.2
51	2	16	50	950	0.5	2.2	24.1
32	1.25	12	100	700	0.6	2.5	12.7
25	1	8	100	910	0.6	2.0	13.2
19	0.75	6	100	1,120	0.6	1.5	12.2

behind their gill covers; (2) they become wedged around their bodies as far as the dorsal fins (However, barramundi can escape the net after being wedged by flaring their gill covers and snapping the mesh.); and (3) they become entangled by mouth parts, fins, or spines. Escapement by breaking the mesh and capture by the third method are not fully accounted for in the calculation procedure described below.

The Holt method was employed to calculate net selectivity (Sparre and Venema 1998). Natural logs of the numbers of fish caught per length-group by two different mesh sizes were assumed to be linearly related to the fish length expressed as the midpoint of the length-class:

$$\log_e\left(\frac{C_{mesh_1}}{C_{mesh_2}}\right) = a + b \times L_{midpoint}.$$

A linear regression of $\log_e(C_{mesh_1}/C_{mesh_2})$ against $L_{midpoint}$ for the length categories with overlapping frequencies produced estimates of a and b (Jennings et al. 2001).

For gilling and wedging, a bell-shaped model described the selection curve:

$$S_L = e^{\left(-\frac{(L-L_o)^2}{2s^2}\right)},$$

where S_L is the probability of a fish of length L getting caught in the net (i.e., selected). The parameter L_o is the optimum length for being caught, and s is the standard deviation of the normal distribution. Model parameters L_o and s were determined using length-frequency data from the 102-mm-mesh and 152-mm-mesh nets. The smaller mesh nets (51 and 19–25–32) caught consistently fewer barramundi than the next larger net and so were not used in the analysis (Milton et al. 1998). The selection factor (SF) was estimated from the following equation:

$$SF = \frac{L_o}{mesh\ size}.$$

The selection curve for the 152-mm-mesh net was used to adjust the data for the catch in that net. The adjusted catch provided an estimate of the true length-frequency distribution of barramundi in larger length-classes ($L_{midpoint}$ = 365–995 mm). Prior to adjustment, the raw length-frequency data were smoothed by calculating a running average with $n = 3$. Then the smoothed length frequencies were divided by the probability of capture for each length-class (Sparre and

Venema 1998). The unbiased results were used in subsequent calculations of total mortality and egg production.

Total Mortality

The exponential decay equation used in population dynamics models describes the total mortality rate for a cohort. Symbolized as Z, the parameter is commonly estimated by employing a length-converted catch curve methodology (Sparre and Venema 1998). To use estimated numbers at length for a given year instead of a true cohort, it is assumed that recruitment has been constant in the past. In addition, the calculated mortality rate is assumed to be constant across all age-classes analyzed (Trenkel and Rochet 2003).

In our calculations, length was converted to age using the von Bertalanffy equation solved for age:

$$t = t_0 - \frac{1}{K}\log_e\left(\frac{L_\infty - L_t}{L_\infty}\right),$$

where K is the Brody growth coefficient, L_∞ is average length at infinity, and L_t is the length at age t. Published length-at-age relationships from five riverine estuaries in tropical, northern Australia (Davis and Kirkwood 1984) were averaged and used to convert length to age. The values used in the calculations were

$t_0 = -1.366$, $K = 0.1428$, and $L_\infty = 1,425.6$ mm.

Converting length values to age involves calculating the amount of time (Δt) it takes a fish to grow from the starting point in a length-class (L_1) to the end point (L_2). Through regression analysis, a and Z parameters were derived from the linearized exponential decay equation

$$\log_e\left(\frac{C_{L_1,L_2}}{\Delta t_{L_1,L_2}}\right) = a - Z \times t_{L_{midpoint_{L_1,L_2}}}.$$

Total mortality (Z) is the sum of fishing mortality (F) and natural mortality (M). Natural mortality for later stage juveniles and adult fishes can be estimated using the Pauly equation (1980; as cited in Froese and Pauly 2003)

$$\log_e M = -0.152 - 0.279 \times \log_e L_\infty + 0.6543 \times$$
$$\log_e K + 0.463 \times \log_e T.$$

In the Pauly equation, L_∞ and K are the von Bertalanffy growth parameters for the stock, and T is the mean annual water temperature for the water body in °C.

Egg Production

Barramundi mature first as males, beginning at approximately 530 mm (2 years; Davis 1986). By the time they reach 680 mm (3 years), about 25% have become mature females. By 920 mm (6 years), theoretically all individuals should have achieved maturity as females. Thus, annual egg production for the stock varies as a function of the relative proportion of males and females and fecundity. A spreadsheet-based model was used to estimate the relative annual egg production based on unbiased barramundi length frequencies for each estuary. The expected sex ratios by length-class were estimated from published population parameters (Davis 1986). The length-sex relationship developed is graphically presented in Figure 3. Female fecundity increases exponentially with size and was estimated from the equation

$$F = 0.3089e^{0.0035L}.$$

Results

Indicators Based on Total Catch

Catch and biomass were significantly greater in each closed system compared to its open counterpart within a region (Figure 4a, b). Total catch averaged 2.8 times greater in the closed systems, and total biomass averaged 3.5 times greater (paired t-tests, respectively: $t = 12.20$, $P < 0.01$, $n = 3$; $t = 6.50$, $P = 0.02$, $n = 3$).

For the length-frequency distributions, abundances within size-classes were significantly reduced in open systems (northern pair: $t = 3.50$, $P < 0.001$, $n = 29$; middle region pair: $t = 4.45$, $P < 0.0001$, $n = 30$; southern pair: $t = 3.49$, $P < 0.01$, $n = 30$) (Figure 5). Finally, although L_{bar} was slightly greater for the closed system in each pair of estuaries, the paired t-test indicated that the difference between the means was not significant ($t = 1.32$, $P = 0.32$, $n = 3$) (Figure 4c).

Indicators Derived from Length-Based Analyses

Gill Net Selectivity

From the analysis, the common standard deviation was $s = 103.4$ and the selection factor was SF = 4.86. When SF was multiplied by each mesh size, L_o in the 102-mm-mesh net was 496 mm and was 739 mm in the 152-mm-mesh net. These parameters were used to calculate the unbiased population levels for length-classes L using the selection curve

$$S_L = e^{\left(-\frac{(L-739.1)^2}{2 \times 102.4^2} \right)}.$$

For the 152-mm-mesh net, the selectivity curve follows the same general trend indicated by the raw length-frequency data indicating a good model fit (Figure 6a) using the selection curve and the true length-frequency distribution of larger length-classes ($L_{midpoint} = 365$–995 mm) (Figure 6b).

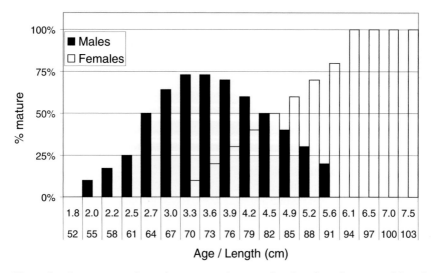

Figure 3.—Average maturity and sex conversion rates by size-class (bottom scale) and age-class (top scale) for barramundi (adapted from Davis 1986).

Total Mortality

On average, length-converted catch curves had steeper slopes for the systems open to fishing (mean $Z = 1.22$) than for the systems closed to fishing (mean $Z = 0.64$). For each pair of systems, Z was greater for the open estuary ($t = 6.04$, $P < 0.05$, $n = 3$) (Table 2; Figure 7). Using Pauly's formula and 2-year mean water temperature of 26.5°C (this study), expected natural mortality was 0.28. Fishing mortality and exploitation rate (F/Z) for each pair of estuaries were consistently lower for the closed systems in each pair (Table 2).

Egg Production

In all estuaries except the Russell River (northern region, closed), the greatest percentage of the barramundi encountering 152-mm-mesh nets was immature (Figure 8a). As derived, of the mature fish in each system,

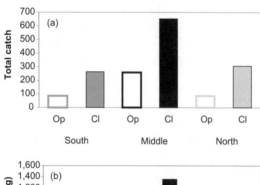

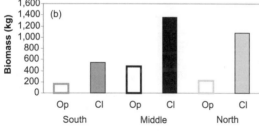

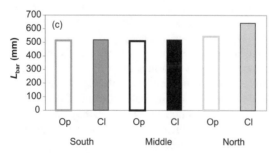

Figure 4.—Overall data (total catch, a; total biomass, b; and average length [L_{bar}], c) for barramundi by system for all nets combined. Op = open to commercial fishing; Cl = closed.

fewer mature males occurred in open systems ($t = 7.17$, $P < 0.03$, $n = 3$) (Figure 8b). The relatively lower abundance of females in any of the systems open to fishing was particularly striking ($t = 9.08$, $P < 0.01$, $n = 3$). The production of eggs in the closed systems was 21 times the production in the open systems ($t = 6.11$, $P = 0.03$, $n = 3$) (Table 3; Figure 8c).

Discussion

Based on comparisons of relative abundance and biomass, estuarine commercial fishing closures were highly effective in conserving barramundi stocks in estuaries closed to commercial fishing. This effect was observed even though recreational fishing was allowed in closed systems. Length-converted catch curve analysis quantified relative levels of fishing mortality attributable to both commercial fishing (ranging from 0.61 to 1.03 in open systems) and recreational fishing (ranging from 0.07 to 0.43 in closed systems). Estimated annual egg production was influenced by differences in sex ratios and age structure between open and closed systems, apparently due to commercial fishing. Length-based indicators applied to nondestructive sampling data from catch-and-release surveys were useful tools in evaluating the effectiveness of MPAs in conserving populations of targeted species.

A major problem with the length-based methodology (as with all population dynamics models) is the estimation of natural mortality rates. One recommendation would be to close at least one remote system along the northeastern Queensland coast to all forms of fishing. Monitoring populations in fully no-take systems would provide an invaluable opportunity to ascertain critical fishery management information such as natural mortality rates for barramundi populations under pristine conditions. In addition, more detailed surveys in fished and no-take estuaries should (1) employ a more finely scaled range of net mesh sizes (e.g., in 1-cm increments) ranging up to 215 mm, and (2) include some sacrifice of the catch to provide subsamples for further analysis of age growth and reproductive parameter estimation. An excellent precedent has been published for coral trout *Plectropomus leopardus* on coral reefs of the Great Barrier Reef (see Russ et al. 1998).

Indicators Based on Total Catch

The lower numbers and biomass netted by the research teams in the estuaries open to commercial fishing can

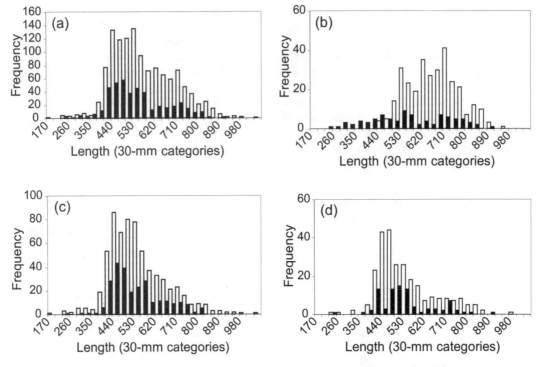

Figure 5.—Length frequency of barramundi caught in all nets combined in (a) all regions; (b) northern pair; (c) middle pair; and (d) southern pair. White bars = closed; black bars = open.

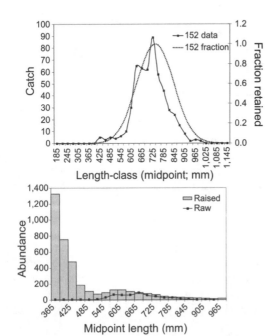

Figure 6.—Analysis (selection curve, top; and adjusted length frequency, bottom) of barramundi caught in largest mesh research nets (152 mm stretched mesh) for all estuaries combined.

Table 2.—Fishing mortality (F) and exploitation rate (F/Z) estimated from length-based analysis of biased corrected catch samples.

Estuary	F		F/Z	
	Open	Closed	Open	Closed
South	1.03	0.27	0.79	0.50
Middle	0.86	0.43	0.76	0.61
North	0.61	0.07	0.69	0.21

be readily attributed to removal of substantial numbers of larger barramundi by commercial fishing. Legal size limits for barramundi restrict the catch to fish greater than 580 mm but less than 1,200 mm TL. However, effects of fishing in terms of relative catch were apparent in size-classes below 580 mm in all systems open to commercial net fishing. These results indicate that recruitment potential in the open systems may be reduced, thereby leading to lower numbers in all size-classes (discussed further below).

In contrast to the effect of fishing on total catch and biomass, L_{bar} did not differ between open and closed systems. Because "plate-sized" barramundi are targeted by fishers due to consumer preference, larger

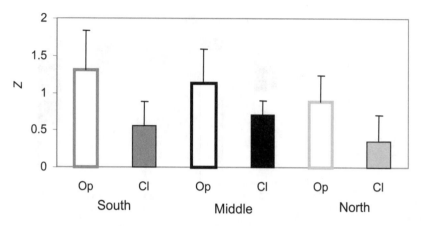

Figure 7.—Total mortality (Z) of barramundi based on length-converted catch curves for each of the six estuaries. Error bars are 95% confidence intervals. Cl = closed; Op = open.

individuals were not entirely removed by commercial net fishermen in the open systems. As a result, the L_{bar} function, which represents average relative length for individuals in a stock, did not reflect differences in relative catch between open and closed systems. Thus, even though the research teams caught greater numbers of larger barramundi in closed systems, overall average catch length did not differ from commercially fished estuaries. Clearly, L_{bar} may not necessarily be a useful indicator of MPA effectiveness.

Indicators Derived from Length-Based Analyses

Optimum length for barramundi caught in our 152-mm-mesh research nets was 739 mm, with the normal curve providing a good approximation of the catching characteristics of this net. A much lower L_o, 556 mm TL, was calculated for barramundi caught in the same mesh size in the Fly River of Papua New Guinea (Milton et al. 1998). A comparison of the hanging ratios may explain these discrepancies. While our research nets were designed with a hanging ratio of 0.40, those nets used in the Fly River had a hanging ratio of 0.67. Thus, our research net had a higher probability of catching fish by entangling (Sparre and Venema 1998). In addition, the mesh in our nets would have remained more elongated when deployed, thereby catching larger fish. The hanging ratio used in our study was typical of the commercial fishery in the northeastern Queensland region.

Total mortality rates derived for the open (commercially fished) systems were greater than in closed systems because of the effects of gill-net fishing. After subtracting the estimated natural mortality rate (M

= 0.28), fishing mortality rates F in the open estuaries increased from north to south (north $F = 0.61$, middle $F = 0.86$, south $F = 1.03$). Commercial landings were relatively greater in the central and southern parts of the study area, as reported by fishers in their logbooks (Queensland Department of Primary Industries 2003).

Mortality rates in the estuaries closed to commercial fishing derive from natural causes and recreational fishing. The lowest rate ($F = 0.07$) occurred in the northern-most closed system, where recreational fishing is dominated by tourist-based guided tour operators known to employ catch-and release angling practices. The highest recreational fishing rate ($F = 0.43$) occurred near the major urban center of Townsville (2003 population of 137,000), where recreational fishing is a popular pastime for residents.

Male fish were likely to be taken in the commercially fished systems before changing sex. Thus, extremely low numbers of females were sampled in the systems open to commercial fishing. Consequently, egg production levels were estimated to be very low in the open systems. These factors may contribute to reduced recruitment from resident breeding populations, explaining the relatively lower catches observed for all size-classes, including those below the legal limit. These findings have implications for management of the fishery in terms of establishing legal size limits. Our 152-mm-mesh research nets, similar to nets used by most regional commercial fishermen, were highly effective at capturing barramundi in the 580–815-mm length-classes (>30% retained, Figure 6a). A standard of sustainability in fisheries management is to maintain legal limits above the length correspond-

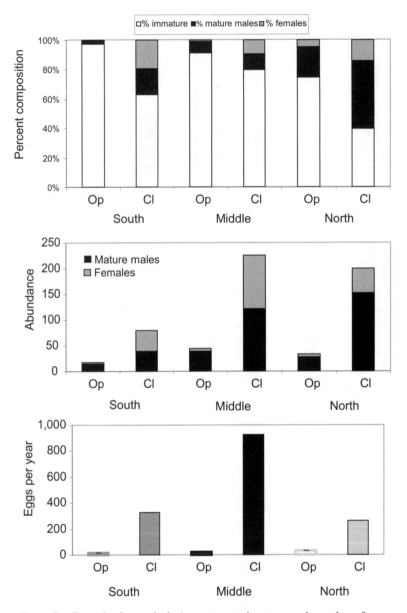

Figure 8.—Reproduction analysis (percent maturity, top panel; number of mature males and females, middle panel; and annual egg production in hundred thousands, bottom panel) for barramundi by estuary.

ing with the age at 50% maturity L_{m50} (Trenkel and Rochet 2003). Unfortunately, at the legal limit of 580 mm, only 17% of the barramundi encountering the 152-mm-mesh nets would be mature (all males, see Figure 3). Furthermore, L_{m50} would be 650 mm TL, a length at which all of these mature individuals would still be males. For protandrous hermaphrodites, establishing a different standard would seem appropriate.

For example, 50% of the mature individuals would be male and 50% female at approximately 820 mm TL or age 4.5 years (Figure 3). Given the selection factor of 4.86, a minimum mesh size of 169 mm would be more advantageous to a sustained fishery. Although this might be more biologically appropriate, barramundi caught may not be as marketable (i.e., larger than "plate sized"). This discrepancy between the best

Table 3.—Summary of egg production calculations. Egg production values are in millions.

Estuary	Region (fishing status	Raw catch	Max age in catch	Abundance	% mature males	% females	Egg production
				Unbiased estimates			
Nobbies	South (open)	19	5.2	622	2%	1%	18
Yellow Gin	South (closed)	55	7.0	215	18%	19%	327
Barrattas	Middle (open)	50	4.2	481	8%	1%	22
Haughton	Middle (closed)	169	7.5	1,112	11%	9%	927
Hull	North (open)	37	5.2	135	21%	5%	32
Russell	North (closed)	198	6.1	333	46%	14%	259
All	Open	106	5.2	1,238	7%	1%	72
All	Closed	422	7.5	1,660	17%	9%	1,513
All	Both	528	7.5	2,898	13%	6%	1,585

legal size from population dynamics versus marketability perspectives presents a dilemma for fisheries management.

If these relatively low egg production estimates in the open systems hold true, then the question arises: What is the source of recruitment to stocks in the open systems? We suggest that estuaries closed to commercial fishing, where mature barramundi are conserved, probably function as sources of recruits to neighboring fished systems (e.g., within < 100 km). Sustainability of the barramundi fishery in this region may substantially rely on recruits originating in systems closed to commercial fishing. This replenishment process in one of main benefits theoretically attributed to marine reserves on reefs (Russ 2002; Halpern 2003). Our results serve as empirical evidence supporting proposed design principles for incorporating marine reserve networks into fishery management programs (Botsford et al. 2003). For hermaphroditic species with a limited dispersal range and special habitat requirements in estuaries, the benefit of marine reserves may be especially significant—if not essential—for maintaining sustainable fisheries in neighboring systems.

Acknowledgments

The Australian Fisheries Research and Development Corporation funded this project (FRDC number 97/206). We express our appreciation to the following colleagues at the Australian Institute of Marine Science and Queensland Department of Primary Industries: A. Tobin, R. Garrett, N. Gribble, P. Dixon, R. Partridge, S. Boyle, A. Caldwell, A. Blair, C. Truscott, K. Ray, I. Ashworth, M. Cappo, D. James, T. McKenna, P. Speare, M. Vaughan, and D. Fenner.

References

Bohnsack, J. A. 1993. Marine reserves: they enhance fisheries, reduce conflicts, and protect resources. Oceanus 1993:63–71.

Botsford, L. W., F. Micheli, and A. Hastings. 2003. Principles for the design of marine reserves. Ecological Applications 13(supplement):S25–S31.

Davis, T. L. O. 1986. Biology of wildstock *Lates calcarifer* in northern Australia. Pages 22–29 *in* J. W. Copland and D. L. Grey, editors. Management of wild and cultured sea bass/barramundi (*Lates calcarifer*). Australian Centre for International Agricultural Research, Brisbane.

Davis, T. L. O., and G. P. Kirkwood. 1984. Age and growth studies on barramundi, *Lates calcarifer* (Bloch), in northern Australia. Australian Journal of Marine and Freshwater Research 35:673–689.

Froese, R., and D. Pauly, editors. 2003. FishBase. Available: *www.fishbase.org* (October 2003).

Grey, D. L. 1987. An overview of *Lates calcarifer* in Australia and Asia. Pages 15–21 *in* J. W. Copland and D. L. Grey, editors. Management of wild and cultured sea bass/barramundi (*Lates calcarifer*). Australian Centre for International Agricultural Research, Brisbane.

Halliday, I. A., J. A. Ley, A. J. Tobin, R. N. Garrett, N. A. Gribble, and D. A. Mayer. 2001. The effects of net fishing: addressing biodiversity and bycatch issues in Queensland inshore waters. Queensland Department of Primary Industries, Deception Bay, Queensland, Australia.

Halpern B. S. 2003. The impact of marine reserves: do reserves work and does reserve size matter? Ecological Applications 13(supplement):S117–S137.

Jennings, S., M. J. Kaiser, and J. D. Reynolds. 2001. Marine fisheries ecology. Blackwell Scientific Publications, London.

Kailola, P. J., M. J. Williams, P. C. Stewart, R. E. Reichelt, A. McNee, and C. Grieve. 1993. Australian fisheries resources. Fisheries Re-

236 LEY AND HALLIDAY

sources and Development Corporation, Canberra, Australia.

King, M. 1995. Fisheries biology, assessment and management. Blackwell Scientific Publications, London.

Ley, J. A., and I. A. Halliday. In press. Role of fishing closures and habitat in conserving regional estuarine biodiversity: a case study in northern Queensland, Australia. Proceedings of the world congress on aquatic protected areas. Australian Society for Fish Biology, Heidelberg, Australia.

Ley, J. A., I. A. Halliday, A. J. Tobin, R. N. Garrett, and N. A. Gribble. 2002. Ecosystem effects of fishing closures in mangrove estuaries of tropical Australia. Marine Ecology Progress Series 245:223–238.

Milton, D. A., D. Die, C. Tenakanai, and S. Swales. 1998. Selectivity for barramundi (*Lates calcarifer*) in the Fly River, Papua New Guinea: implications for managing gill-net fisheries on protandrous fishes. Marine and Freshwater Research 49:499–506.

Queensland Department of Primary Industries. 2003. CFISH database. Queensland Department of Primary Industries, Deception Bay, Queensland, Australia.

Rochet, M.-J., and V. M. Trenkel. 2003. Which community indicators can measure the impact of fishing? A review and proposals. Canadian Journal of Fisheries and Aquatic Sciences 60:86–99.

Russ, G. R. 2002. Yet another review of marine reserves as reef fishery management tools. Chapter 19 *in* P. Sale, editor. Coral reef fishes. Elsevier, New York.

Russ, G. R., D. C. Lou, J. B. Higgs, and B. P. Ferreira. 1998. Mortality rate of a cohort of the coral trout, *Plectropomus leopardus*, in zones of the Great Barrier Reef Marine Park closed to fishing. Marine and Freshwater Research 49:507–511.

Russell, D. J., and R. N. Garrett. 1988. Movements of juvenile barramundi, *Lates calcarifer* (Bloch), in north-eastern Queensland. Australian Journal of Marine and Freshwater Research 39:117–123.

Salini, J., and J. Shaklee. 1987. Stock structure of Australian and Papua New Guinean barramundi (*Lates calcarifer*). Pages 30–34 *in* J. W. Copland and D. L. Grey, editors. Management of wild and cultured sea bass/barramundi (*Lates calcarifer*). Australian Centre for International Agricultural Research, Brisbane.

Sparre, P., and S. C. Venema. 1998. Introduction to tropical fish stock assessment, part 1. FAO Fisheries Technical Paper 306.1.

Trenkel, V. M., and M.-J. Rochet. 2003. Performance of indicators derived from abundance estimates for detecting the impact of fishing on a fish community. Canadian Journal of Fisheries and Aquatic Sciences 60:67–85.

Williams, L. E. 1997. Queensland's fisheries resources: current condition and recent trends, 1988–1995. Queensland Department of Primary Industries, Brisbane, Australia.

American Fisheries Society Symposium 42:237, 2004

Abstract Only

Effects of Fishing on the Benthic Habitat and Fauna of Seamounts on the Chatham Rise, New Zealand

MALCOLM R. CLARK,[1] ASHLEY A. ROWDEN, AND STEVE O'SHEA

National Institute of Water and Atmospheric Research, New Zealand,
Private Bag 14-901, Kilbirnie, Wellington, New Zealand 6003

Abstract.—Major deepwater trawl fisheries occur for orange roughy *Hoplostethus atlanticus* on seamounts in New Zealand waters. These seamounts are often small, and trawling can be concentrated in a very localized area. Seamount habitat is thought to be productive but also fragile, and there is growing concern about effects of fishing on biodiversity and ecosystem productivity. This has prompted research to examine the nature and extent of deepwater trawling impact on seamount habitat in New Zealand. Results are presented from a recent survey where video and still imagery were applied to classify benthic habitat and an epibenthic sled was used to sample the deepwater fauna. The study took place on the Chatham Rise, where a group of eight seamounts in close proximity allowed for a spatially unconfounded comparison of replicated fished and unfished seamounts. Commercial fisheries data were analyzed to determine the amount of trawling on each. Similarities within, and differences between, fished and unfished seamounts were identified for distribution of trawl gear modification of habitat, extent of live coral, macroinvertebrate assemblage composition, taxonomic distinctness, and size spectra . This study provided information to help plan management strategies and develop effective management practices to allow both conservation and exploitation of seamounts. In May 2001, 19 seamounts throughout the New Zealand region, including several features on the Chatham Rise, were closed to bottom trawling as a precautionary measure.

---

[1] E-mail: m.clark@niwa.co.nz

American Fisheries Society Symposium 42:239, 2004

Abstract Only

Evaluation of Areas Closed to Fishing on the Continental Shelf off Nova Scotia

Kenneth T. Frank,[1] James E. Simon, and Robert K. Mohn

Bedford Institute of Oceanography, Fisheries and Oceans Canada,
Post Office Box 1006, Dartmouth, Nova Scotia B2Y 4A2, Canada

Abstract.—Areas closed to fishing, either on a seasonal or year-round basis, have been used as a regulatory measure since the 1970s in the Northwest Atlantic. Generally, such areas coincide with spawning or juvenile nursery areas and have been widely accepted by fishermen as having intrinsic biological value. We briefly review the rationale and performance of area closures directed at various fisheries, both finfish and invertebrates, on the Scotian Shelf. An analysis is presented of the year-round closed area for haddock *Melanogrammus aeglefinus*, encompassing two major offshore banks (total area of ~13,700 km^2), that was established in 1987 on the central Scotian Shelf. The management objective was to protect juvenile haddock and allow the stock to rebuild following several years of excessive discarding of undersized haddock. We evaluate changes in abundance, growth, and survival rate of juvenile haddock before and after the closure. In addition, the potential for larval spillover (seeding) due to physical processes is examined. At present, the higher concentrations of small haddock do not correspond to the closed area boundaries, and industry, while still in support of the closure, is calling for a revision of the boundaries.

[1] E-mail: FrankK@mar.dfo-mpo.gc.ca

American Fisheries Society Symposium 42:241, 2004

Abstract Only

The Use of a Refuge Area in the Restoration of Lake Trout in Parry Sound, Lake Huron

DAVID M. REID[1]

Upper Great Lakes Management Unit, Lake Huron Office, Ontario Ministry of Natural Resources, 1450 Seventh Avenue, East, Owen Sound, Ontario N4K 2Z1, Canada

DAVID M. ANDERSON AND BRYAN A. HENDERSON

Lake Huron Fisheries Research, Ontario Ministry of Natural Resources, 1450 Seventh Avenue, East, Owen Sound, Ontario N4K 2Z1, Canada

Abstract.—Parry Sound contains the only successfully rehabilitated population of lake trout *Salvelinus namaycush* in the four lower Great Lakes. This population fell to extremely low levels in the 1960s, probably due to parasitism by sea lamprey *Petromyzon marinus*. Since 1988, the relative abundance of wild spawning lake trout has increased in Parry Sound as a result of a series of key management actions (Reid et al. 2001). These actions included stocking yearling lake trout derived from the remnant stock (beginning in 1981) and stringent angler exploitation controls, including the creation of a 1,061-ha refuge in 1987. The boundaries of this protected area were established to reduce angling mortality of wild lake trout. Biotelemetry studies conducted in 1989 and 1990 led to the protected area being increased to 1,908 ha in 1990. Although the specific role that the refuge played in rehabilitating Parry Sound lake trout is difficult to assess, prior to establishing the refuge, harvest levels of wild lake trout exceeded sustainable levels and there was limited evidence of natural reproduction. From 1994 to 2001, tagging studies conducted in the refuge area showed 84% of recaptured lake trout were originally tagged within the refuge, indicating high fidelity to refuge spawning shoals between years. Differences in seasonal movements into Georgian Bay and thiamine levels (likely related to diet) between fish that spawn in the refuge area versus non-refuge fish further support the effectiveness of the refuge in protecting a relatively isolated portion of the population. The refuge-protected wild fish may have played a vital role in attracting stocked fish to suitable spawning areas. By 1997, the Parry Sound lake trout population was deemed successfully rehabilitated, and stocking was discontinued. Our data indicates that the lake trout refuge protected a significant portion of the remnant lake trout population by reducing angler-induced mortality rates and assisted in the successful rehabilitation of the population.

Reference

Reid, D. M., D. M. Anderson, and B. A. Henderson. 2001. Restoration of lake trout in Parry Sound, Lake Huron. North American Journal of Fisheries Management 21:156–169.

---

[1] E-mail: david.m.reid@mnr.gov.on.ca

American Fisheries Society Symposium 42:243–253, 2004

Dramatic Increase in the Relative Abundance of Large Male Dungeness Crabs *Cancer magister* following Closure of Commercial Fishing in Glacier Bay, Alaska

S. James Taggart[1]

U.S. Geological Survey, Alaska Science Center,
3100 National Park Road, Juneau, Alaska 99801, USA

Thomas C. Shirley[2]

Juneau Center, School of Fisheries and Ocean Sciences, University of Alaska Fairbanks,
11120 Glacier Highway, Juneau, Alaska 99801, USA

Charles E. O'Clair[3]

National Marine Fisheries Service, Auke Bay Laboratory,
11305 Glacier Highway, Juneau, Alaska 99801, USA

Jennifer Mondragon[4]

U.S. Geological Survey, Alaska Science Center,
3100 National Park Road, Juneau, Alaska 99801, USA

Abstract.—The size structure of the population of the Dungeness crab *Cancer magister* was studied at six sites in or near Glacier Bay, Alaska, before and after the closure of commercial fishing. Seven years of preclosure and 4 years of postclosure data are presented. After the closure of Glacier Bay to commercial fishing, the number and size of legal-sized male Dungeness crabs increased dramatically at the experimental sites. Female and sublegal-sized male crabs, the portions of the population not directly targeted by commercial fishing, did not increase in size or abundance following the closure. There was not a large shift in the size-abundance distribution of male crabs at the control site that is still open to commercial fishing. Marine protected areas are being widely promoted as effective tools for managing fisheries while simultaneously meeting marine conservation goals and maintaining marine biodiversity. Our data demonstrate that the size of male Dungeness crabs can markedly increase in a marine reserve, which supports the concept that marine reserves could help maintain genetic diversity in Dungeness crabs and other crab species subjected to size-limit fisheries and possibly increase the fertility of females.

Introduction

Declining fish and invertebrate stocks around the world are creating concerns about the long-term sustainability of many fisheries (Jackson et al. 2001; Stergiou 2002; Myers and Worm 2003). Fisheries in Alaska are not immune to these declines, and crustacean fisheries, in particular, are prone to serial depletion and collapse (Orensanz et al. 1998). The Alaska fishery for Dungeness crabs *Cancer magister* began in the southeastern portion of the state in 1916 and subsequently expanded to Prince William Sound, Cook Inlet, and Kodiak (Orensanz et al. 1998). The fishery is open during specified seasons and is limited to male crabs with a carapace width greater than 165 mm (Koeneman

[1] Corresponding author: jim_taggart@usgs.gov
[2] E-mail: tom.shirley@uaf.edu
[3] E-mail: cnroclair@rockisland.com; present address: 1290 Three Meadows Lane, Friday Harbor, Washington 98250, USA.
[4] E-mail: jennifer_mondragon@usgs.gov

1985). In southeastern Alaska, the Dungeness crab harvest has been characterized by large fluctuations on both annual and decadal scales. In other parts of the state, including Kodiak, Cook Inlet, and Prince William Sound, the fisheries declined during the 1980s and 1990s and have not shown significant rebounds (Orensanz et al. 1998).

Marine protected areas are being widely promoted as effective tools for managing fisheries while simultaneously meeting marine conservation goals and maintaining marine biodiversity (National Research Council 2001; Palumbi 2002). The positive effects of marine reserves on the size of individuals and increases in density, biomass, and diversity have been demonstrated in numerous studies of fish and invertebrate populations in both temperate and tropical ecosystems (for review, see Halpern 2003). Almost all of these studies, however, have focused on either coral reef or rocky-reef habitats. Data on the effectiveness of subarctic marine reserves or reserves in soft-bottom habitats are lacking.

Controlled experiments testing the impact of human exploitation on the population structure of marine species are rare (Underwood 1994, 1995) and even more unusual for crustaceans (Kelly et al. 2000). Closures of crustacean fisheries are usually prompted by major declines in the abundance of the harvested species, resulting in the collapse of those fisheries (Kruse 1993; Orensanz et al. 1998). Such closures normally remain in effect only until there is evidence that the fished stocks are rebounding; so there are limited opportunities to compare changes in the structure of crustacean populations in a closed area with comparable nearby populations still being exploited.

In 1991, the National Park Service proposed closing commercial fishing in Glacier Bay National Park, Alaska (U.S. Department of the Interior 1991). Intense negotiations among the stakeholders (National Park Service, environmental organizations, commercial fishing organizations, and the State of Alaska) ensued and were ended by Congressional action in 1998 (U.S. Congress 1998). The legislation created one of North America's largest marine reserves by closing commercial fishing in Glacier Bay. Although the closure included a phase-out period for some of the fisheries, the Dungeness crab fishery was closed immediately.

In anticipation of commercial fishery closures, we initiated a study in 1992 to document changes in the population structure of Dungeness crabs. Study sites were selected inside and outside the proposed closure areas. We collected 7 years of preclosure and 4 years of postclosure data. The many years of data provide powerful documentation of changes in population structure following the creation of the marine reserve.

Methods

Study Area

Our study area was located at the northern end of the southeastern Alaskan panhandle in and adjacent to Glacier Bay. The study area included six sites: North Beardslee Islands (58°33′N 135°54′W), South Beardslee Islands (58°33′N 135°53′W), Berg Bay (58°31′N 136°13′W), Bartlett Cove (58°27′N 135°53′W), Gustavus Flats (58°23′N 135°43′W), and Secret Bay (58°29′N 135°54′W; Figure 1). All study sites were located within Glacier Bay National Park and Preserve with the exception of Gustavus Flats, which was located adjacent to the park boundary in Icy Strait.

Glacier Bay is a large (1,312 km²), glacial fjord system with high sedimentation rates of clay-silt particles from streams and tidewater glaciers (Cowan et al. 1988). The maximum depth is approximately 450 m, and the tides are mixed semidiurnal with a maximum range of approximately 7.5 m. The primarily unconsolidated rocky coastline is highly convoluted, creating numerous small bays characterized by muddy bottoms that also commonly include sand, pebble, cobble, and shell substrates.

Selection of Study Sites

In 1991, the National Park Service proposed regulations to close commercial fishing in Glacier Bay National Park after a 7-year phase-out period (U.S. Department of the Interior 1991). At the time, the most likely scenario was that commercial fishing would initially close in the Park Wilderness waters (Wilderness Act of 1964, Public Law 88-577, 78 Stat. 890, 88th Congress, 3 September 1964), followed by closure of the rest of Glacier Bay. We initially selected five study sites; two sites were located in the Beardslee Islands wilderness area and two were located in non-wilderness waters. A fifth site was selected outside of the park to increase the odds that at least one of our study sites would remain open to commercial fishing.

By 1997, the Dungeness crab fishery had still not been closed, and, at that time, it appeared that portions of the wilderness waters would remain open to commercial fishing. To increase the probability that at

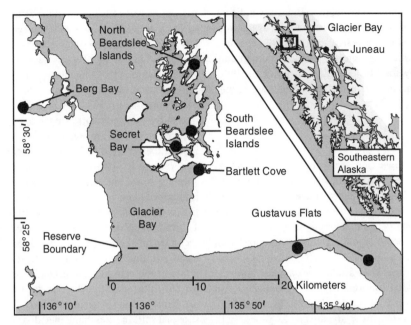

Figure 1.—Map of the study area showing the experimental sites (Bartlett Cove, Secret Bay, South Beardslee Islands, North Beardslee Islands, and Berg Bay) inside Glacier Bay and the control site (Gustavus Flats) outside the bay.

least two study sites would be located in areas that closed to commercial fishing, we added an additional site, Secret Bay, located in the southern portion of the Beardslee Islands wilderness area. When the closure decision was finalized in 1998 (U.S. Congress 1998), all of Glacier Bay immediately closed to commercial Dungeness crab fishing. The only study site that remained open to fishing was located outside of Glacier Bay (Gustavus Flats).

It is important to clarify that sport fishing and personal use fishing are allowed in Glacier Bay. Thus, it would be incorrect to classify Glacier Bay as a no-take marine reserve. However, only two of the study sites, Bartlett Cove and Gustavus Flats, are accessible via a boat ramp and a road from park headquarters and the town of Gustavus; the potential population of people participating in sport harvest is low (the population of Gustavus is about 400 people).

Sampling Procedures

Sampling was conducted from 1992 to 2002 during September when the commercial fishery is closed (the summer fishery is from 15 June to 15 August and the winter fishery is from 1 October to 30 November). Crabs were sampled with commercial crab pots (0.91

m in diameter, 0.36 m tall, with 5-cm-wire mesh). Escape rings were sealed with webbing on each pot to retain smaller crabs. Pots were baited with hanging bait (comprised of chum salmon *Oncorhynchus keta*, Pacific cod *Gadus macrocephalus*, or Pacific halibut *Hippoglossus stenolepis*) and bait jars (filled with herring and squid). Pots were soaked for 24 h. Within each study site, we set 25 pots in shallow water (0–9 m) and 25 in deeper water (10–25 m). Pot locations within the study area were placed in prime Dungeness crab habitat using knowledge from a local fisherman. The pots were set in strings parallel to shore at intervals of approximately 100 m. Each year, we attempted to place the pots in the same locations using a global positioning system (GPS; PLGR+96, Rockwell Collins, Cedar Rapids, Iowa). We estimate that the pots were set within 20 m from the original waypoints. Water depth (standardized to mean lower low water), set and retrieval time, and GPS location were recorded for each pot. Water temperature and salinity profiles were measured at each study site during each sampling period with a profiling conductivity-temperature-depth probe (SBE-19 SEA-CAT, Sea-Bird Electronics, Bellevue, Washington).

As the pots were retrieved, we counted and identified all organisms. We recorded the sex, carapace width, shell condition, reproductive condition, and appendage damage for all crabs. Carapace width was measured to the nearest mm immediately anterior to the 10th anterolateral spine with vernier calipers (Shirley and Shirley 1988; Shirley et al. 1996). All organisms were returned to the water where they were captured.

Data Analysis

We plotted average catch-per-pot as a time series, split by females, legal-sized males, and sublegal-sized males. We calculated product-moment correlation coefficients between year and average catch for the years before and the years after the fishery closure. Individual pots were the sampling unit for the correlation coefficient calculations.

For the experimental sites, we compared the size abundance distributions between preclosure and postclosure years. We also calculated the size-abundance distributions by study site and by year at both the experimental sites and the control site. Size distributions were grouped into 5-mm size-classes.

For each study site and year, we compared the relationship between average catch-per-pot and the average size for both males and females. Individual crabs were the sampling unit for size, and pot was the sampling unit for average catch. The number of pots sometimes deviated from 50 when a pot was lost or when the degradable cotton string securing the pot lid broke (range = 44–50 pots). We quantified the relationship before and after the fishery closure with a product-moment correlation coefficient. We examined the response of individual sites to closure of the fishery by plotting the data as a time series.

We used StatView (SAS Institute Inc., Cary, North Carolina) and Access (Microsoft Corp., Redmond, Washington) for all statistical tests and calculations.

Results

During the preclosure phase of the study (1992–1998), there was a significant decline in the average catch-per-pot for all size-classes and sex-classes (legal-sized males, sublegal-sized males, and females) in both the five experimental study sites and the one control study site (Figure 2). The experimental and control sites responded differently to the commercial fishery closure (Table 1). At the control site, the number of females caught per pot continued to decline but the average catch

per pot increased for both sublegal males and legal males. In the experimental sites, sublegal males did not have a significant trend while females continued to decline; in contrast, the average catch per pot of legal-sized males increased dramatically (Figure 2A).

After the closure of commercial fishing, a dramatic shift occurred in the size-abundance distribution of the male population at the experimental sites (Figure 3). During the preclosure phase of the study, the number of male crabs over 165 mm (legal size) was relatively small compared to the number of sublegal-sized males. After the fishery closure, the number of male crabs over 165 mm began increasing, and by 2000, the number of crabs larger than 170 mm exceeded the highest abundance we had recorded during any of the 7 preclosure years. This trend continued and in each subsequent year the number and size of male crabs increased. In contrast, neither males at the control sites (Figure 4) nor females at the experimental sites (Figure 3) have shown a shift in size-abundance distribution toward larger-sized crabs.

Before the closure of the commercial fishery, there was a significant negative relationship ($R = -0.563$, $P = 0.0009$) between average size of male crabs per pot and average catch of male crabs per pot (Figure 5). After the fishery closure, this relationship reversed and became significantly positive ($R = 0.685$, $P = 0.0009$). The response at individual sites, however, was highly variable (Figure 6). The average catch per pot and the average size of crabs increased at the South Beardslee Island study site beyond the range of values we had observed before the closure (Figure 6A). The average catch per pot and the average size dramatically increased at the Secret Bay study site, although there were only 2 years of preclosure data for this comparison (Figure 6C). The average size of the Bartlett Cove crabs also increased beyond values we had previously observed, but the average catch was within the range of values we had measured during the preclosure study (Figure 6E). In contrast to the other experimental sites, North Beardslee Islands and Berg Bay did not exhibit a strong pattern after the fishery closure (Figures 6B, D). The postclosure sizes and abundances remained at the lower end of the range of values we had observed before the closure. At the control site, Gustavus Flats, the average catch of male crabs increased between 2001 and 2002, but all of the postclosure values were within the range we had observed prior to the closure (Figure 6F).

No significant relationship existed between average size of female crabs per pot and the average

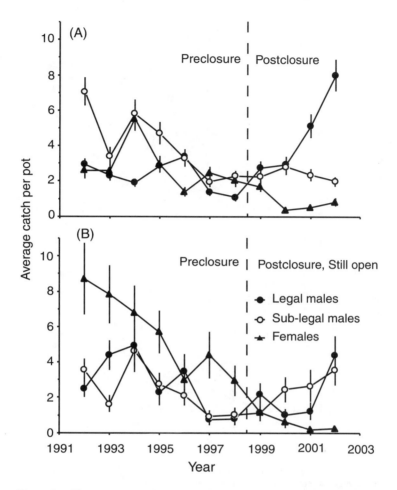

Figure 2.—The average catch per pot of legal-sized male crabs (≥165 mm), sublegal-sized males, and females from 1992 through 2002 at (A) the five experimental sites and (B) the control site. Vertical bars are 95% confidence intervals. The dashed line indicates when the commercial fishery for Dungeness crabs closed.

catch per pot before ($R = -0.333$, $P = 0.07$) or after ($R = 0.316$, $P = 0.177$) the fishery closure (Figures 5C, D). Some variation occurred among sites (Figures 6G, H, I, J, K, L), however, none of the sites had increases in either catch or size after the closure.

Discussion

After the closure of Glacier Bay to commercial fishing, the number and size of legal-sized male Dungeness crabs increased dramatically at the experimental

Table 1.—Correlation coefficient between average catch and year (split before and after the closure of the commercial fishery) for legal-sized males (>= 165 mm), sublegal-sized males, and females.

Category	Experimental sites		Control site	
	Preclosure	Postclosure	Preclosure	Postclosure, still open
Legal males	−0.205 ($P < 0.0001$)	0.364 ($P < 0.0001$)	−0.304 ($P < 0.0001$)	0.272 ($P < 0.0001$)
Sublegal males	−0.328 ($P < 0.0001$)	−0.520 ($P = 0.1041$)	−0.300 ($P < 0.0001$)	0.297 ($P < 0.0001$)
Females	−0.123 ($P < 0.0001$)	−0.157 ($P < 0.0001$)	−0.367 ($P < 0.0001$)	−0.296 ($P < 0.0001$)

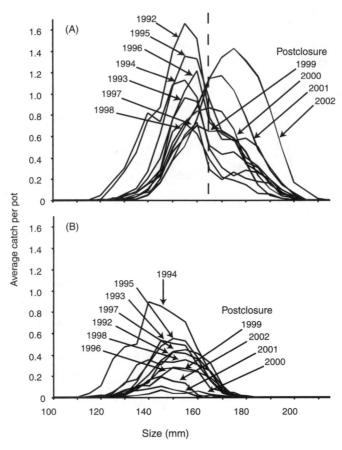

Figure 3.—Average size abundance distribution for (A) males and (B) females at all the experimental sites combined. Each line represents a year from 1992 to 2002. The dashed line represents the legal size limit (165 mm) for male Dungeness crabs.

sites (Figures 2, 3). The shift in the size-abundance distribution toward larger crabs has continued in the postclosure years.

The key question is whether the shift in the abundance distribution of male crabs was caused by a release from commercial fishing mortality; several lines of evidence support this hypothesis. First, there were large changes in abundance and considerable differences among study sites during the 7 preclosure years; however, we never observed high numbers of males larger than the legal size limit at any of the study sites before the closure. Second, female and sublegal male crabs, the portions of the population not directly targeted by commercial fishing, did not increase in size or abundance following the closure. Thus there is no

evidence that a strong recruitment event occurred and caused the sudden increase in large males.

While commercial fishing was occurring there was a significant negative correlation between average size and average catch per pot (Figure 5). This relationship switched to a positive correlation after the closure, indicating that commercial fishing influenced the initial relationship. For females, no significant correlation existed between average size and average catch per pot before or after the commercial fishing closure. Since females are not harvested, they serve as a within-site control and provide further support that fishing mortality drove the preclosure male correlations. Finally, although the number of legal males did increase in the control site between 2001 and 2002, there was

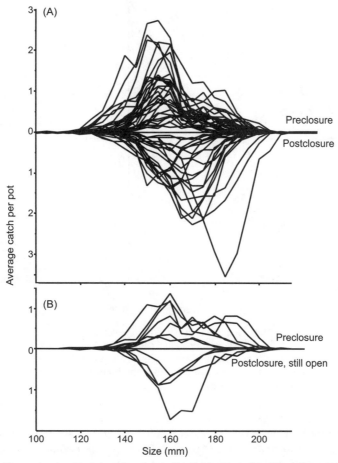

Figure 4.—Average size abundance distribution for male Dungeness crab at the (A) experimental sites and (B) control site. The distributions above the zero line represent data from each site in each year prior to the closure of the fishery. The distributions below the zero line show data from each site in each year after the fishery closure.

not a large shift in the size-abundance distribution of male crabs at this site (Figure 4).

High variation occurred among the study sites inside the park (where commercial fishing was closed) before and after the closure, and the dramatic shift in the size-abundance distribution (Figures 3, 4) was driven by three of the five study sites: South Beardslee Island (Figure 6A), Secret Bay (Figure 6C), and Bartlett Cove (Figure 6E). The variation among sites demonstrates the importance of multiple study sites both inside and outside the reserve when testing reserve effectiveness (Underwood 1994). In addition to not having multiple control sites due to the evolving nature of the proposed closures, our single control was compromised for an additional reason. When Congress passed the legislation that closed fishing in the

park, the commercial fishermen who held permits for Glacier Bay were bought out and their permits were retired. Most of the Dungeness crab fishermen who fished in the park also fished outside of the park at Gustavus Flats. Consequently, commercial fishing at Gustavus Flats was reduced until new fishermen recruited to the area.

In the presence of commercial fishing, we did not find areas that had both high crab abundance and large mean size. At Bartlett Cove, crabs were abundant but the average size of male crabs was small (Figure 6E); at Berg Bay, the average size of crabs was large but the average catch per pot was consistently low (Figure 6D). In the absence of commercial fishing, several sites exhibited high abundance and large mean size. In other studies, as the exploitation rate for

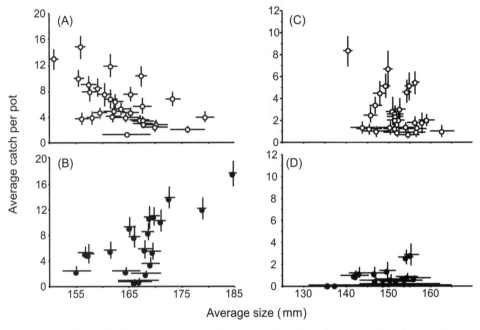

Figure 5.—Relationship between average catch per pot of males and average size of males for each site in each year (A) before the fishery closure and (B) after the fishery closure. Relationship between average catch per pot of females and average size of females for each site in each year (C) before the fishery closure and (D) after the fishery closure. Horizontal and vertical bars are 95% confidence intervals.

a size-limited fishery increased, the mean size of the individuals in the fished population decreased (Abbe and Stagg 1996). Based on this relationship, we hypothesize that the sites with higher catch per pot attracted intense commercial fishing effort, which resulted in the maximum size converging on the legal size limit, and this caused the observed lower mean sizes. Conversely, sites with large mean size and low average abundance, such as Berg Bay, did not change markedly after the closure. This low-abundance area may not have been attractive to commercial fishing; therefore, fishing effort was low. Thus, closing commercial fishing did not result in a large change.

In this paper, we have presented strong evidence that the size structure of male Dungeness crabs changed in the absence of commercial fishing. The increase in the number of large male crabs in the reserve is striking and could have other important population effects. One possible outcome of the population structure changes, for example, is an increase in female fertility. In the preclosure phase of the study, no female Dungeness crabs larger than 179 mm were ovigerous, and it is possible that this is due to the limited availability of large male crabs (Swiney et al.

2003). In Dungeness crabs, (Smith and Jamieson 1991) blue crabs *Callinectes sapidus* (Kendall et al. 2001, 2002), Tanner crabs *Chionoecetes bairdi* (Paul 1984), and snow crabs *Chionoecetes opilio* (Sainte-Marie et al. 2002), a population with more females than males or small-sized males can have lower quality or availability of sperm resources. A reduction in the abundance and size of males due to commercial harvesting may decrease fertility for females. Marine reserves may, therefore, increase the fertility of females where males are more abundant or larger than in commercially fished locations. Since Dungeness crabs have meroplanktonic larvae that are able to disperse during the pelagic phase (Epifanio 1988; McConnaughey et al. 1992), higher female fertility inside marine reserves could also result in an increase in larval export to areas outside of the reserve (Shanks et al. 2003).

In addition to reductions in fertility, fisheries that remove most of the large individuals from a population can select against genotypes that promote fast growth (Reznick et al. 1990; Conover and Munch 2002), and slower growth can reduce productivity of fisheries (Conover and Munch 2002). If reserves protect adult animals, they have also protected the oppor-

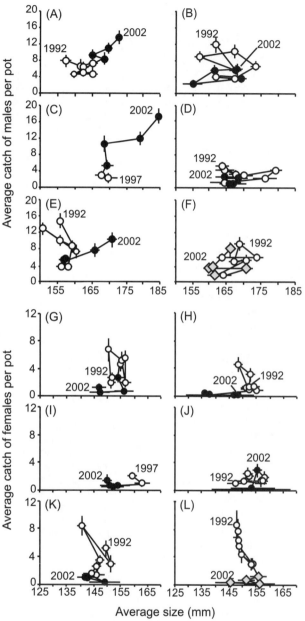

Figure 6.—Relationship between average catch per pot of males and average size of males for each year in (A) South Beardslee Islands, (B) North Beardslee Islands, (C) Secret Bay, (D) Berg Bay, (E) Bartlett Cove, and (F) Gustavus Flats. Relationship between average catch per pot of females and average size of females for each year in (G) South Beardslee Islands, (H) North Beardslee Islands, (I) Secret Bay, (J) Berg Bay, (K) Bartlett Cove, and (L) Gustavus Flats. Open circles are samples collected before the fishery closure, black circles are samples collected after the fishery closure, and gray diamonds are samples from the control site after the fishery closure. Horizontal and vertical bars are 95% confidence intervals. The first and last samples are labeled and the lines between the data points indicate the chronology.

252   TAGGART ET AL.

tunity for adults to grow to a larger size. Large males
have higher reproductive success than do small males,
so the genetic consequences of commercial fishing
could potentially be mitigated by strategically located
marine reserves (Trexler and Travis 2000; National
Research Council 2001). Data that we have presented
demonstrate that the size of male Dungeness crabs
can markedly increase in a marine reserve, which supports the concept that marine reserves could help maintain genetic diversity in Dungeness crabs and other
crab species subjected to size-limit fisheries.

Acknowledgments

This long-term study was made possible by the support of a large number of people. J. de La Bruere made
the field work efficient and enjoyable through his expert ability to operate the R/V *Alaskan Gyre*. We thank
A. Andrews and J. Nielsen for large efforts in data
management and analysis. L. Herron helped produce
the graphics. E. Mathews, A. Andrews, and J. Nielsen
provided critical review of the manuscript. G. Bishop,
C. Dezan, B. Foster, E. Hooge, P. Hooge, J. Luthy,
C. Schroth, and L. Solomon each participated in the
project for several years. D. Hart, E. Leder, W. Park,
K. Scheding, and K. Swiney conducted thesis research
that was integral to aspects of this project. We thank
T. Lee and E. Knudsen for their continued support of
long-term studies at Glacier Bay National Park and
Preserve. We especially thank the large number of
unnamed graduate students, faculty, and state and federal agency researchers, over 70 people total, who gen-
erously donated their time and efforts to this long-term
project. This project was funded by the U.S. Geological Survey and the National Park Service.

References

Abbe, G. R., and C. Stagg. 1996. Trends in blue crab
(*Callinectes sapidus* Rathbun) catches near
Calvert Cliffs, Maryland, from 1968 to 1995 and
their relationship to the Maryland commercial fishery. Journal of Shellfish Research 15:751–758.
Conover, D. O., and S. B. Munch. 2002. Sustaining
fisheries yields over evolutionary time scales. Science 297:94–96.
Cowan, E. A., R. D. Powell, and N. D. Smith. 1988.
Rainstorm-induced event sedimentation at the
tidewater front of a temperate glacier. Geology
16:409–412.
Epifanio, C. E. 1988. Transport of invertebrate larvae between estuaries and the continental shelf.

Pages 104–114 *in* M. P. Weinstein, editor. Larval
fish and shellfish transport through inlets. Ameri-
can Fisheries Society, Symposium 3, Bethesda,
Maryland.
Halpern, B. S. 2003. The impact of marine reserves:
do reserves work and does reserve size matter?
Ecological Applications 13:S117–S137.
Jackson, J. B. C., M. X. Kirby, W. H. Berger, K. A.
Bjorndal, L. W. Botsford, B. J. Bourque, R. H.
Bradbury, R. Cooke, J. Erlandson, J. A. Estes, T.
P. Hughes, S. Kidwell, C. B. Lange, H. S. Lenihan,
J. M. Pandolfi, C. H. Peterson, R. S. Steneck, M.
J. Tegner, and R. R. Warner. 2001. Historical over-
fishing and the recent collapse of coastal ecosys-
tems. Science 293:629–638.
Kelly, S., D. Scott, A. B. MacDiarmid, and R. C.
Babcock. 2000. Spiny lobster, *Jasus edwardsii*,
recovery in New Zealand marine reserves. Bio-
logical Conservation 92(3):359–369.
Kendall, M. S., D. L. Wolcott, T. G. Wolcott, and A. H.
Hines. 2001. Reproductive potential of individual
male blue crabs, *Callinectes sapidus*, in a fished
population: depletion and recovery of sperm number and seminal fluid. Canadian Journal of Fisheries and Aquatic Sciences 58:1168–1177.
Kendall, M. S., D. L. Wolcott, T. G. Wolcott, and A. H.
Hines. 2002. Influence of male size and mating
history on sperm content of ejaculates of the blue
crab *Callinectes sapidus*. Marine Ecology Progress
Series 230:235–240.
Koeneman, T. M. 1985. A brief review of the commercial fisheries for *Cancer magister* in southeast
Alaska and Yakutat waters, with emphasis on recent seasons. University of Alaska Sea Grant Report 85-3:61–76.
Kruse, G. H. 1993. Biological perspectives on crab
management in Alaska. Alaska Sea Grant College
Program Report 93-02: 355–384.
McConnaughey, R. A., D. A. Armstrong, B. M. Hickey,
and D. R. Gunderson. 1992. Juvenile Dungeness
crab (*Cancer magister*) recruitment variability and
oceanic transport during the pelagic larval phase.
Canadian Journal of Fisheries and Aquatic Sciences 49(10):2028–2044.
Myers, R. A., and B. Worm. 2003. Rapid worldwide
depletion of predatory fish communities. Nature
(London) 423:280–283.
National Research Council. 2001. Marine protected
areas: tools for sustaining ocean ecosystems. National Academy Press, Washington D.C.
Orensanz, J. M., J. Armstrong, D. Armstrong, and R.
Hilborn. 1998. Crustacean resources are vulnerable to serial depletion—the multifaceted decline
of crab and shrimp fisheries in the greater Gulf of
Alaska. Reviews in Fish Biology and Fisheries
8:117–176.
Palumbi, S. R. 2002. Marine reserves: a tool for eco-
system management and conservation. Pew
Oceans Commission, Arlington, Virginia.

Paul, A. J. 1984. Mating frequency and viability of stored sperm in the Tanner crab *Chionoecetes bairdi* (Decapoda, Majidae). Journal of Crustacean Biology 4:375–381.

Reznick, D. A., H. Bryga, and J. A. Endler. 1990. Experimentally induced life-history evolution in a natural population. Nature (London) 346:357–359.

Sainte-Marie, B., J.-M. Sevigny, and M. Carpentier. 2002. Interannual variability of sperm reserves and fecundity of primiparous females of the snow crab (*Chionoecetes opilio*) in relation to sex ratio. Canadian Journal of Fisheries and Aquatic Sciences 59:1932–1940.

Shanks, A. L., B. A. Grantham, and M. H. Carr. 2003. Propagule dispersal distance and the size and spacing of marine reserves. Ecological Applications 13:S159–S169.

Shirley, S. M., and T. C. Shirley. 1988. Appendage injury in Dungeness crabs, *Cancer magister*, in southeastern Alaska. U.S. National Marine Fisheries Service Fishery Bulletin 86:156–160.

Shirley, T. C., G. Bishop, C. E. O'Clair, S. J. Taggart, and J. L. Bodkin. 1996. Sea otter predation on Dungeness crabs in Glacier Bay, Alaska. Alaska Sea Grant College Program Report 96-02:563–576.

Smith, B. D., and G. S. Jamieson. 1991. Possible consequences of intensive fishing for males on the mating opportunities of Dungeness crabs. Transactions of the American Fisheries Society 120:650–653.

Stergiou, K. I. 2002. Overfishing, tropicalization of fish stocks, uncertainty and ecosystem management: resharpening Ockham's razor. Fisheries Research 55:1–9.

Swiney, K. M., T. C. Shirley, S. J. Taggart, and C. E. O'Clair. 2003. Dungeness crab, *Cancer magister*, do not extrude eggs annually in southeastern Alaska: an in situ study. Journal of Crustacean Biology 23:280–288.

Trexler, J. C., and J. Travis. 2000. Can marine protected areas restore and conserve stock attributes of reef fishes? Bulletin of Marine Science 66:853–873.

U.S. Congress. 1998. Omnibus consolidated and emergency supplemental Appropriations Act for FY 1999. Federal Register 63:Section 123(signed 21 October 1998):268–270.

U.S. Department of the Interior. 1991. Glacier Bay National Park, Alaska: fishing regulations. Federal Register 56:Section 13(5 August 1991):37262–37265.

Underwood, A. J. 1994. On beyond BACI: sampling designs that might reliably detect environmental disturbances. Ecological Applications 4:3–15.

Underwood, A. J. 1995. Ecological research and (and research into) environmental management. Ecological Applications 5:232–247.

Part VI:
Economic and Legal Issues

American Fisheries Society Symposium 42:257, 2004

Abstract Only

Full article in press in *Marine Resource Economics.*

Incorporating Information and Expectations in Fishermen's Spatial Decisions

RITA E. CURTIS[1]

National Marine Fisheries Service, Office of Science and Technology,
1315 East-West Highway, Number 12752, Silver Spring, Maryland 20910, USA

KENNETH E. MCCONNELL[2]

Department of Agricultural and Resource Economics, University of Maryland,
3218 Symons Hall, College Park, Maryland 20742, USA

Abstract.—Applied economic analyses conducted to date of fishermen's spatial decisions have primarily used random utility models of location choice. A common characteristic of these studies is that they typically assume that fishermen have current information on catch rates at all fishing sites in the fishery, which implies a high degree of information sharing among fishermen while at sea. Using data from the Hawaii longline fishery, this paper tests this hypothesis, analyzing whether varying assumptions on information available to fishermen for basing spatial choices affects predictions regarding those decisions.

[1] E-mail: rita.curtis@noaa.gov
[2] E-mail: kmcconnell@arec.umd.edu

American Fisheries Society Symposium 42:259–265, 2004

The Economics of Protected Areas in Marine Fisheries Management: An Overview of the Issues

Susan Hanna[1]

Department of Agricultural and Resource Economics, Oregon State University,
Corvallis, Oregon 97331-3601, USA

Abstract.—Marine protected areas (MPAs) are under increasing consideration as a fishery management tool but remain controversial. The design, implementation, and management of MPAs necessarily involve human behavior and institutions, but these components are often overlooked. Social and economic elements permeate all stages of MPA development and implementation: the design, site selection, implementation, monitoring, and enforcement. This paper presents an overview of seven key economic elements of MPAs: economic uncertainty, design processes, distributional effects, institutional placement, economic impacts, management costs, and enforcement costs. The paper assesses the extent to which these elements are receiving explicit consideration in current practice and the degree to which they are contributing to controversy in MPA implementation.

Introduction

Marine protected areas (MPAs) are increasingly promoted as a tool to protect marine ecosystems and, in fishery management, to help rebuild overexploited stocks and provide a buffer against management error (Allison et al. 1998). Executive Order 13158 (U.S. Office of the Federal Register 2000) directs federal agencies to develop a national system of MPAs by strengthening established MPAs and establishing new ones. Under the auspices of the North American Free Trade Agreement, the Commission for Environmental Cooperation (CEC) is developing a Marine Protected Area Action Plan for North America (CEC 2002). The American Fisheries Society policy statement on the management of Pacific rockfish *Sebastes* spp. also supports the establishment of a system of MPAs to protect habitat and promote rockfish population recovery (Parker et al. 2000).

The promotion of MPAs reflects concern over the status of fish populations and the quality of habitat. In fishery management, MPAs represent a precautionary approach in which protective actions are taken proactively to ensure sustainability of fishery resources and associated ecosystems (Restrepo et al.

1999; FAO 2000). Lauck et al. (1998) see MPAs as contributors to the precautionary approach through hedging against uncertainty.

Several international ocean agreements are directed toward the protection of marine ecosystems through fishery management. These include the Rome Consensus on World Fisheries, the Code of Conduct on Responsible Fisheries, and the Kyoto Declaration adopted at the Conference on the Sustainable Contribution of Fisheries to Food Security. The agreements include precautionary guidelines for reducing overfishing, reducing bycatch and discards, reducing fishing capacity, strengthening governance, and strengthening the scientific basis for ecosystem management (FAO 2000).

In its 1999 report to Congress, the National Marine Fisheries Service Ecosystems Principles Advisory Panel recommended MPAs as part of an ecosystem-based approach to fishery management (NMFS 1999). The panel also noted that existing governance systems create incentives that are incompatible with ecosystem-level goals and recommended that that the United States explore ecosystem-based approaches through fishery ecosystem plans. Ecosystem management is intended to provide a vehicle to address economic and ecological interactions at multiple scales for the protection of ecosystem structure and function (Langton and Haedrich 1997).

---

[1] E-mail: susan.hanna@oregonstate.edu

Marine protected areas have received rapid acceptance at the policy level of federal fisheries, but there is a long way between general policy statements and actual—to say nothing of effective—implementation. Marine protected areas remain extremely controversial in fisheries. Proponents of MPAs often focus on fixing ecological problems and do not address, or do not know how to address, the human dimensions of marine ecosystems.

Marine protected areas are often perceived to be ecological management tools. Although MPAs may be directed toward achieving ecological goals, their design, implementation, and management involve human concerns. As such, they necessarily incorporate human behavior and institutions. There are important economic, cultural, and institutional elements that influence how MPAs can be structured, how they function, and ultimately, how successful they will be in meeting their objectives. Neglecting these elements, human issues are often considered only in terms of social or economic impacts to be assessed after MPA design and siting has been established "scientifically." Thus, design proceeds without reflecting the context within which MPAs will be implemented. By excluding consumptive use, MPAs appear to be independent of people who are engaged in consumptive activities. But in their development and their impacts, MPAs directly involve consumptive (commercial and recreational) fishery users.

Far from being relegated to impacts, social and economic elements permeate all stages of MPA development and implementation: the design, site selection, implementation, monitoring, and enforcement. Ignoring these elements will lead to MPAs of weak design and ineffective performance. This paper presents an overview of seven key economic elements of MPAs, assessing the extent to which they are receiving explicit consideration in current practice and the degree to which they are contributing to controversy in MPA implementation.

A Regional Example: Pacific Groundfish Management

The Pacific Groundfish Fishery Management Plan covers 82 species of demersal fish, nine of which are now designated as overfished. The overfished stocks and their designation dates are lingcod *Ophiodon elongatus* (1999), Pacific ocean perch *Sebastes alutus* (1999), bocaccio *S. paucispinis* (1999), canary rockfish *S. pinniger* (2000), cowcod *S. levis* (2000),

darkblotched rockfish *S. crameri* (2001), widow rockfish *S. entomelas* (2001), yelloweye rockfish *S. ruberrimus* (2002), and Pacific hake *Merluccius productus* (also known as whiting; 2002). The Pacific Fishery Management Council (PFMC) is managing these stocks under interim rebuilding plans and is in the process of developing final rebuilding plans (PFMC 2003).

One goal of the PFMCs strategic plan for groundfish (PFMC 2000) is to increase the productivity of West Coast groundfish resources, and to do so, the plan recommends large reductions in fleet capacity across all sectors of the commercial groundfish fishery. Another goal is to use marine reserves as a groundfish fishery management tool that is integrated with other fishery management approaches and contributes to conservation in a measurable way (PFMC 2000a). The council's intent is to use marine reserves as an additional control on fishing mortality, for example, in the protection of cowcod in Southern California. The council has adopted six objectives for groundfish marine reserves (in order of priority): stock rebuilding; biological productivity; economic productivity; insurance; habitat protection; and research and education. The economic productivity objective is focused on achieving long-term production while minimizing short-term negative economic impacts on user groups (PFMC 2000b, 2002).

At the state level, Oregon's Ocean Policy Advisory Council (OPAC) began a review of marine protected areas in 2000, and after extensive consultations with scientists, marine industries, conservationists, and the public, issued a report in 2002 (OPAC 2002). The report recommends that a limited system of marine reserves be established to test their effectiveness in achieving statewide conservation goals and policies and to provide baseline information on marine environmental conditions and species. The report further recommends that marine reserves intended to enhance state or federal fishery management be left to the respective state and federal fishery management agencies.

In its recommendations, OPAC does not specify size, number, or location of sites in a limited system of marine reserves, but instead places emphasis on establishing a coherent process through which to develop these parameters. The intention of OPAC is that the process be based on clear goals and criteria; be flexible; minimize economic effects on existing uses; be based on a full range of physical, natural, and social sciences; promote research; and emphasize enforcement. The process should be a collaborative, community-based process involving all affected parties and should address the six "process principles" for marine reserves adopted by the Oregon Coastal Zone Management Association. These principles are: clear goals, impact analysis, fishing community involvement, achievable enforcement, baseline information monitoring, and adequate funding. A planning process is now in development (OPAC 2002), but some unresolved questions about how Oregon will proceed with the development of marine reserves remain.

The OPAC places a strong emphasis on ensuring that the process of MPA planning and development has transparency and integrity. Industry statements about MPAs further emphasize the importance of developing effective, inclusive processes. In a review of Oregon industry attitudes about MPAs, Anderson (2000) found a high degree of skepticism about the motives and science behind MPAs, about the potential for political manipulation of the MPA development process, and about the extent to which fishermen and their knowledge would be included in the design of MPAs. A general feeling was that the human dimensions of marine reserves would be ignored. Other industry statements on marine reserves emphasize the importance of clear definitions of their purpose, monitoring the extent to which they achieve that purpose, preventing the use of MPAs as indirect allocation mechanisms, and adequate enforcement (Grader and Spain 1999; Waldeck and Buck 1999; Zinn and Buck 2001; R. Moore, statement on marine reserves before U.S. Senate Subcommittee on Oceans and Fisheries, 2000).

Economic Issues Surrounding Marine Protected Areas

Economic elements permeate all stages of MPA development and implementation: design, site selection, implementation, monitoring, and enforcement. Seven key economic aspects of MPAs are discussed here: economic uncertainty, design processes, distributional effects, institutional placement, economic impacts, management costs, and enforcement costs.

The Effect of Economic Uncertainty

A primary economic principle of resource behavior is that uncertainty of tenure will create an incentive to make short-term decisions, and instability in the means of production will lead to investments in ca-

pacity that improve short-term competitive positions. These behaviors result in a collective level of extraction pressure that is unsustainable by the resource, unprofitable to the industry, and non-remunerative to the public owners of the resource (Hanna 1998). Implementation of an MPA within this context of uncertainty will only shift these problems into another area. Until the base incentive problem of the lack of ownership and race for fish is resolved, the ability of fishery management to improve its effectiveness or the ability of MPAs to enhance that effectiveness is problematic. Policy makers will continue to face the same overcapacity problems that have enabled the current momentum for expanding the system of MPAs (Sanchirico 2000).

Building an Effective Design Process

The ecological basis of MPA design (NRC 2000) has received much attention. Much less attention has been given to the process used in the design of MPAs, and the design process has proven to be an area of vulnerability in MPA development. Objectives for protected areas are ultimately representations of societal, not scientific, values (Agardy 2000). Although the design of an MPA affects user groups most directly, development processes may not include those interests until after the design is developed. We know from fishery management processes that there can be clear cost-effectiveness advantages to active user group participation in the development of regulations (Hanna 1995a, 1995b). And in development of MPAs, there are similar advantages: participation provides an educational process to user groups; vests them in the development process; and, in the information user groups bring to the process, provides a better informed, and often more cost-effective, result to managers (Fenton et al. 2002). Participation of user groups is also essential to the formation of a common set of objectives, without which the sustainability of an MPA will be in question. Design processes have been a controversial element of MPAs, and the ensuing political actions against MPAs can produce high levels of transactions costs.

Distributional Effects

Following the principles of effective participatory processes can help keep transactions costs within acceptable bounds. But the extent to which the content of one MPA design process can generalize to another is limited by the specificity of the fisheries context

(Lindholm and Auster 2002). An important component of that context is the economic structure of the
fishery and the distribution over space, scale, and time
of benefits earned from the fishery. That distribution
reflects the history of a fishery, and historic "shares"
are rigorously guarded. A defining characteristic of
MPA development is the expectation that users with
accustomed shares in a given area will give up the
benefits from that area in exchange for highly uncertain future returns. To the fishing industry, MPAs often look like an attempt to reallocate accustomed uses
of resources to other interests, and many perceive
MPAs as being highly subject to political manipulation (Waldeck and Buck 1999; Zinn and Buck 2001).
The allocatable nature of MPAs is a major element of
their controversy.

An important distributional issue with MPAs is
that the benefits are diffuse while costs are concentrated. Benefits and costs stem from the value held in
de facto property rights to a fishery in a particular area.
Loss of these rights is a cost that is exacerbated by a
sense of inequity when the creation of an MPA benefits one group at the expense of another (NRC 2000).
Some opportunity to mitigate distributional issues associated with MPAs could be found by choosing areas with the lowest opportunity cost to the fishery, for
example, overfished areas that are under rebuilding
plans required under the Magnuson-Stevens Fishery
Conservation and Management Act (MSFCMA;
NRC 2000; Sanchirico and Wilen 2000). When MPA
creation creates large losses to affected user groups,
opportunities for compensation can be explored. Although some might balk at the idea of compensating
users of a public resource for the loss of access to
which they have no formal rights, Rettig (1994) lists
several pragmatic advantages to compensation. They
may lead to better (uncontested) outcomes, reduce
long-run transactions costs, and even force the government to consider the opportunity costs of actions
more carefully.

Institutional Placement Effects

Fisheries and the people who work in them vary widely
in economic, cultural, and political characteristics. So
too does the institutional environment within which fisheries are managed. Although all federal U.S. fisheries
are managed under the authority of the MSFCMA as
implemented by the Department of Commerce through
the system of regional fishery management councils,
there is considerable variation in the specifics of management at the interregional and intraregional level. Fish-

eries may be managed by individual states, interstate
fisheries commissions, or regional fishery management
councils. Different suites of regulations apply to different species, gear types, and vessels. Property rights range
from nonexistent to tradable quota shares. This variation forms the institutional context for a given area, a
context that is critical in determining how the benefits
and costs of MPAs are distributed over time and space
(Sanchirico et al. 2002).

Implementation of MPAs that reflects the institutional context is most likely to be cost-effective, but
an MPA layered over the existing context without any
particular accommodation to it is likely to create new
costs and conflicts. An MPA developed independent
of a fishery and its institutional context will likely el-
evate costs for both.

Beyond developing and implementing an
MPA, achieving and sustaining its objectives may
depend critically on that part of the institutional
environment that fundamentally influences human
behavior: the system of property rights. If the fishery is de facto open access, the race for fish by an
overcapitalized fleet will erode the enhancement
benefits of the MPA. Rights-based management,
accompanied by a decrease in capacity, could make
an MPA easier to implement and sustain (Arnason
2000; NRC 2000).

Economic Impacts

Economic impacts of MPAs are the economic element
most frequently considered during MPA development,
although not usually until the end of the design process. They are also the elements most feared by industry and are often identified as the reason MPAs
should not be implemented. The impacts of an MPA,
and the relative magnitude of its benefits and costs,
will be affected by its objectives, size, location, and
allowed uses (Thomson 1998; Sanchirico and Wilen
2000). Impacts result from the spatial adjustments that
take place in response to the removal of an ocean area
from use and are felt in spatial and temporal distributions of effort, landings, and management.

One obvious economic impact of an MPA is the
displacement of fishing effort from the reserve area to
another area. Poorly sited MPAs may increase the risk
to those species subjected to the new higher-intensity
levels of effort, reducing the sustainable value of the
system of fisheries they impact (Holland 2000).
Changing the distribution of fishing effort across space,
species, or time as a result of MPA implementation
also creates costs to the system of data collection and

analysis. The current system is not oriented around spatial accounting, and the potential impacts of MPAs on data and models that currently support management should be carefully considered (Holland 2000).

Management Costs

Marine protected areas create both benefits and costs to management. Benefits include the potential for increased scientific knowledge of ecosystems, hedges against uncertain stock assessments, and educational opportunities (Sanchirico et al. 2002). Costs include those associated with data collection, performance monitoring, and enforcement. These costs are additional to routine management costs, the magnitude of which are affected by the efficiency with which it is conducted. T. McClurg (presentation from New Zealand Seafood Industry Commission Conference, 2002) notes that it is commonly assumed that a major barrier to more effective management is the poor state of knowledge about the marine environment and that as knowledge expands, levels of certainty will improve, and the quality of management will increase. But he concludes that what is really needed to produce more effective (and cost-effective) fishery management is a deeper understanding of the process of management (McClurg, presentation, 2002).

The efficiency of management is related not only to its effectiveness but also to the economic benefits that an MPA can provide. Arnason (2000) demonstrates that if the fishery management system is inefficient and there is no realistic expectation of improvement, it is difficult to see an economic justification for an MPA. This is because in an inefficiently managed fishery, economic rents are normally zero, and marine reserves induce an initial loss. The costs of imposing, adjusting, and enforcing MPAs further reduce the net economic contribution of fisheries.

Enforcement Costs

Enforcement is the hidden cost of fishery management to those who develop new management approaches. As new and controversial claimants to ocean space, MPAs carry substantial implications for costs in enforcing compliance with boundaries and rules of use. The cost of ineffective enforcement can be large, not only in preventing the realization of MPA benefits but also in the erosion of general rule compliance among a broader group external to the MPA.

Promoting compliance is an important reason to ensure that the MPA development process is as inclu-

sive and transparent as possible. Ultimately, the goal of a participatory process is to find plans that have common ground, in which user groups can gain from MPAs or in which MPA costs to the industry are relatively low (Sanchirico and Wilen 2000). When those most affected by MPA design are not given the opportunity to participate until a much later stage, they will have a direct incentive to undermine the effort (NRC 2000).

Proulx (1998) notes that estimates of resources needed to enforce MPAs are almost always too low. He argues that unrealistic estimates are based on three false assumptions: that lawful users will provide sufficient policing, that cooperating agencies will contribute to enforcement, and that aerial and surface patrols transiting the area will be frequent. He further states that the classic enforcement formula for maintaining a closed area has never succeeded. This formula is based on a reliance on an enforcement presence frequent enough to deter violations, severe penalties, and sanctions of violations. He recommends that rather than thinking only of traditional enforcement resources for MPAs, managers give more consideration to nontraditional solutions to compliance.

One approach would be to recognize that the population of users closest to the MPA is the one that bears the greatest impacts and to tailor enforcement efforts to that group. Proulx (1998) recommends a detailed threat analysis be performed in advance of a decision to create an MPA. This assessment would identify all sources of potential incursions into the marine reserve, be realistic about the existence of willful violators, identify cultural variations in ideas about using the marine environment, and seek to eliminate motivations for violations. The threat analysis, community-oriented policing techniques, and educational programs that Proulx (1998) recommends all carry costs in addition to those for existing enforcement programs.

Conclusion

Although finding a balance between environmental and economic objectives is a core purpose of fishery management, the integration of these into management planning for marine fisheries has been limited. Maintaining business competitiveness is often considered to be the concern of individual fishing and processing firms rather than of management. Achieving management efficiency while managing for environmental objectives has also received little consideration. This is particularly true in the case of MPAs, where the objective is to remove areas *from* fisheries, and the idea of integrating them with fishery management may seem foreign. However, integration holds the best prospects for efficient as well as effective implementation for both MPAs and fishery management.

Developing and managing MPAs are areas of vulnerability and controversy in the economic issues outlined above. Building MPA development and management processes that integrate economic and ecological elements is sorely needed if MPAs are to be effectively integrated into fisheries management. The integration could also help modernize the lingering management philosophy that defines marine ecosystem problems as ecological, proceeds without economic performance indicators, and applies inappropriate incentives.

This paper has presented an overview of seven key economic elements of MPAs in fisheries, assessing the extent to which they are receiving explicit consideration in current practice. The degree to which human dimensions of fisheries are anticipated, understood, and integrated in the development of MPAs is an important determinant of the ultimate effectiveness of MPAs in fisheries.

There are many driving forces behind the implementation of MPAs in the United States. The strength of these forces ensures that systems of marine protected areas will be implemented. The question is whether they will be implemented effectively or carelessly. Effective implementation of MPAs will not happen without processes that successfully involve all affected economic interests in their development. As well, they will not be sustained without well-specified management processes for monitoring, evaluation, and enforcement.

Finally, it is worth reiterating that while MPAs may create a buffer of safety for fish stock protection, they are not a solution to the fundamental problem of fishery management. That problem is the perverse behavioral incentive created by the absence of property rights. Like many other fishery management tools, MPAs treat the symptoms of the incentive problem (e.g., overexploitation of fish stocks) but do not address the underlying causes of those symptoms, such as the race for fish and fleet overcapacity. Until the basic institutional problem of fishery management is resolved, the contribution of MPAs to fishery management will be limited by the same problems that underlie current recommendations for their establishment.

Acknowledgments

This paper was funded by the National Oceanic and Atmospheric Administration Office of Sea Grant and Extramural Programs under grant number NA16RG1039 (project number RCF-10) and by appropriations made by the Oregon state legislature. The views expressed herein do not necessarily reflect the views of any of those organizations. The comments of two anonymous reviewers are appreciated.

References

Agardy, T. 2000. Information needs for marine protected areas: scientific and societal. Bulletin of Marine Science 66:876–888.

Allison, G. W., J. Lubchenco, and M. H. Carr. 1998. Marine reserves are necessary but not sufficient for marine conservation. Ecological Applications 8(Supplement 1):S79–S92.

Anderson, L. 2000. Marine reserves in Oregon: analysis of industry perceptions and strategies for improved communications. Report prepared for Environmental Defense, Oakland, California.

Arnason, R. 2000. Marine reserves—is there an economic justification? Pages 19–30 *in* R. Sumaila, editor. Economics of marine protected areas. The Fisheries Centre, University of British Columbia, Vancouver.

CEC (Commission for Environmental Cooperation). 2002. North American agenda for action: 2003–2005. A three-year program plan for the Commission for Environmental Cooperation of North America. CEC, Montreal.

FAO (Food and Agriculture Organization of the United Nations). 2000. The state of world fisheries and aquaculture, 2000. FAO, Rome.

Fenton, D. G., P. Macnab, J. Simms, and D. Duggan. 2002. Developing marine protected area programs in Atlantic Canada—a summary of community involvement and discussions to date. Pages 1413–1426 *in* S. Bondrop-Nielsen and N. W. P. Munro, editors. Managing protected areas in a changing world, proceedings of the fourth international conference on science and management of protected areas, 14–19 May 2000, Wolfville, Nova Scotia, Canada. Science and Management of Protected Areas Association, Wolfville, Nova Scotia.

Grader, Z., and G. Spain. 1999. Marine reserves: friend or foe? Pacific Coast Federation of Fishermen's Associations, Fishermen's News (February), Eugene, Oregon.

Hanna, S. 1995a. User participation and fishery management performance within the Pacific Fishery

Management Council. Ocean and Coastal Management 28(1–3):23–44.

Hanna, S. 1995b. Efficiencies of user participation in natural resource management. Pages 59–68 *in* S. Hanna and M. Munasinghe, editors. Property rights and the environment: social and ecological issues. World Bank, Washington, D.C.

Hanna, S. 1998. Institutions for marine ecosystems: economic incentives and fishery management. Ecological Applications 8(Supplement):170–174.

Holland, D. 2000. Integrating marine protected areas into dynamic spatial models of fish and fishermen. Pages 93–98 *in* R. Sumaila, editor. Economics of marine protected areas. The Fisheries Centre, University of British Columbia, Vancouver.

Langton, R. W., and R. L. Haedrich. 1997. Ecosystem-based management. Pages 153–158 *in* J. Boreman, B. S. Nakashima, J. A. Wilson, and R. L. Kendall, editors. Northwest Atlantic groundfish: perspectives on a fishery collapse. American Fisheries Society, Bethesda, Maryland.

Lauck, T. C., C. W. Clark, M. Mangel, and G. R. Munro. 1998. Implementing the precautionary principle in fisheries management through marine reserves. Ecological Applications 8(Supplement 1):S72–S78.

Lindholm, J. and P. J. Auster. 2002 Marine protected area design: toward a generalized life history approach. Pages 1126–1136 *in* S. Bondrop-Nielsen and N. W. P. Munro, editors. Managing protected areas in a changing world, proceedings of the fourth international conference on science and management of protected areas, 14–19 May 2000, Wolfville, Nova Scotia, Canada. Science and Management of Protected Areas Association, Wolfville, Nova Scotia.

NMFS (National Marine Fisheries Service). 1999. Ecosystem based fishery management. A report to Congress by the Ecosystems Principles Advisory Panel as mandated by the Sustainable Fisheries Act amendments to the Magnuson-Stevens Fishery Conservation and Management Act 1996. NMFS, Washington, D.C.

NRC (National Research Council). 2000. Marine protected areas: tools for sustaining ocean ecosystems. National Academy Press, Washington, D.C.

OPAC (Oregon Ocean Policy Advisory Council). 2002. Oregon and marine reserves. Report and recommendations to the Governor, August 16. OPAC, Portland, Oregon.

Parker, S. J., S. A. Berkeley, J. T. Golden, D. R. Gunderson, J. Heifetz, M. A. Hixon, R. Larson, B. M. Leaman, M. S. Love, J. A. Musick, V. M. O'Connell, S. Ralston, H. J. Weeks, and M. M.

Yoklavich. 2000. Marine stocks at risk: management of Pacific rockfish. American Fisheries Society policy statement 31d. Fisheries 25(3):22–29.

PFMC (Pacific Fishery Management Council). 2000a. Transition to sustainability: groundfish fishery strategic plan, October. PFMC, Portland, Oregon.

PFMC (Pacific Fishery Management Council). 2000b. Marine reserves to supplement management of West Coast groundfish resources: draft report, July. PFMC, Portland, Oregon.

PFMC (Pacific Fishery Management Council). 2002. Information sheet: marine reserves. PFMC, Portland, Oregon.

PFMC (Pacific Fishery Management Council). 2003. Information sheet: questions and answers about rebuilding plans. PFMC, Portland, Oregon.

Proulx, E. 1998. The role of law enforcement in the creation and management of marine reserves. NOAA Technical Memorandum NOAA-TM-NMFS-SWFSC-255:74–77.

Restrepo, V. R., P. M. Mace, and F. M. Serchuk. 1999. The precautionary approach: a new paradigm or business as usual? NOAA Technical Memorandum NMFS-F/SPO-41:61–70.

Rettig, R. B. 1994. Who should preserve the marine environment? Marine Resource Economics 9:87–94.

Sanchirico, J. N. 2000. Marine protected areas as fishery policy: a discussion of potential costs and benefits. Resources for the Future, Discussion paper 00-23, Washington, D.C.

Sanchirico, J. N., K. A. Cochran, and P. M. Emerson. 2002. Marine protected areas: social and economic implications. Resources for the Future, Discussion paper 02-26, Washington, D.C.

Sanchirico, J. N., and J. E. Wilen. 2000. The impacts of marine reserves in limited entry fisheries. Pages 212–222 *in* R. Sumaila, editor. 2000. Economics of marine protected areas. The Fisheries Centre, University of British Columbia, Vancouver.

Thomson, C. 1998. Evaluating marine harvest refugia: an economic perspective. NOAA Technical Memorandum NOAA-TM-NMFS-SWFSC-255:78–83.

U.S. Office of the Federal Register. 2000. Executive Order 13158. Federal Register 65:105(31 May 2000):34909.

Waldeck, D. A., and E. H. Buck. 1999. The Magnuson-Stevens Fishery Conservation and Management Act: reauthorization issues for the 106th Congress. Congressional Research Service, Report RL30215, Washington, D.C.

Zinn, J., and E. H. Buck. 2001. Marine protected areas: an overview. Congressional Research Service, Report February 8, 2001, Washington, D.C.

American Fisheries Society Symposium 42:267–273, 2004

Taking a Risk: Legal Challenges and Issues for No-Take Zones

KRISTEN M. FLETCHER[1]

Roger Williams University School of Law, 10 Metacom Avenue,
Bristol, Rhode Island 02809, USA

Abstract.—Preserving public lands through restrictive uses is not a new concept in resource management; it has been recognized in U.S. law for over 100 years. Generally, however, these efforts have been directed to areas above the high water mark. In recent years, restricting access or harvest in marine areas has increased as fisheries managers try to facilitate the rebuilding of fish stocks. With the increase in use of marine protected areas and their "controversial progeny," the no-take marine reserve, questions regarding the legal underpinnings of these zones—the authority necessary to create and manage them—as well as other potential legal challenges have arisen. This paper focuses on three potential legal challenges to marine protected areas and no-take zones, including the authority to create the zone, the environmental review required, and private property-related challenges.

Introduction

In recent years, as fisheries managers struggle with traditional quota limitations and seasonal restrictions to facilitate the rebuilding of fish stocks, the use of marine protected areas (MPAs) and their "controversial progeny" (Brax 2002), the no-take marine reserve, is increasing in use both domestically and internationally. The no-take zones preclude fishing activity on some or all species with a goal of protecting habitat, rebuilding stocks, and enhancing fishery yield, and sometimes prohibiting fishing and the removal or disturbance of any living or nonliving marine resource. Preserving public lands through restrictive uses is not a new concept in resource management; it has been recognized in U.S. law for over 100 years. Generally, however, these efforts have been directed to areas above the high water mark.

With the increase in use, questions regarding the legal underpinnings of no-take zones, the authority necessary to create and manage no-take zones, as well as other potential legal challenges have arisen. This paper focuses on three potential legal challenges to no-take marine zones, including (1) the authority to create the zone, (2) the environmental review required, and (3) property-related challenges.

Marine Protected Area Authority

While the terms "marine reserves" and "marine protected areas" have been used synonymously, the terms can actually refer to marine areas that serve different functions and are governed by distinct regulations. For instance, one MPA may restrict certain types of fishing gear while another may prohibit fishing altogether. Categories of protected areas can range from strictly protected wilderness areas to multiple-use areas. Such areas are often proposed as components of fisheries management to enhance the long-term sustainable exploitation of fishery resources or to rebuild depleted stocks and protect particularly delicate areas or previously exploited areas. For example, fishery management councils have proposed closing areas to assist in the rebuilding of particular species of managed fish.

Federal and state governments use MPAs to preserve habitat or manage valuable marine ecosystems but must have a legal right or authority to preclude certain uses or activities. At a state level, depending upon the state agency overseeing the creation of an area and the location of the area, enabling legislation may be necessary to establish the authority to create and manage an MPA. While some states may need to adopt statutes, in other cases, legislation may exist that grants authority to particular agencies or subagencies (see California Marine Life Protection Act, California Fish and Game, section 2856–2863). At a federal

[1] E-mail: kfletcher@rwu.edu

level, authority to create an MPA exists under a number of statutes resulting in overlapping jurisdiction by federal agencies.

In an attempt to connect these efforts, President Clinton signed Executive Order 13158 in May 2000, calling for the creation of a comprehensive system of MPAs, defining an MPA as "any area of the marine environment reserved by Federal, State, territorial, tribal, or local laws or regulations to provide lasting protection for part or all of the natural and cultural resources therein," potentially including many sites in state waters (U.S. Office of the Federal Register 2000). The order found that an "expanded and strengthened comprehensive system of marine protected areas throughout the marine environment would enhance the conservation of our Nation's natural and cultural marine heritage and the ecologically and economically sustainable use of the marine environment for future generations" (U.S. Office of the Federal Register 2000). The order directs the Department of Commerce and Department of the Interior, in consultation with other federal agencies, to develop a national system of MPAs. In addition, federal agencies must consult regarding the impacts of their activities on MPAs and avoid causing harm to MPAs (sections 3–5). The federal agencies and authorities are provided in several major statutes.

National Marine Sanctuaries Act

The National Marine Sanctuaries program was created in 1972 as part of the Marine Protection, Research, and Sanctuaries Act (MPRSA, U.S. Code,[2] title 16, sections 1401–1445). The purpose of the program is to identify marine areas of special national or international significance due to their resource or human-use values. The statute also provides authority for comprehensive management of the areas where existing regulatory authority is inadequate to assure coordinated conservation and management. National or international significance is determined by assessment of the area's natural resource and ecological qualities, including its contribution to biological productivity, maintenance of ecosystem structure, maintenance of ecologically or commercially important or threatened species or species assemblages, maintenance of critical habitat of endangered species, and

the biogeographic representation of the site (MPRSA, section 1433[b][1][A]).

The act particularly identifies "the importance of maintaining natural biological communities... and protect[ing]... natural habitats, populations, and ecological processes" (MPRSA, section 1431[b][3]). Designation of a marine area as a sanctuary, does not prohibit, in itself, all development but does require special use permits from the Department of Commerce to authorize specific activities that are compatible with the purposes of the sanctuary (MPRSA, section 1440).

Through 1988 and 1992 amendments, the National Oceanic and Atmospheric Administration (NOAA) was given authority to review federal agency actions that may affect a sanctuary resource, and enforcement and liability provisions were added that give sanctuary designation and sanctuary management plans greater authority. The amendments provide that it is unlawful to: (1) destroy, cause the loss of, or injure any sanctuary resource managed under law or regulations for that sanctuary; and (2) possess, sell, offer for sale, purchase, import, export, deliver, carry, transport, or ship by any means any sanctuary resource taken in violation of this section (MPRSA, section 1436).

A "sanctuary resource" is "any living or nonliving resource... that contributes to the...value of the sanctuary" (MPRSA, section 1432[8]). The amendments create a rebuttable presumption that all sanctuary resources on board a vessel were taken in violation of the act or regulations (MPRSA, section 1437[e][4]). Enforcement authorities are granted broad powers to board, search, and seize vessels and impose penalties of up to US$100,000 per violation per day (MPRSA, section 1437). In addition, persons damaging or injuring any sanctuary resources are liable for response costs and damages, with retention of damage awards for restoration work (MPRSA, section 1442). The act also allows NOAA to review any federal agency action that might impact a sanctuary resource and requires NOAA to review and revise sanctuary management plans every 5 years (MPRSA, sections 1434[d]–1434[e]).

There are currently 14 designated sanctuaries. The designation process for marine sanctuaries has been streamlined, but Congress and the executive branch have also accelerated the process by designating or ordering the designation of certain sanctuaries. Because designation of large areas affects numerous user groups, conflict management has become an important part of program development and implementation.

---

[2] U.S. Code available at *www4.law.cornell.edu/uscode/* (April 2004).

National marine sanctuaries are part of the larger national system of marine protected areas, as mandated under Executive Order 13158. The many different categories of marine protected areas may correspond with the purposes set forth at the time of establishment or may relate to the entity in charge of managing the area. For example, national marine sanctuaries are included as one type of marine protected area already existing under federal law. Zones within the borders of a national marine sanctuary may be created and regulated in different fashions (i.e., particular activities, including fishing, biodiversity conservation, recreation, and oil and gas exploration, may be assigned to sectioned-off areas of the map, and each area may then have its own rules and regulations attached). Some national parks are also considered marine protected areas; they may actually be located within the boundaries of a national marine sanctuary; for example, Channel Islands National Park is within Channel Islands National Marine Sanctuary. State governments may establish marine protected areas in the 4.827-km (3-mi) limit of the state territorial waters and may work with the federal government to manage larger marine protected areas that span both federal and state waters.

Magnuson-Stevens Act

The Magnuson–Stevens Act (U.S. Code, title 16, sections 1801–1883) was passed in 1976 as the Fishery Conservation and Management Act to conserve and manage the fishery resources found off the U.S. coasts and to prevent foreign fishing within 321.8 km (200 mi) of the U.S. coast. The act set up the management scheme that is still used in U.S. fisheries management.

Reacting to heavy fishing by foreign vessels off U.S. coasts, Congress passed the Fishery Conservation and Management Act and claimed sovereign rights and exclusive fishery management authority over all fishery resources within 321.8 km (200 mi) of the coast. It was renamed the Magnuson Fishery Conservation and Management Act in 1980 and the Magnuson–Stevens Fishery Conservation and Management Act (FCMA) in 1996.

The FCMA established national standards for fishery conservation and management in U.S. waters. The FCMA established eight regional fishery management councils composed of state officials with fishery management responsibilities, the regional administrators of the National Marine Fisheries Service, and individuals appointed by the Secretary of Commerce who are knowledgeable regarding the conservation and management, or the commercial or recreational harvest, of the fishery resources of the particular geographical area. Additional nonvoting members include representatives from the U.S. Coast Guard, U.S. Fish and Wildlife Service, Department of State, and regional interstate fisheries commissions (Atlantic States Marine Fisheries Commission, Gulf States Marine Fisheries Commission, and Pacific States Marine Fisheries Commission).

The councils are responsible for preparing and amending fishery management plans for each fishery under their authority that requires conservation and management. The plans must follow the National Standards set out in the act. The National Standards are included in the text of the act (U.S. Code, title 16, section 1851).

Fishery management plans, submitted to the Secretary of Commerce for approval, describe the fisheries and contain necessary and appropriate conservation and management measures, and once approved, the Secretary of Commerce promulgates implementing regulations. The Secretary of Commerce may prepare secretarial fishery management plans if the appropriate council fails to develop such a plan.

The Sustainable Fisheries Act, adopted in 1996, reauthorized and amended the Magnuson–Stevens Act by adding three new National Standards (for a total of 10) and essential fish habitat (EFH) provisions. Fishery management plans have now been amended to include the essential fish habitat for each managed species and proposed protection and conservation measures for the habitat.

Antiquities Act

The Antiquities Act (U.S. Code, title 16, sections 431–433) was originally passed in response to vandalism occurring at the Casa Grande ruins in Arizona. The purpose of the act was to protect objects of historic or scientific interest on public lands. It authorizes the President to designate historic landmarks and structures as national monuments and provides penalties for people who damage these historic sites. Presidents have used the Antiquities Act for decades to set aside valuable public lands, including submerged lands. In general, U.S. courts interpret the President's discretion broadly.

The act has two main components: (1) a criminal enforcement component, which provides for the prosecution of persons who appropriate, excavate, injure, or destroy any historic or prehistoric ruin or monument, or any object of antiquity on lands owned or controlled by the United States and (2) a component that authorizes a permit for the examination of ruins and archaeological sites and the gathering of objects of antiquity on lands owned or controlled by the United States.

Historic properties, including shipwrecks, located on public lands were first protected using the Antiquities Act, followed by the Archaeological Resources Protection Act (U.S. Code, title 16, sections 470gg–470ll) and expansion of the National Historic Preservation Act (U.S. Code, title 16, sections 470–470v-1) in 1980. In the marine environment, where the United States has ownership or control of the submerged lands in or on which submerged cultural resources are located, the Antiquities Act permitting provision can be used to regulate salvage. It is possible, however, that its reach may be limited to regulating salvage only in marine protected areas in which the United States has the authority to protect submerged cultural resources.

Coastal Zone Management Act

The Coastal Zone Management Act (U.S. Code, title 16, sections 1451–1464) directs states to provide an inventory and designation of areas that contain significant resources and to develop standards to protect them. States can accomplish this through specific state laws such as coastal wetlands protection laws or through designating sites as areas of concern. State programs must also address issues of shoreline management, beach access and land acquisition, and ocean management. In developing the program, states must strive to resolve conflicts among competing uses and, in some cases, acquire title or other interest in land or waters when necessary to achieve conformance with the management program.

Environmental Review

The National Environmental Policy Act (NEPA; U.S. Code, title 42, sections 4321–4270d) is the national statute that requires federal agencies to conduct an environmental review of certain actions. Enacted in 1969 with the inspiring goal to "create and maintain conditions under which man and nature can exist in productive harmony, and fulfill the social, economic, and other requirements of present and future generations of Americans" (U.S. Code, title 42, section 4331[a]), NEPA "sets forth a ringing and vague statement of purposes" (Rodgers 1994: 801). This vagueness grew into a powerful tool for environmentalists seeking to judicially challenge federal actions that ignored potential environmental impacts.

Aside from its statements of policy objectives, NEPA's "action-forcing" mechanism is in section 102, which requires all federal agencies to include a detailed statement of the environmental impact of a major federal action significantly affecting the human environment. A "major" federal action is one which requires substantial planning, time, resources, or expenditure that the federal agency proposes or permits. Through Environmental Assessment (EA) and Environmental Impact Statement (EIS) reviews, agencies are forced to consider environmental impacts before action is taken. In addition, NEPA mandates coordination and collaboration among federal agencies. Specifically, "prior to making any detailed statement, the responsible Federal official shall consult with and obtain the comments of any Federal agency which has jurisdiction by law or special expertise with respect to any environmental impact involved" (U.S. Code, title 42, section 4332). This includes the U.S. Fish and Wildlife Service for freshwater and anadromous species and the National Marine Fisheries Service for marine and anadromous species.

For practical purposes, this is where NEPA's mandates end. The Supreme Court has declared that NEPA's reach is procedural rather than substantive, and NEPA cannot "mandate particular results but only prescribe the necessary process" (*Robertson v. Methow Valley Citizens Council*, 490 U.S. 332, 350 [1989]; also see *Vermont Yankee Nuclear Power Corp. v. Natural Resources Defense Council*, 435 U.S. 519, 548 [1978], and *Kleppe v. Sierra Club*, 427 U.S. 390 [1976][3]). The court stated that "once an agency has made a decision subject to NEPA's procedural requirements, the only role for a court is to ensure that the agency has considered the environmental consequences; it cannot interject itself within the area of discretion of the executive as to the choice of the action to be taken." (*Robertson v. Methow Valley Citizens Council*, 410). Thus, once a federal agency has completed the "detailed statement" that NEPA requires, it may then continue its proposed

---

[3] Legal notation is used for citing court cases in this paper. The citation lists the case name, volume number, reporter, page numbers, and year of decision. Supreme Court decisions are published in the United States Supreme Court Reports (U.S.), while lower federal court decisions are published in the Federal Reporter (F.) or Federal Supplement (F. Supp.). "Slip op." refers to a slip opinion not published in an official reporter. P = Pacific Reporter.

activity. Essentially, NEPA offers a procedural challenge that "merely prohibits uninformed—rather than unwise—agency actions" (*Robertson v. Methow Valley Citizens Council*, 332, 351).

In general, the establishment of a National Marine Sanctuary under the National Marine Sanctuary Act (U.S. Code, title 16, sections 1431–1445), the designation of Essential Fish Habitat under the Magnuson–Stevens Act, or the designation of a monument under the Antiquities Act triggers NEPA section 102 and the need for an EA or EIS. Litigation related to EFH may provide a primer. The first EFH-related lawsuit charged that the EFH Amendments from the Gulf of Mexico, New England, Caribbean, Pacific, and North Pacific were unlawfully prepared and approved in reliance on inadequate environmental analyses and in violation of the specific requirements of the Magnuson–Stevens Act. The Earthjustice Legal Defense Fund challenged the continued fishing activities, which they asserted produce adverse effects on EFH. These activities specifically include shrimp trawling in the Gulf of Mexico, bottom trawling off New England, and bottom trawling off the Pacific coast (Second Amended Complaint for Declaratory, Mandatory, and Injunctive Relief, *Earthjustice Legal Defense Fund v. National Oceanic and Atmospheric Administration, National Marine Fisheries Service* [and Regional Councils] [1999], unpublished, on file with authors).

A recent decision regarding the application of NEPA in the Exclusive Economic Zone (EEZ) may also affect the application of NEPA on the creation of offshore MPAs. In response to a Navy challenge of the application of NEPA in the EEZ, the U.S. District Court for the Central District of California held that, although the Navy need not subject its entire Littoral Warfare Advanced Development Program to environmental review, NEPA does apply "to federal actions which may affect the environment in the [Exclusive Economic Zone]" (*Natural Resources Defense Council v. United States Department of the Navy*, No. CV-01–07781 Slip op. at 21; Central District of California Court, September 19, 2002). The Navy's Littoral Warfare Advanced Development Program, initiated in 1996, oversees the sea testing of experimental anti-submarine technologies. As noted in the opinion, "the purpose of the sea tests is to provide a robust, "real world" environment for the testing and demonstration of anti-submarine warfare technologies that the Navy may want to acquire" (*Natural Resources Defense Council v. United States Department of the Navy*, 2). Concerned for the welfare of cetacean populations, the Natural Resources Defense Council filed suit, seek-

ing to enjoin the Navy from conducting further sea tests until it completed environmental studies as required by NEPA. The Navy argued that its activities in the EEZ are not subject to environmental review under NEPA. While the EEZ is not strictly considered part of the territory of the United States, the federal government does have certain "sovereign rights" within the area "for the purposes of exploring, exploiting, conserving, and managing natural resources" (NRDC at 20 [citing *Environmental Defense Fund, Inc. v. Massey*, 986 F. 2d 528, 534 (D.C. Circuit Court, 1994]). Furthermore, regarding natural resource conservation and management, "the United States does have substantial, if not exclusive, legislative control of the EEZ" (*Natural Resources Defense Council v. United States Department of the Navy*, 21). As a result, the court held "that NEPA applies to federal actions which may affect the environment in the EEZ" (*Natural Resources Defense Council v. United States Department of the Navy*, 21).

Property Right Challenges

While several statutes provide the necessary authority to designate and review the environmental impacts of MPAs, precluding certain activities within an MPA may still be challenged as going too far. The "takings clause" of the 5th Amendment prohibits the government from taking private property for public use without just compensation (U.S. Constitution, 5th amendment). When there has been a permanent physical invasion of land by the government, it is generally incontrovertible that there has been a taking of private property requiring compensation (*Loretto v. Teleprompter Manhattan CATV Corp.*, 458 U.S. 419 [1982]). But in addition to instances of physical invasion or government confiscation of property, a government action such as limiting the beneficial use of personal property may be recognized as a taking if it goes too far (*Pennsylvania Coal Co. v. Mahon*, 260 U.S. 393, 415 [1922]). Recognizing that "government hardly could go on if to some extent values incident to property could not be diminished without paying for every such change in the general law... when it reaches a certain magnitude, in most if not all cases, there must be an exercise of eminent domain and compensation to sustain the act." (*Pennsylvania Coal Co. v. Mahon*, 413).

Recent U.S. Supreme Court cases have placed emphasis on the economic impact of the regulation on the property owner and the degree to which the owner's distinct investment-backed expectations have been frustrated (*Penn Central Transportation Co. v. City of New York*, 438 U.S. 104 [1978]). While no set formula exists for determining when a government limitation on property amounts to a regulatory taking, the Supreme Court has found that when a property owner has lost all economically beneficial use of the property, a taking has occurred (*Lucas v. South Carolina Coastal Council*, 505 U.S. 1003, 1019 [1992]). In the seminal case, *Lucas v. South Carolina Coastal Council* (1992), developer David Lucas sued South Carolina for denying him the right to build residential homes on two waterfront lots on a South Carolina barrier island, the Isle of Palms.[4] The Coastal Council denied his building permit under authority of the Beachfront Management Act (South Carolina Code Annotated, section 48-39-250), which limited development behind an erosion line (known as the "setback line"), effectively prohibiting Lucas from building any structures on the property. The court found that a taking had occurred since the legislature's actions deprived the land of all of its economic viability.[5] South Carolina bought the land from Lucas for over US$1.5 million.

---

[4] Tibbetts (1997) discusses the current issue between regulators in South Carolina and North Carolina, who prohibit the building of seawalls, and beachfront property owners, who claim that seawalls are the only method of saving their property from falling into the sea due to extreme erosion. Tibbetts references the *Lucas v. South Carolina Coastal Council* decision and explains:

> Legal experts doubt that the U.S. Supreme Court would even consider the *Lucas* case if it were presented now, because beach erosion has eaten away the land in question, just as state regulators predicted. The ocean has eroded the buildable land on the undeveloped lot and threatened a home on the developed lot formerly owned by Lucas. In the *Lucas* decision, the Court relied on the traditional legal assumption that land is unchanging, says R. J. Lyman, an attorney with the Massachusetts Office of Environmental Affairs. Nature, however, shows that land forms, especially beachfront property, are in flux. The Court saw the Lucas lots in a "snapshot" taken when the beachfront was usually wide, he says, "but that snapshot was not representative of the moving picture" of oceanfront property that can erode and disappear. (Tibbetts 1997)

David Lucas stated that the high land on the Isle of Palms lots will return, likely to remain "erosion-free for several years" (Tibbetts 1997).

For those cases that fall in between a physical invasion and a total loss of property value, the Supreme Court has ruled that takings inquiries should be made on a case-by-case basis using factors such as character of the governmental action and economic impact of the regulation to determine whether a regulatory scheme effects a taking of property (*Penn Central Transportation Co. v. City of New York*, 124). From another beachfront case (*Nollan v. California Coastal Commission*, 483 U.S. 825 [1987]), the Supreme Court found that the state cannot condition a permit to build without showing a legitimate state purpose for the governmental mandate. Furthermore, if the state has both a legitimate interest, such as preventing erosion or protecting a flood plain, and the exactions bear a relationship to the impact of the proposed development, then the state's requirements will not be considered a taking (*Dolan v. City of Tigard*, 512 U.S. 374 [1994]; for a complete analysis of the case, see Freis and Reyniak 1996).

There are several potential challenges from property owners to the designation and management of an MPA. First, as noted above, coastal property owners may challenge an MPA if they own property within an MPA and can show a physical taking of their property (i.e., their property been included in an MPA and either has been physically taken from them or they have lost all economically beneficial use of the land) or prove that the government action requires them to bear an unfair burden more appropriately born by the public.

---

[5] The court recognized an exception to the application of the *Lucas* analysis when the government bases the regulation on the "background principles of the State's law of property and nuisance" (*Lucas v. South Carolina Coastal Council*, 1029). The Oregon Supreme Court applied the *Lucas v. South Carolina Coastal Council* analysis in *Stevens v. City of Cannon Beach* (*Stevens v. City of Cannon Beach*, 845 P. 2d 449, Oregon Supreme Court, 1993, review by U.S. Supreme Court denied, 510 U.S. 1207 [1994]), where the landowner sued the city of Cannon Beach and the state of Oregon for denials of permits to construct a seawall on the dry sand portion of the plaintiff's beachfront lot. The Oregon Supreme Court found that the plaintiffs had no property interest in developing the dry sand portion of their property, applying the nuisance exception of the *Lucas v. South Carolina Coastal Council* decision and upholding the applicability of the Oregon Beach Bill (Oregon Revised Statutes, section 390.605).

However, how do private property rights challenges relate to the right to fish or, in general, extract resources from MPAs? For fishing permit holders, the private property right challenges available under current law are limited. The holder of a permit, even a rights-based individual transferable quota (ITQ) or individual fishing quota, does not have the same "rights" as those claimed by a landowner. The permit is legally viewed as a license to fish depending upon season, total allowable catch, and other restrictions. The ITQ represents, generally, a right to harvest a certain number of fish in a specific fishery. Neither the ITQ nor license represents the right to fish in a particular area, and it is generally held that such mechanisms do not create permanent property rights but only revocable harvesting privileges (see *Sea Watch International v. Mosbacher*, 762 F. Supp. 370, 376 [District Court of D.C., 1991] that found the surf clam–ocean quahog ITQ regulations were such revocable harvesting privileges). Such privileges are not protected by the Fifth Amendment's takings provision (see Code of Federal Regulations, title 50, section 679.40[f] [1996].

In conclusion, potential legal challenges exist, but the nature of MPAs as part of the public trust and generally open to use limit the challenges on the basis of private property rights. While other private challenges may exist, such as constitutional challenges under the Equal Protection Clause, or legal claims challenging the governmental authority to create reserves or the process that is used to establish them, MPAs will likely be challenged with regard to the question of authority to create an MPA and limit others' access as well as the impacts of setting aside such lands.

Acknowledgments

This research was sponsored in part by the National Sea Grant Law Center, University of Mississippi School of Law, and Roger Williams University School of Law.

References

Brax, J. 2002. Zoning the oceans: using the National Marine Sanctuaries Act and the Antiquities Act to establish marine protection areas and marine reserves in America. Ecology Law Quarterly 29:71.

Freis, J. H., Jr., and S. V. Reyniak. 1996. Putting takings back into the Fifth Amendment: land use planning after *Dolan v. City of Tigard*. Columbia Journal of Environmental Law 21:103.

Rodgers, W. 1994. Environmental law. West, St. Paul, Minnesota.

Tibbetts, J. 1997. Beachfront battles over seawalls. Coastal Heritage 12:3.

U.S. Office of the Federal Register. 2000. Executive Order 13158. Federal Register 65:105(31 May 2000):34909.

Appendices

Appendix 1: Selected Bibliography

The following is a list of literature citations relating to aquatic protected areas.

General (Biodiversity, Conservation, and More)

Agardy, M. T. 1994. Advances in marine conservation: the role of marine protected areas. Trends in Ecological Evolution 9:267–270.

Agardy, T., editor. 1995. The science of conservation in the coastal zone: new insights on how to design, implement, and monitor marine protected areas. International Union for the Conservation of Nature and Natural Resources, Gland, Switzerland.

Agardy, T. 1997. Marine protected areas and ocean conservation. Academic Press and R. G. Landes Co., Austin, Texas.

Agardy, T. 1999. Creating havens for marine life. Issues in Science and Technology 16(1):37–44. Available: *www.nap.edu/issues/16.1/agardy.htm* (January 2004).

Agardy, T. 2000. Key steps taken to preserve the U.S.'s marine heritage. Issues in Science and Technology 17(1):26. Available: *www.nap.edu/issues/17.1/update.htm* (January 2004).

Alder, J. 1996. Have tropical marine protected areas worked? An initial analysis of their success. Coastal Management 24(2):97–114.

Allison, G. W. 1998. Marine reserves are necessary but not sufficient for marine conservation. Ecological Applications 8(1):79–92.

Atkinson, J., P. Brooks, A. Chatwin, and P. Shelly. 2000. The wild sea: saving our marine heritage. Conservation Law Foundation, Boston.

Auster, P. J., K. Joy, and P. C. Valentine. 2001. Fish species and community distributions as proxies for seafloor habitat distributions: the Stellwagen Bank National Marine Sanctuary example (Northwest Atlantic, Gulf of Maine). Environmental Biology of Fishes 60(4):331–346.

Batisse, M. 1990. Development and implementation of the biosphere reserve concept and its applicability to coastal regions. Environmental Conservation 17(2):111–116.

Batisse, M. 1997. Biosphere reserves: a challenge for biodiversity conservation and regional development. Environment 39(5):7–15, 31–33.

Brailovskaya, T. 1998. Obstacles to protecting marine biodiversity through marine wilderness preservation: examples from the New England region. Conservation Biology 12:1236–1240.

Brunckhorst, D. J., editor. 1994. Marine protected areas and biosphere reserves: towards a new paradigm. Australian Nature Conservation Agency, Canberra. Available: *www.environment.gov.au/marine/mpa/nrsmpa/paradigm/index.html* (January 2004).

Carr, M. H. 2000. Marine protected areas: challenges and opportunities for understanding and conserving coastal marine ecosystems. Environmental Conservation 27(2):106–109.

Crosby, M. P. 1994. Opportunities for nonindigenous species research and monitoring in NOAA's National Estuarine Research Reserves. Pages 69–78 *in* Nonindigenous estuarine and marine organism conference, Seattle, WA. Proceedings of the conference and workshop, April 1993. National Oceanic and Atmospheric Administration, Washington, D.C.

Crosby, M. P., K. Geenen, D. Laffoley, C. Mondor, and G. O'Sullivan. 1997. Proceedings of the second international symposium and workshop on marine and coastal protected areas: integrating science and management. National Oceanic and Atmospheric Administration, Silver Spring, Maryland.

Crosby, M. P., and H. M. Golde. 1993. A review and synthesis of the first decade of research in the National Estuarine Research Reserve System. NOAA Technical Memorandum 26.

Dyer, M., and M. Holland. 1991. The biosphere reserve concept: needs for a network design. BioScience 41(5):319–325.

Eichbaum, W. M., M. P. Crosby, M. T. Agardy, and S. A. Laskin. The role of marine and coastal protected areas in the conservation and sustainable use of biological diversity. Oceanography 9(1):60–70.

Environment Conservation Council. 2000. Marine, coastal and estuarine investigation: final report. Environment Conservation Council, East Melbourne, Victoria, Australia. Available: *www.nre.vic.gov.au/ecc* (January 2004).

Faye, D. 1999. Marine protection. Ecos 98:17–21.

Fisheries Society of the British Isles. 2001. Briefing paper 1: marine protected areas in the North Sea. Fisheries Society of the British Isles, Granta Information Systems, Cambridge, UK. Available: *www.le.ac.uk/biology/fsbi/fsbi.pdf* (January 2004).

Florida State University and Mote Marine Laboratory. 1998. Essential fish habitat and marine reserves:

proceedings of the second William R. and Lenore Mote international symposium in fisheries ecology. Bulletin of Marine Science Special Issue 6(3):525–1010.

Francour, P., J. G. Harmelin, D. Pollard, and S. Sartoretto. 2001. A review of marine protected areas in the northwestern Mediterranean region: siting, usage, zonation and management. Aquatic Conservation: Marine and Freshwater Ecosystems 11(3):155–188.

Garcia-Charton, J. A., I. D. Williams, A. Perez-Ruzafa, M. Milazzo, R. Chemello, C. Marcos, M.-S. Kitsos, A. Koukouras, and S. Riggio. 2000. Evaluating the ecological effects of Mediterranean marine protected areas: habitat, scale and the natural variability of ecosystems. Environmental Conservation 27(2):159–178.

Goni, R, N. V. C Polunin, and S. Planes. 2000. The Mediterranean: marine protected areas and the recovery of a large marine ecosystem. Environmental Conservation 27(2):95–97.

Hixon, M. A., P. D. Boersma, M. L. Hunter, Jr., F. Micheli, E. A. Norse, H. P. Possingham, and P. V. R. Snelgrove. 2001. Oceans at risk: research priorities in marine conservation biology. Pages 125–154 in M. E. Soulee and G. H. Orians, editors. Conservation biology: research priorities for the next decade. Island Press, Washington, D.C.

Jamieson, G. S., and C. O. Levings. 2001. Marine protected areas in Canada—implications for both conservation and fisheries management. Canadian Journal of Fisheries and Aquatic Sciences 58:138–156.

Jegalian, K. 1999. Plan would protect New England coast. Science 284(5412):237–242.

Jones, P. J. S. 1999. Marine nature reserves in Britain: past lessons, current status and future issues. Marine Policy 23(4–5):375–396.

Junaes, F. 2001. Mediterranean marine protected areas. Trends in Ecology and Evolution 16(4):169–170.

Kelleher, G. 1997. Marine protected areas: are they necessary? Connect: UNESCO International Science, Technology and Environmental Education Newsletter 22(3/4):11–12.

Kelleher, G., C. Bleakley, and S. Wells, editors. 1995. A global representative system of marine protected areas, volume 1. The Great Barrier Reef Marine Park Authority, The World Bank, and IUCN–The World Conservation Union, Washington, D.C.

Kenchington, R. A., and M. T. Agardy. 1990. Achieving marine conservation through biosphere reserve planning and management. Environmental Conservation 17(1):39–44.

Levinton, J. S. 1995. Marine biology: function, biodiversity, ecology. Oxford University Press, New York.

McNeil, S. E. 1994. The selection and design of marine protected areas: Australia as case study. Biodiversity and Conservation 3:586–605.

Mills, C. E., and J. T. Carlton. 1998. Rationale for a system of international reserves for the open ocean. Conservation Biology 12(1):244–247.

Mumby, P. J. 2001. Beta and habitat diversity in marine systems: a new approach to measurement, scaling and interpretation. Oecologia 128(2):274–280.

National Research Council. 2000. Marine protected areas: tools for sustaining ocean ecosystems. National Academy Press, Washington, D.C. Available: www.nap.edu/books/0309072867/html/ (January 2004).

Natural Resources Defense Council and The Ocean Conservancy. 2001. A citizen's guide to the Marine Life Protection Act. Available: www.cencal.org/MLPAguidebook.pdf (January 2004).

New England Aquarium and The Pew Charitable Trusts. 2001. Oceans for the future: the making of marine protected areas. New England Aquarium, Boston.

Palumbi, S. R. 2000. The ecology of marine protected areas. In M. Bertness, editor. Marine ecology: the new synthesis. Sinauer Press, Sunderland, Massachusetts.

Parker, S., and M. Munawar, editors. 2001. Ecology, culture, and conservation of a protected area: Fathom Five National Marine Park, Canada. Backhuys Publishers, Leiden, The Netherlands.

Pew Ocean Commission. 2002. Commission activities: Dr. Kennel testimony to the U.S. Commission on Ocean Policy. Prepared for Pew Ocean Commission, San Pedro, California. Available: www.pewoceans.org/activities/2002/04/24/ activities_25957.asp (January 2004; see www.pewoceans.org/articles/2002/04/24/ pr_25961.asp for other testimony.)

Pitcher, T. J. 2001. Fisheries managed to rebuild ecosystems? Reconstructing the past to salvage the future. Ecological Applications 11(2):601–617.

Roberts, C., and J. Hawkins. 2000. Fully protected marine reserves: a guide. World Wildlife Fund, Washington, D.C.

Salmona, P., and D. Verardi. 2001. The marine protected area of Portofino, Italy: a difficult balance. Ocean and Coastal Management 44(1–2):39–60.

Siebig, S. 1997. The benefits of marine protected areas: Bazaruto, Mozambique. Connect: UNESCO International Science, Technology, and Environmental Education Newsletter 22(3/4):12–13. Available: www.unesco.org/education/educprog/ste/pdf_files/ connect/connect97-3.pdf (January 2004).

Sobel, J. 1993. Conserving biological diversity through marine protected areas. Oceanus 36:19–26.

Steele, J. H. 1974. The structure of marine ecosystems. Harvard University Press, Cambridge, Massachusetts.

Thorsell, J., R. Ferster Levy, and T. Sigaty. 1997. A global overview of wetland and marine protected areas on the world heritage list. IUCN–The World Conservation Union and The World Conservation Monitoring Centre, Gland, Switzerland. Available: *www.unep-wcmc.org/wh/reviews/wetlands/* (January 2004).

Ticco, P. C. 1995. The use of marine protected areas to preserve and enhance marine biological diversity: a case study approach. Coastal Management 23:309–314.

Zinn, J., and E. H. Buck. 2001. Marine protected areas: an overview. Congressional Research Service Report made available to the public by the National Council for Science and the Environment, Washington, D.C. Available: *http://cnie.org/NLE/CRSreports/Marine/mar-39.cfm* (January 2004).

Coral Reefs

Bohnsack, J. A., B. Causey, M. P. Crosby, R. B. Griffis, M. A. Hixon, T. F. Hourigan, K. H. Koltes, J. E. Maragos, A. Simons, and J. T. Tilmant. 2002. A rationale for minimum 20–30% no-take protection. Pages 615–620 in M. K. Kasim Moosa, S. Soemodihardjo, A. Nontji, A. Soegiarto, K. Romimohtarto, Sukarno, and Suharsono, editors. Proceedings of the Ninth International Coral Reef Symposium, Bali, Indonesia, October 23–27 2000. Ministry of Environment, Indonesian Institute of Sciences and International Society for Reef Studies, Jakarta.

Crosby, M. P., G. Brighous, and M. Pichon. 2002. Priorities and strategies for addressing natural and anthropogenic threats to coral reefs in Pacific Island Nations. Ocean and Coastal Management 45:121–137.

Loya, Y., S. M. Al-Moghrabi, M. Ilan, and M. P. Crosby. 1999. The Red Sea Marine Peace Park coral reef benthic communities: ecology and biology monitoring program. Pages 239–250 in J. E. Maragos and R. Grober-Dunsmore, editors. Proceedings of the Hawaii Coral Reef Monitoring Workshop, June 9–11, 1998, Honolulu, Hawaii. Division of Aquatic Resources, Department of Land and Natural Resources, Honolulu, Hawaii.

Maragos, J. E., M. P. Crosby, and J. McManus. 1996. Coral reefs and biodiversity: a critical and threatened relationship. Oceanography 9:83–99.

Masica, M. A. 2001. Designing effective coral reef marine protected areas. a synthesis report based on presentations given at the 9th International Coral Reef Symposium, Bali, Indonesia, October 2000. IUCN World Commission on Protected Areas—Marine. Available: *http://wcpa.iucn.org/pubs/pdfs/ICRSreport.pdf* (January 2004).

McClanahan, T. R., and R. Arthur. 2001. The effect of marine reserves and habitat on populations of east African coral reef fishes. Ecological Applications 11(2):559–569.

McClanahan, T. R., and S. Mangi. Spillover of exploitable fishes from a marine park and its effect on the adjacent fishery. Ecological Applications 10(6):1792–1805.

McClanahan, T. R., M. McField, M. Huitric, K. Bergman, E. Sala, M. Nystrom, I. Nordemar, T. Elfwing, and N. A. Muthiga. 2001. Responses of algae, corals and fish to the reduction of macroalgae in fished and unfished patch reefs of Glovers Reef Atoll, Belize. Coral Reefs 19(4)367–379.

National Park Service. 1998. Coral reefs under National Parks Service jurisdiction: overview of areas, protection, and management issues. Water Resources Division, U.S. Department of the Interior, Washington, D.C.

Roberts, C. M., and J. P. Hawkins. 1997. How small can a marine reserve be and still be effective? Coral Reefs 16(3):150.

Rouphael, A. B., and G. J. Inglis. "Take only photographs and leave only footprints"?: an experimental study of the impacts of underwater photographers on coral reef dive sites. Biological Conservation 100(3):281–287.

Salm, R. V., and S. L. Coles, editors. 2001. Coral bleaching and marine protected areas: proceedings of the workshop on mitigation coral bleaching impact through MPA design, Bishop Museum, Honolulu, Hawaii, May 29–31, 2001. Asia Pacific Coastal Marine Program, Report 0102, Nature Conservancy, Honolulu, Hawaii.

Cultural Resources

Robinson, J. 2000. A guide to historical U.S. coast survey data significant to cultural resource management in the national marine sanctuaries. Marine Chart Division, Office of the Coast Survey, National Ocean Service, National Oceanic and Atmospheric Administration, Silver Spring, Maryland.

Economics

Agardy, M. T. 1993. Accommodating ecotourism in multiple use planning of coastal and marine protected areas. Ocean and Coastal Management 20:219–239.

Davis, D., and C. Tisdell. 1995. Recreational SCUBA diving and carrying capacity in marine protected areas. Ocean and Coastal Management 226:19–40.

Dixon, J. A. 1993. Economics of marine protected areas. Oceanus 36:35–40.

Hoagland, P., K. Yoshiaki, and J. Broadus. 1995. A methodological review of net benefit evaluation for marine reserves. World Bank, Environment Department, Washington, D.C.

Holland, D., and R. Brazee. 1996. Marine reserves for fisheries and management. Marine Resource Economics 11(3):157–171.

Holland, D. S. 2000. A bioeconomic model of marine sanctuaries on Georges Bank. Canadian Journal of Fisheries and Aquatic Sciences 57:1307–1319.

Pezzey, J., C. V. Callum, M. Roberts, and B. T. Urdal. 2000. A simple bioeconomic model of a marine reserve. Ecological Economics 33(1):77–91.

Sumaila, U. R., and J. Alder, editors. 2001. Economics of marine protected areas: papers, discussions and issues: a conference held at the UBC Fisheries Centre, July 2000. University of British Columbia, Fisheries Centre, Fisheries Centre Research Reports 9:8, Vancouver. Available: *www.fisheries.ubc.ca/publications/reports/report9_8.php* (January 2004).

Williams, I. D., and N. V. C. Polunin. 2000. Differences between protected and unprotected reefs of the western Caribbean in attributes preferred by dive tourists. Environmental Conservation 27(4):382–391.

Fisheries and Marine Reserves

Adams, S., B. D. Mapstone, G. R. Russ, and C. R. Davies. 2000. Geographic variation in the sex ratio, sex specific size, and age structure of *Plectropomus leopardus* (Serranidae) between reefs open and closed to fishing on the Great Barrier Reef. Canadian Journal of Fisheries and Aquatic Sciences 57(7):1448–1458.

Agardy, M. T. 1994. Closed areas: a tool to complement other forms of fisheries management. *In* Limiting access to marine fisheries: keeping the focus on conservation. Center for Marine Conservation and World Wildlife Fund, Washington, D.C.

Apostolaki, P., E. J. Milner-Gulland, G. Kirkwood, and M. McAllister. 2002. Modelling the effects of establishing a marine reserve for mobile fish species. Canadian Journal of Fisheries and Aquatic Sciences 59(3):405–415.

Auster, P. J., and R. J. Malatesta. 1995. Assessing the role of non-extractive reserves for enhancing harvested populations in temperate and boreal marine systems. Pages 82–89 *in* Marine protected areas and sustainable fisheries. Science and Management of Protected Areas Association, Wolfville, Nova Scotia.

Auster, P. J., C. Michalopoulos, P. C. Valentine, and R. J. Malatesta. 1998. Delineating and monitoring habitat management units in a temperate deepwater marine protected area. Pages 169–185 *in* N. W. P. Munro and J. H. M. Willison, editors. Linking protected areas with working landscapes, conserving biodiversity. Proceedings of the Third International Conference on Science and Management of Protected Areas, 12–16 May 1997. Science and Management of Protected Areas Association, Wolfville, Nova Scotia.

Auster, P. J., and N. L. Shackell. 1997. Fishery reserves. Pages 159–166 *in* J. G. Boreman, B. S. Nakashima, H. W. Powles, J. A. Wilson, and R. L. Kendall, editors. Northwest Atlantic groundfish: perspectives on a fishery collapse. American Fisheries Society, Bethesda, Maryland.

Auster, P. J., and N. L. Shackell. 2000. Marine protected areas for the temperate and boreal Northwest Atlantic: the potential for sustainable fisheries and conservation of biodiversity. Northeastern Naturalist 7:419–434.

Barr, B. W. 1995. The U.S. National Marine Sanctuary Program and its role in preserving sustainable fisheries. Pages 165–173 *in* N. L. Shackell and J. H. M. Willison, editors. Marine protected areas and sustainable fisheries. Science and Management of Protected Areas Association, Wolfville, Novia Scotia.

Beck, M. W., K. L. Heck, K. W. Able, D. L. Childers, D. B. Eggleston, B. M. Gillanders, B. Halpern, C. G. Hays, K. Hoshino, T. J. Mnello, R. J. Orth, P. F. Sheridan, and M. R. Weinstein. 2001. The identification, conservation, and management of estuarine and marine nurseries for fish and invertebrates. BioScience 51(8):633–641.

Bohnsack, J. A. 1993. Marine reserves: they enhance fisheries, reduce conflicts, and protect resources. Oceanus 36:63–71.

Bohnsack, J. A. 1996. Marine reserves, zoning, and the future of fishery management. Fisheries 21(9):14–16.

Bohnsack, J. A. 1998. Application of marine reserves to reef fisheries management. Australian Journal of Ecology 23:298–304.

Bohnsack, J. A. 1998. Reef fish response to divers in two "no-take" marine reserves in Hawaii. Reef Encounter 23:22–24.

Buck, E. 1995. Summaries of major laws implemented by the National Marine Fisheries Service. National Council for Science and the Environment. Available: *http://cnie.org/NLE/CRSreports/legislative/leg-11.cfm* (January 2004).

Cadrin, S. X., A. B. Howe, S. J. Correia, and T. P. Currier. 1995. Evaluating the effects of two coastal mobile gear fishing closures on finfish abundance off Cape Cod. North American Journal of Fisheries Management 15:300–315.

Clapp, D. F., R. D. Clark, and J. S. Diana. 1990. Range, activity, and habitat of large, free-ranging Brown Trout in a Michigan stream. Transactions of the American Fisheries Society 119:1022–1034

Collins, M. K., T. J. Smith, W. E. Jenkins, and M. R. Denson. 2002. Small marine reserves may increase escapement of red drum. Fisheries 27(2):20–24.

Dayton, P., S. Thrush, T. Agardy, and R. Hoffman. 1995. Environmental effects of marine fishing.

Aquatic Conservation: Marine and Freshwater Ecosystems 5:1–28.

Delaney, J. M. 2001. Marine reserve design in Florida's Tortugas. Earth System Monitor 11(3):12–13. Available: *www.nodc.noaa.gov/General/NODCPubs/ESM/ESM_MAR2001vol11no3.pdf* (January 2004).

Dugan, J. E. 1993. Applications of marine refugia to coastal fisheries management. Canadian Journal of Fisheries and Aquatic Sciences 50:2029–2042.

Fogarty, M. J. 1999. Essential habitat, marine reserves and fishery management. Trends in Ecology and Evolution 14(4):133–134.

Fogarty, M. J., J. A. Bohnsack, and P. K. Dayton. 2000. Marine reserves and resource management. *In* C. Sheppard, editor. Seas at the millennium: an environmental evaluation. Pergamon, New York.

Gladstone, W. 2002. The potential value of indicator groups in the selection of marine reserves. Biological Conservation 104(2):211–220.

Gulf of Mexico Fishery Management Council. 1999. Marine reserves technical document: a scoping document for the Gulf of Mexico. Gulf of Mexico Fishery Management Council, Tampa, Florida. Available: *www.gulfcouncil.org/downloads/MR-TechScope.pdf* (January 2004).

Hall, S. J. 1999. Marine protected areas. Pages 230–240 *in* The effects of fishing on marine ecosystems and communities. Blackwell Science, Oxford, UK.

Halpern, B. S., and R. R. Warner. 2002. Marine reserves have rapid and lasting effects. Ecology Letters 5(3):361–366.

Hancock, D. A., D. C. Smith, A. Grant, and J. P. Beumer, editors. 1997. Developing and sustaining world fisheries resources: the state of science and management. Second World Fisheries Congress. Australian Society for Fish Biology, Brisbane.

Hastings, A., and L. W. Botsford. 1999. Equivalence in yield from marine reserves and traditional fisheries management. Science 284:1537–1538.

Hayes, M. C., L. F. Gates, and S. A. Hirsch. 1997. Multiple catches of smallmouth bass in a special regulation fishery. North American Journal of Fisheries Management 17:182–187.

Hyrenbach, K. D., K. A. Forney, and P. K. Dayton. 2000. Marine protected areas and ocean basin management. Aquatic Conservation: Marine and Freshwater Ecosystems 10:437–458.

Jackson, J. B. C. 2001. What was natural in the coastal oceans? Proceedings of the National Academy of Sciences of the United States of America 98(10):5411–5418. Available: *www.pnas.org/cgi/content/full/98/10/5411* (January 2004).

Jamieson, G. S., and J. Lessard. 2000. Marine protected areas and fishery closures in British Columbia. Canadian Special Publication of Fisheries and Aquatic Sciences 131.

Jennings, S. 2000. Patterns and prediction of population recovery in marine reserves. Reviews in Fish Biology and Fisheries 10(2):209–231.

Jennings, S., and M. Kaiser. 1998. The effects of fishing on marine ecosystems. Advances in Marine Biology 34:201–352.

Johnson, D. R., N. A. Funicellli, and J. A. Bohnsack. 1999. Effectiveness of an existing estuarine no-take fish sanctuary within the Kennedy Space Center, Florida. North American Journal of Fisheries Management 19:436–453.

Jurado-Molina, J., and P. Livingston. 2002. Multispecies perspectives on the Bering Sea groundfish fisheries management regime. North American Journal of Fisheries Management 22:1164–1175.

Koenig, C. C., F. C. Coleman, C. B. Grimes, G. R. Fitzhugh, K. M. Scanlon, C. T. Gledhill, and M. Grace. 2000. Protection of fish spawning habitat for the conservation of warm-temperate reef-fish fisheries of shelf-edge reefs of Florida. Bulletin of Marine Science 66(3):593–616.

Kramer, D. L., and M. R. Chapman. 1999. Implications of fish home range size and relocation for marine reserve function. Environmental Biology of Fishes 55:65–79.

Lauck, T., C. Clark, M. Mangel, and G. Munro. 1998. Implementing the precautionary principle in fisheries management through marine reserves. Ecological Applications 8(Supplement):S72–S78.

Lindeman, K. C., R. Pugliese, and G. T. Waugh. 2000. Developmental patterns within a multispecies reef fishery: management applications for essential fish habitats and protected areas. Bulletin of Marine Science 66(3):929–956.

Lindholm, J., P. J. Auster, M. Ruth, and L. Kaufman. 2001. Juvenile fish responses to variations in seafloor habitats: modeling the effects of fishing and implications for the design of marine protected areas. Conservation Biology 15:424–437.

Lindholm, J. M., P. J. Auster, M. Ruth, and L. Kaufman. 1998. A modeling approach to the design of marine refugia for fishery management. Pages 138–150 *in* N. L. Shackell and J. H. M. Willison, editors. Linking protected areas with working landscapes, conserving biodiversity, proceedings of the Third International Conference on Science and Management of Protected Areas, 12–16 May 1997. Science and Management of Protected Areas Association, Wolfville, Novia Scotia.

Lipicus, R., R. Seitz, W. Goldsborough, M. Montane, and W. Stockhausen. 2001. A deep-water marine dispersal corridor for adult female blue crabs in Chesapeake Bay. *In* G. H. Kruse, N. Bez, A. Booth, M. W. Dorn, S. Hills, R. Hills, editors. Spatial processes and management of marine populations. Alaska Sea Grant, AK-SG-01-92, Fairbanks.

Lipcius, R., W. Stockhausen, and D. Eggleston. 2001. Marine reserves for Caribbean spiny lobster: empirical evaluation and theoretical metapopulation recruitement dynamics. Marine and Freshwater Research 52:1589–1598.

Lizaso, J. L. S., R. Goni, O. Renones, G. Charton, R. Galzin, J. T. Bayle, P. S. Jerez, A. P. Ruzafa, and A. A. Ramos. 2000. Density dependence in marine protected populations: a review. Environmental Conservation 27(2):144–158.

Malakoff, D. 2002. Picturing the perfect reserve. Science 296(5566):245.

Manriquez, P. H., and J. C. Castilla. 2001. Significance of marine protected areas in central Chile as seeding grounds for the gastropod *Concholepas concholepas*. Marine Ecology Progress Series 215:201–211.

McArdle, D. A. 1997. California marine protected areas. California Sea Grant College System, Publication T-039, University of California, La Jolla. Available: *www-csgc.ucsd.edu/EXTENSION/ CAFisheries/mpas.html* (January 2004).

McClanahan, T. R., and S. Mangi. 2001. The effect of a closed area and beach seine exclusion on coral reef fish catches. Fisheries Management and Ecology 8(2):107–121.

McNeill, S. E. 1994. The selection and design of marine protected areas: Australia as a case study. Biodiversity and Conservation 3:586–605.

Murray, M. R. 1998. The status of marine protected areas in Puget Sound. Puget Water Quality Action Team, Puget Sound/Georgia Basin Environmental Report Series 8, Olympia, Washington. Available: *www.wa.gov/puget_sound/shared/ download.html* (January 2004).

Murray, S. N., R. F. Ambrose, J. A. Bohnsack, L.W. Botsford, M. H. Carr, G. E. Davis, P. K. Dayton, D. Gotshall, D. R. Gunderson, M. A. Hixon, J. Lubchenco, M. Mangel, A. MacCall, D. A. McArdle, J. C. Ogden, J. Roughgarden, R. M. Starr, M. J. Tegner, and M. M. Yoklavich. 1999. No-take reserve networks: sustaining fishery populations and marine ecosystems. Fisheries 24(11):11–25.

National Marine Fisheries Service. 2001. Generic amendment addressing the establishment of the Tortugas Marine Reserves in the fishery management plans of the Gulf of Mexico available for public review and comment. National Marine Fisheries Service, Southeast Regional Office, Southeast Fishery Bulletin NR01-011, St. Petersburg, Florida. Available: *http://caldera.sero.nmfs.gov/ fishery/newsbull.001/news001.htm* (January 2004).

Paddack, M. J., and J. A. Estes. 2000. Kelp forest fish populations in marine reserves and adjacent exploited areas in central California. Ecological Applications 10(3):855–870.

Palsson, W. 2001. Marine refuges offer haven for Puget Sound fish. Fish and Wildlife Science, An Online Science Magazine from the Washington Department of Fish and Wildlife. Available: *http:// wdfw.wa.gov/science/articles/marine_sanctuary/ index.html* (January 2004)

Palumbi, S. 2001. MPA perspective: genetics, marine disposal distances, and the design of marine reserve networks. MPA News 2(8):5–6. Available: *http://depts.washington.edu/mpanews/MPA17.htm* (January 2004).

Parks, N. 2002. Rapid rewards of marine reserves. Science NOW. Available (with subscription): *http:// sciencenow.sciencemag.org/cgi/content/full/2002/ 516/3* (January 2004).

Pinnegar, J. K., N. V. C. Polunin, P. Francour, F. Badalementi, R. Chemello, M.-L. Harmelin-Vivien, B. Hereu, M. Milazzo, M. Zabala, G. D'Anna, and C. Pipitone. 2000. Trophic cascades in benthic marine ecosystems: lessons for fisheries and protected-area management. Environmental Conservation 27:179–200.

Pitcher, T., editor. 1997. The design and monitoring of marine reserves. University of British Columbia, Fisheries Centre, Fisheries Centre Research Reports 5(1), Vancouver. Available: *www.fisheries.ubc.ca/publications/reports/ report5_1.php* (January 2004).

Pitcher, T .J. 2001. Fisheries managed to rebuild ecosystems? Reconstructing the past to salvage the future. Ecological Applications 11(2):601–617.

Pitcher, T. J., R. Watson, N. Haggan, S. Guenette, R. Sylvie, R. Kennish, D. Sumaila, D. Cook, K. Wilson, and A. Leung. 2000. Marine reserves and the restoration of fisheries and marine ecosystems in the South China Sea. Bulletin of Marine Science 66(3):527–534.

Planes, S., R. Galzin, A. Garcia-Rubies, R. Goni, J. G. Harmelin, L. LeDireach, P. Lenfant, and A. Quetglas. 2000. Effects of marine protected areas on recruitment processes with special reference to Mediterranean littoral ecosystems. Environmental Conservation 27:1–18.

Rieman, B., and J. Clayton. 1997. Wildfire and native fish: issues of forest health and conservation of sensitive species. Fisheries 22(11):6–15.

Roberts, C. 2000. Selecting marine reserves and nonindigenous species. Bulletin of Marine Science 66(3):581–592.

Roberts, C., and N. Polunin. 1993. Marine reserves: simple solutions to managing complex fisheries? Ambio 22(6):364–368.

Roberts, C. M., J. A. Bohnsack, F. Gell, J. P. Hawkins, and F. Goodridge. Effects of marine reserves on adjacent fisheries. Science 294:1920–1923.

Roberts, C. M., B. Halpern, S. R. Palumbi, and R. R. Warner. 2001. Designing marine reserve networks: why small, isolated protected areas are not enough. Conservation Biology In Practice 2(3). Available: *http://cbinpractice.org/inpractice/article23DES.cfm* (January 2004).

Rodrigues, A. S. L., and K. J. Gaston. 2002. Optimisation in reserve selection procedures-why not? Biological Conservation 107(1):123–129.

Rogers-Bennett, L., and J. S. Pearse. 2001. Indirect benefits of marine protected areas for juvenile abalone. Conservation Biology 15(3)642–647.

Rosenberg, A., T. E. Bigford, S. Leathery, R. L. Hill, and K. Bickers. 2000. Ecosystem approaches to fishery management through essential fish habitat. Bulletin of Marine Science 66(3):535–542.

Rowe, S. 2001. Movement and harvesting mortality of American lobsters (*Homarus*) tagged inside and outside no-take reserves in Bonavista Bay, Newfoundland. Canadian Journal of Fisheries and Aquatic Sciences 58:1336–1346.

Russ, G., and A. Alcala. 1997. Do marine reserves export adult fish biomass? Evidence from Apo Island, Central Philippines. Marine Ecology Progress Series 132:1–9.

Sanchez Lizaso, J. L., R. Goni, O. Renones, J. A. Garcia-Charton, R. Galzin, J. T. Bayle, P. Sanchez-Jerez, A. Perez Ruzafa, and A. A. Ramos. 2000. Density dependence in marine protected populations: a review. Environmental Conservation 27:144–158.

Seitz, R., R. Lipicus, W. Stockhausen, and M. Montane. 2001. Efficacy of blue crab spawning sanctuaries in Chesapeake Bay. In G. H. Kruse, N. Bez, A. Booth, M. W. Dorn, S. Hills, and R. Hills, editors. Spatial processes and management of marine populations. Alaska Sea Grant, AK-SG-01-92, Fairbanks.

Shipp, R. L. 2002. No-take marine protected areas (MPAs) as a fishery management tool, a pragmatic perspective. Report to the FishAmerica Foundation, Alexandria, Virginia. Available: *www.asafishing.org/images/gais_shipp.pdf* (January 2004).

Simberloff, D. 2000. No reserve is an island: marine reserves and indigenous species. Bulletin of Marine Science 66(3):567–580.

Sobel, J., and C. Dahlgren. 2002. Marine reserves: a guide to science, design, and use. Island Press, Washington, D.C.

Southeast Fisheries Science Center and American Fisheries Society. 1995. Review of the use of marine fishery reserves in the U.S. southeastern Atlantic: proceedings of a symposium at the American Fisheries Society 125th Annual Meeting, 28–29 August, 1995, Tampa, Florida. National Oceanic and Atmospheric Administration, National Marine Fisheries Service, Southeast Fisheries Science Center, Miami.

Starr, R. M., and K. A. Johnson. 1998. Goal oriented marine reserves: a zoogeographic approach. In O. R. Magoon, H. Converse, B. Baird, and M. Miller-Henson, editors. Conference proceedings of California and the World Ocean '97: taking a look at California's ocean resources: an agenda for the future. American Society of Civil Engineers, Reston, Virginia.

Stockhausen, W., and R. Lipcius. 2001. Single large or several small marine reserves for the Caribbean spiny lobster? Marine and Freshwater Research 52:1605–1614.

Stockhausen, W., R. Lipcius, and B. Hickey. 2000. Joint effects of larval dispersal, population regulation, marine reserve design and exploitation on production and recruitment in the Caribbean spiny lobster. Bulletin of Marine Science 66(3):957–990.

The Ocean Conservancy. 2001. Marine and coastal protected areas in the United States Gulf of Maine region. Available: *www.oceanconservancy.org/dynamic/learn/publications/publications.htm* (January 2004).

Trexler, J. C., and J. Travis. 2000. Can marine protected areas restore and conserve stock attributes of reef fishes? Bulletin of Marine Science 66(3):853–873.

Vacchi, M., and others, editors. 1999. Fish visual census in marine protected areas: proceedings of the international workshop held at Ustica, Italy, 26–28 June 1997. Palermo: Societa siciliana di scienze naturali: Dipartimento di biologia animale dell'Universita di Palermo, Palermo, Italy.

Walters, C. 2000. Impacts of dispersal, ecological interactions, and fishing effort dynamics on efficacy of marine protected areas: how large should protected areas be? Bulletin of Marine Science 66(3):745–757.

Wing, K. 2001. Keeping oceans wild: how marine reserves protect our living seas. Natural Resources Defense Council, New York. Available: *www.nrdc.org/water/oceans/kow/kowinx.asp* (January 2004).

Legislation, Regulations, and Hearings

Sinclair, L. 2000. Protecting oceans and coastlines. Safety and Health 162(3):26–27.

Spadi, F. 2000. Navigation in marine protected areas: national and international law. Ocean Development and International Law 31(3):285–303.U.S. Office of the Federal Register. 2000. Executive Order on marine protected areas, Executive Order 13158. Federal Register 65:105(31 May 2000):34909–34911. Available: *http://mpa.gov/executive_order/execordermpa.pdf* (January 2004).

U.S. House of Representatives. Committee on Resources. Subcommittee on Fisheries Conservation, Wildlife, and Oceans. Testimony of Scott B. Gudes, Acting Under Secretary for Oceans and Atmosphere, Deputy Under Secretary for Oceans and Atmosphere, National Oceanic and Atmospheric Administration, U.S. Department of Commerce, on Chesapeake Bay oyster restoration, management, and research. 107th Congress, 1st session, October 22, 2001. Available: *www.legislative.noaa.gov/Testimony/102201gudes.html* (March 2004).

U.S. House of Representatives. House Resources Committee. Subcommittee on Fisheries Conservation, Wildlife, and Oceans. Written testimony of Timothy R. E. Keeney, Deputy Assistant Secretary for Oceans and Atmosphere, National Oceanic and Atmospheric Administration, U.S. Department of Commerce, on marine protected areas. 107th Congress, 2nd session, May 23, 2002. Available: *www.legislative.noaa.gov/Testimony/ 052302keeney.pdf* (March 2004).

U.S. House of Representatives. Subcommittee on Commerce, Justice, State, and Judiciary Appropriations. President's FY 2003 NOAA budget request before the subcommittee. 107th Congress, 2nd session, April 10, 2002. Available: *www.legislative.noaa.gov* (March 2004).

U.S. Office of the Federal Register. 2000. Executive Order on marine protected areas. Executive Order 13158. Federal Register 65:105(31 May 2000):34909–34911.

Management and Policy

Baker, J. L. 2000. Guide to marine protected areas. Prepared under contract to the Coast and Marine Section, Environment Protection Agency, Department for Environment and Heritage, South Australia, Adelaide. Available: *www.environment.sa.gov.au/ coasts/pdfs/mpa1.pdf* (March 2004).

Ballantine, W. Marine reserves—the need for networks. New Zealand Journal of Marine and Freshwater Research 25:115–116.

Barr, B., and J. Lindholm. 2000. Conservation of the sea using lessons from the land. George Wright Forum 17(3).

Barr, B. W. 2000. Establishing effective marine protected areas networks. *In* S. Bondrop-Nielsen and N. W. P. Munro, editors. 2002. Managing Protected Areas in a Changing World, Proceedings of the Fourth International Conference on Science and Management of Protected Areas, 14–19 May 2000. Science and Management of Protected Areas Association, Wolfville, Nova Scotia.

Boersma, P., and J. Parrish. 1999. Limiting abuse: marine protected areas, a limited solution. Ecological Economics 31:287–304.

Borrini-Feyerabend, G. 1998. Managing marine protected areas in partnership with communities. *In* Partnership for conservation: report of the regional workshop on marine protected areas, tourism, and communities. International Union for Conservation of Nature, Eastern Africa Regional Office, Nairobi, Kenya.

Botsford, L.W., A. Hastings, and S. D. Gaines. 2001. Dependence of sustainability on the configuration of marine reserves and larval dispersal distance. Ecology Letters 4(2):144–150.

Brodie, J. E., and I. McPhail. 1995. Science and management: the Great Barrier Reef Marine Park Authority experience. *In* Proceedings of the Second International Symposium and Workshop on Marine and Coastal Areas: integrating science and management. Office of Ocean and Coastal Resource Management, NOAA, and the U.S. Man and the Biosphere Reserve Program, Washington, D.C.

Brody, S. 1996. Marine protected areas in the Gulf of Maine: a survey of marine users and other interested parties. Maine State Planning Office and Gulf of Maine Council on the Marine Environment, Augusta. Available: *www.gulfofmaine.org/library/ mpas/survey.pdf* (March 2004).

Chiappone, M., editor. 2001. Fisheries investigations and management implications in marine protected areas of the Caribbean. Part 1 of 3 *in* Series on science tools for marine park management. The Nature Conservancy, Arlington, Virginia.

Chiappone, M., editor. 2001. Water quality conservation in marine protected areas: a case study of Parque Nacional del Este, Dominican Republic. Part 2 of 3 *in* Series on science tools for marine park management. The Nature Conservancy, Arlington, Virginia.

Conservation Law Foundation. 2001. Conservation coast to coast: comparing state-level action on marine protected areas in the Gulf of Maine, California, and Washington. Conservation Law Foundation, Boston. Available: *www.clf.org* (March 2004).

Craik, W. 1992. The Great Barrier Reef Marine Park: its establishment, development, and current status. Marine Pollution Bulletin 25(5–8):122–133.

Craik, W. 1996. The Great Barrier Reef Marine Park, Australia: a model for regional management. Natural Areas Journal 16(4):344–353.

Crosby, M. P. 1994. A proposed approach for studying ecological and socio-economic impacts of alternative access management strategies in marine protected areas. Pages 45–65 *in* D. J. Brunkhorst, editor. Marine protected areas and biosphere reserves: towards a new paradigm. Australian Nature Conservation Agency, Canberra.

Crosby, M. P. 1997. Moving towards a new paradigm for interactions among scientists, managers and the public in marine and coastal protected areas. Pages 10–24 *in* M. P. Crosby, D. Laffoley, C. Mondor, G. O'Sullivan, and K. Geenen, editors. Proceeding of the second international symposium and workshop on marine and coastal protected areas, July 1995. Office of Ocean and Coastal Resource Management, National Oceanic and Atmospheric Administration, Silver Spring, Maryland.

Crosby, M. P., A. Abu-Hilal, A. Al-Homoud, J. Erez, and R. Ortal. 2000. Interactions among scientists, managers and the public in defining research priorities and management strategies for marine and coastal resources: Is the Red Sea Marine Peace Park a new paradigm? Water, Air and Soil Pollution 123:581-594.

Crosby, M. P., B. Al-Bashir, M. Badran, S. Dweiri, R. Ortal, M. Ottolenghi, and A. Perevolotsky. 2002. The Red Sea Marine Peace Park: early lessons learned from a unique trans-boundary cooperative research, monitoring and management program. *In* Proceedings of the fourth conference on the protected areas of East Asia—benefits beyond boundaries in East Asia, March 18–23, 2002. Yangmingshan National Park, Taipei, Taiwan.

Crosby, M. P., and A. D. Beck. 1995. Management-oriented research in National Estuarine Research Reserves, with examples of fisheries-focused studies. Natural Areas Journal 15:12–20.

Crosby, M. P., K.S. Geenen, and R. Bohne. 2000. Alternative access management strategies for marine and coastal protected areas: a reference manual for their development and assessment. U.S. Man and the Biosphere Program, Washington, D.C.

Driml, S. 1994. Protection for profit: economic and financial values of the Great Barrier Reef World Heritage Area and other protected areas: a report to the Great Barrier Reef Marine Park Authority. Great Barrier Reef Marine Park Authority, Research Publication 35, Townsville, Queensland, Australia.

Eichbaum, W. M., and T. Agardy, 1995. The role of marine protected areas in comprehensive marine governance. *In* Proceedings of the Second International Symposium and Workshop on Marine and Coastal Protected Areas: integrating science and management. Office of Ocean and Coastal Resource Management, National Oceanic and Atmospheric Administration, and the U.S. Man and the Biosphere Program, Washington, D.C.

Foster, N., and M. H. Lemay, editors. 1986. Managing marine protected areas: an action plan. *In* International marine protected area management seminar, June 1–12, 1986. Marine and Estuarine Management Division, National Ocean Service, National Oceanic and Atmospheric Administration, Washington, D.C.

Fujita, R. M. 2002. Marine protected areas for protecting biological diversity, ecosystem integrity, and sustainable fisheries. Testimony of environmental defense concerning marine protected areas presented to the U.S. Commission on Ocean Policy, April 19, 2002, Los Angeles, California. Available: *www.environmentaldefense.org/documents/1992_Ocean%20Commission%2Ehtm* (January 2004).

Gilman, E. L. 1997. Community-based and multiple-purpose protected areas: a model to select and manage protected areas with lessons from the Pacific Islands. Coastal Management 25:59–91.

Gjerde, K. M. 2001. High seas marine protected areas—participant report of the Expert Workshop on Managing Risks to Biodiversity and the Environment on the High Seas, including tools such as marine protected areas: scientific requirements and legal aspects. International Journal of Marine and Coastal Law 16(3):515–528.

Gubbay, S. 1995. Marine protected areas: principles and techniques for management. Chapman and Hall, London.

Jones, P. J. S. 1994. A review and analysis of the objectives of marine nature reserves. Ocean and Coastal Management 24:179–198.

Kelleher, G. 1999. Guidelines for marine protected areas. IUCN–The World Conservation Union, Gland, Switzerland.

Kelleher, G., and R. Kenchington. 1991. Guidelines for establishing marine protected areas. IUCN–The World Conservation Union, Gland, Switzerland.

Kenchington, R. A. 1990. Managing marine environments. Taylor and Francis, New York.

McManus, J. W., C. van Zwol, L. R. Garces, and D. Sadacharan, editors. 1998. A framework for future training in marine and coastal protected area management. International Center for Living Aquatic Resources Management (ICLARM), Manila, and the Coastal Zone Management Centre, The Hague, The Netherlands.

Morin, T., M. C., J. Schubel, D. Shaw, and J. Pederson. 2002. Marine protected areas: a discussion with stakeholders in the Gulf of Maine, summer and fall 2001. New England Aquarium and the MIT Sea Grant College Program, Boston.

Muchiri, S. M., and V. Kalehe. 1999. Management of marine resources by zoning: the case of Kisite/Mpunguti marine protected area of Kenya's south coast. *In* Proceedings of the Conference on Advances on Marine Sciences in Zanzibar, Tanzania, 28 June–1 July 1999. Institute of Marine Sciences, Dar es Salaam University, Zanzibar, Tanzania.

National Academy of Public Administration. 2000. Protecting our national marine sanctuaries: a report by the Center for the Economy and the Environment. Center for the Economy and the Environment, Washington, D.C.

National Oceanic and Atmospheric Administration. 1999. Turning to the sea: America's ocean future. National Oceanic and Atmospheric Administration, Office of Public and Constituent Affairs, Washington, D.C. Available: *www.publicaffairs.noaa.gov/pdf/ocean_rpt.pdf* (March 2004).

National Research Council, Committee on Marine Governance Area and Management. 1997. Striking a balance: improving stewardship of marine areas. National Academy Press, Washington, D.C. Available: *http://bob.nap.edu/books/0309063698/html/index.html* (March 2004).

Phillips, A., editor. 2000. Financing protected areas: guidelines for protected area managers. Financing Protected Areas Task Force of the World Commission on Protected Areas (WCPA), and Economics Unit of IUCN, Best Practice Protected Areas Guidelines 5, Gland, Switzerland and Cambridge, UK. Available: *http://iucn.org/themes/wcpa/pubs/pdfs/Financing_PAs.pdf* (January 2004).

Ray, G. C. 1975. Critical marine habitats: definition, description, criteria and guidelines for identification and management. *In* Proceedings of an International Conference on Marine Parks and Reserves, May 12–14, Tokyo, Japan. IUCN–The World Conservation Union, IUCN Publication 37, Gland, Switzerland.

Recksiek, H., and G. Hinchcliff. March 2002. Marine protected areas needs assessment final report. National Oceanic and Atmospheric Administration Coastal Services Center in cooperation with the National MPA Protected Areas Center, National Oceanic and Atmospheric Administration, Charleston, South Carolina. Available: *www.csc.noaa.gov/cms/cls/MPANAFINAL.pdf* (March 2004).

Salm, R. V., J. R. Clark, and E. Siirila. 2000. Marine and coastal protected areas: a guide for planners and managers. IUCN–The World Conservation Union, Gland, Switzerland.

Salvat, B. 1975. Guidelines for the planning and management of marine parks and reserves. *In* Proceedings of an International Conference on Marine Parks and Reserves, May 12–14, 1975, Tokyo, Japan. IUCN–The World Conservation Union, IUCN Publicaton 37, Gland, Switzerland.

Scinto, L., editor. 1999. Strategies for developing and applying marine protected area science in Puget Sound/Georgia Basin, May 17–18, 1999, workshop, Bellingham, Washington. Puget Sound/Georgia Basin International Task Force, Olympia, Washington.

Scovazzi, T., editor. 1999. Marine speciality protected areas: the general aspects and the Mediterranean regional system. Kluwer Law International, International Environmental Law and Policy Series, volume 52, The Hague, The Netherlands, and Boston.

Syms, C., and M. H. Carr. 2001. Marine protected areas: evaluating MPA effectiveness in an uncertain world. Scoping paper presented at the Guidelines for Measuring Management Effectiveness in Marine Protected Areas Workshop, Monterey, California, May 1–3, 2001. North American Commission for Environmental Cooperation, Montreal, Quebec. Available: *www.biology.ucsc.edu/people/carr/Syms/syms_download_page.htm* (March 2004).

United Nations Environment Programme–Caribbean Environment Programme. 2000. Training manual: training of trainers course in marine protected areas management. United Nations Environment Programme–Caribbean Environment Programme, Caribbean Regional Coordinating Unit and Coastal Zone Management Centre. Available: *www.cep.unep.org/issues/MPA%20manual.htm* (March 2004).

Villa, F., L. Tunesi, and T. Agardy. 1992. Zoning marine protected areas through spatial multiple-criteria analysis: the case of the Asinara Island National Marine Reserve of Italy. Conservation Biology 16(2):515.

Socioeconomics

Badalamenti, F., A. A. Ramos, E. Voultsiadou, J. L. Sanchez-Lizaso, G. D'Anna, C. Pipitone, J. Mas, J. A. Ruiz Fernandez, D. Whitmarsh, and S. Riggio. 2000. Cultural and socio-economic impacts of Mediterranean marine protected areas. Environmental Conservation 27(2):110–125.

Brown, K., W. N. Adger, E. Tompkins, P. Bacon, D. Shim, and K. Young. 2001. Trade-off analysis for marine protected area management. Ecological Economics 37(3):417–434.

Bunce, L., K. Gustavson, J. Williams, and M. Miller. 1999. The human side of reef management: a case study analysis of the socioeconomic framework of Montego Bay Marine Park. Coral Reefs 18(4):369–380.

Dobryznski, T. J., and E. E. Nicholson. 2001. An evaluation of the short-term social and economic impacts of marine reserves on user groups in Key West. Nicholas School of the Environment, Duke University, Master's thesis, Durham, North Carolina.

Dobrzynski, T. J., and E. E. Nicholson. 2001. New study looks at the socioeconomic impacts of marine reserves in the Florida Keys. Coastal Society Bulletin 23(1):1, 14–18.

Guénette, S., R. Chuenpagdee, and R. Jones. 2000. Marine protected areas with an emphasis on local communities and indigenous peoples: a review. University of British Columbia, Fisheries Centre, Fisheries Centre Research Reports 8(1), Vancouver. Available: *www.fisheries.ubc.ca/publications/reports/report8_1.php* (January 2004).

Appendix 2: Institutions and Organizations

The following table is a listing of organizations, governmental and nongovernmental, involved in the aquatic protected areas field. This list is by no means complete and is simply meant to provide readers with a place to begin their research. Web site addresses are current as of May 2004; keep in mind that they may change. For more information, visit the federal marine protected areas web site at *www.mpa.gov*.

Table A2.1.—List of organizations in the aquatic protected areas field.

Institution or organization	Involvement	Web site address
	Governmental federal, national, or state programs affecting MPAs	
Atlantic States Marine Fisheries Commission	One of three interstate marine fishery commissions helping to manage and conserve coastal fisheries within the first 3 miles of the nation's coastline. Effective March 7, 2001, the commission approved a ban on horseshoe crab fishing in federal waters off the mouth of Delaware Bay in the newly named Carl N. Schuster, Jr., Horseshoe Crab Reserve.	*www.asmfc.org; www.publicaffairs noaa.gov/releases2001/feb01/noaa 01r104.html*
California Department of Fish and Game	California's Marine Life Protection Act (MLPA) requires the adoption of a Marine Life Protection Program and master plan to improve the system of marine protected areas in California. The act requires the California Department of Fish and Game to develop the plan for developing the statewide network. The web site provides information on the background, goals, ongoing process, and work accomplished of the MLPA.	*www.dfg.ca.gov/mrd/mlpa/ index.html*
Coastal Zone Management Program, Office of Ocean and Coastal Resource Management, National Ocean Service, National Oceanic and Atmospheric Administration	State–federal partnership to manage the nation's coastal resources, including specifically identified Geographic Areas of Particular Concern and Special Coastal Areas.	*www.ocrm.nos.noaa.gov/czm*
Environmental Protection Agency (EPA), Office of Water, Oceans, and Coastal Protection Division, Ocean Discharge Criteria	Executive Order 13158 ordered the EPA, relying upon existing Clean Water Act authorities, to issue new, science-based regulations, as necessary, "to better protect beaches, coasts, and the marine environment from pollution."	*www.epa.gov/owow/oceans/*
Federal Fisheries Management Program, Office of Sustainable Fisheries, National Marine Fisheries Service, National Oceanic and Atmospheric Administration	Manages the sustainable long-term harvest of over 700 commercially and recreationally important marine species under 40 fishery management plans. Protected areas include important spawning sites, Habitat Areas of Particular Concern, gear use and access limitation zones, and closed areas.	*www.nmfs.noaa.gov/*
Fish and California Game Commission	Expects to establish a network of MPAs, including marine reserves, in the next several years. Decisions will be based on California's Marine Life Protection Act, input from the public, review of proposals by the public, and biological, social, and economic information.	*www.dfg.ca.gov/fg_comm/ mlma_mlpa.html; www.dfg.ca.gov/ mrd/channel_islands/*

Table A2.1.—Continued.

Institution or organization	Involvement	Web site address
Fisheries management councils	Eight regional fishery management councils were established under the Magnuson Fishery Conservation and Management Act and the Sustainable Fisheries Act of 1996 to be responsible for the conservation and management of fish stocks from 3 to 200 miles off the coast of the United States. Several of the councils are involved in the process of setting up marine reserves or marine protected areas as tools to manage fisheries.	www.nmfs.noaa.gov/partnerships.htm
New England Fishery Management Council	Has a marine protected area committee; reports of recent meetings available.	www.nefmc.org
Pacific Fishery Management Council	Responsible for the conservation and management of fish stocks from 3 to 200 miles off the coasts of California, Oregon, and Washington. During a two-step process, the council is considering creating marine reserves as a tool to manage West Coast groundfish. Information on the "Marine Reserves Initiative, Phase I Marine Analysis for Public Review," is available on the web site.	www.pcouncil.org/
South Atlantic Fisheries Management Council	Considering the use of MPAs as a fishery management tool emphasizing the snapper group complex. The council has produced a "Marine Protected Areas Scoping Document" and will hold public scoping meetings to gather additional public comments regarding the use of and siting for MPAs.	www.safmc.net; www.safmc.net/library/MPApid02.pdf
Institute for Marine Protected Areas Training and Technical Assistance, Coastal Services Center, National Oceanic and Atmospheric Administration	Provides information, tools, and strategies for the design and effective management of marine protected areas for federal, state, local, tribal, and nongovernmental coastal resource managers, strengthening their ability to protect and enhance the nation's MPAs.	www.csc.noaa.gov/cms/cls/mpa_training.html
National Estuarine Research Reserve System, Office of Ocean and Coastal Resource Management, National Ocean Service, National Oceanic and Atmospheric Administration	Manages a national network of 25 estuarine research reserves in partnership with state governments.	http://nerrs.noaa.gov/
National Marine Sanctuaries Program, Office of Ocean and Coastal Resource Management, National Ocean Service, National Oceanic and Atmospheric Administration	Manages the nation's 13 national marine sanctuaries, which were established to manage and protect marine areas of nationally significant ecological, cultural, recreational, or aesthetic value.	www.sanctuaries.nos.noaa.gov/
Channel Islands Marine Sanctuary	Engaged in a joint process with the California Department of Fish and Game to consider marine reserves in the sanctuary. Included on the web site is the document, "A Recommendation for Marine Protected Areas in the Channel Islands National Marine Sanctuary."	www.cinms.nos.noaa.gov/
Florida Keys National Marine Sanctuary and Tortugas Ecological Reserve	Effective July 1, 2001, a no-take ecological reserve was established in federal waters of the Florida Keys National Marine Sanctuary. Web site addresses what makes the sanctuary special and how it was established.	www.fknms.nos.noaa.gov/tortugas

Table A2.1.—Continued.

Institution or organization	Involvement	Web site address
National Park Service, U.S. Department of the Interior	Manages the nation's system of national parks, many of which include marine areas. At least 39 national parks contain MPAs.	*www.nps.gov; www2.nature.nps.gov/ geology/coastal/index.htm*
Oregon Ocean Policy Advisory Council Working Group on Marine Protected Areas	Provides reports and recommendations to Oregon governor on MPA policies and actions for Oregon. Web site includes a dialogue center and meeting notices.	*www.oregonocean.org*
Office of Protected Resources, National Marine Fisheries Service, National Oceanic and Atmospheric Administration	Evaluates the status of marine species, including sea turtles, cetaceans, pinnipeds, marine and anadromous fish, plants, invertebrates and their habitats, and identifies those in need of increased protection as threatened or endangered under the Endangered Species Act, or depleted under the Marine Mammal Protection Act. The program develops conservation and recovery programs for these species, such as critical habitats.	*www.nmfs.noaa.gov/prot_res/ prot_res.html*
U.S. Fish and Wildlife Service, Department of the Interior	Manages the nation's system of national wildlife refuges, many of which include marine areas.	*www.fws.gov*

Academic

National Center for Ecological Analysis and Synthesis, University of California Santa Barbara	Theory of Marine Reserves Working Group focuses on theoretical basis for design and siting of marine reserves and protected areas. See "Scientific Consensus Statement on Marine Reserves and Marine Protected Areas."	*www.nceas.ucsb.edu; www.nceas.ucsb.edu/Consensus*
Natural Reserve System, University of California, Office of the President	Manages a number of coastal areas for research and study through its Natural Reserve System.	*http://nrs.ucop.edu/*

Research

Woods Hole Oceanographic Institution, Marine Policy Center	One of the research themes is "ocean use and protection, including studies of MPAs, the economics and management of ocean waste disposal and marine pollution, and fisheries management."	*www.whoi.edu/mpcweb/ overview_main.html*

Nongovernmental

American Fisheries Society	Founded in 1870, represents 10,000 fisheries scientists, and promotes fisheries science and research. The Society's Policy Statement #31a, "Protection of Marine Fish Stocks at Risk of Extinction," recommends the use of marine reserves and MPAs.	*www.fisheries.org/html/Public_ Affairs/Policy_Statements/ ps_31a.shtml*

Table A2.1.—Continued.

Institution or organization	Involvement	Web site address
COMPASS (Communication Partnership for Science and the Sea)	Collaboration among SeaWeb, Monterey Bay Aquarium, Island Press, and a board of scientific experts to advance marine conservation science and communicate its knowledge to policymakers, the public, and the media. A current focus is communicating the importance of immediately establishing marine reserves as an important oceans management tool. See the "Scientific Consensus Statement on Marine Reserves and Marine Protected Areas" released at the 2001 AAAS meeting.	*www.compassonline.org*
Conservation International	The Global Marine Program uses a systems approach to identify "priorities for coastal and marine conservation" and to "develop individual solutions that work together synergistically to best conserve whole ecosystem."	*www.conservation.org*
Conservation Law Foundation	Advocates a network of fully protected marine areas in the Gulf of Maine; see also its report "The Wild Sea: Saving our Marine Heritage."	*www.clf.org/hot/20001120.htm*
Cousteau Society	Recommends, with the Marine Conservation Biology Institute, a network of MPAs "to protect the biological diversity and integrity of U.S. waters."	*www.cousteausociety.org*
George Wright Society	Works on behalf of the science and heritage of protected areas. Publishes *The George Wright Forum* and sponsors biennial conferences.	*www.georgewright.org*
Marine Conservation Biology Institute	In a policy statement, "Safeguarding America's Seas: Establishing a National System of Marine Protected Areas, A Call", the institute calls upon the federal government to evaluate marine areas in the U.S. and establish a system of MPAs.	*www.mcbi.org*
Marine Ecosystem Health Program, University of California, Davis, Wildlife Health Center	Helping to develop the science needed by stakeholders to make sound scientific decisions regarding MPAs in the Pacific Northwest region. Key components of their work include funding research on MPAs, sharing research findings, and ensuring that pertinent scientific data are available to managers, policymakers, and concerned citizens.	*http://mehp.vetmed.ucdavis.edu*
The Ocean Conservancy (formerly Center for Marine Conservation)	The Ocean Conservancy is a nonprofit organization dedicated to protecting marine life. Action Item 5 of its "Agenda for the Oceans" urges federal and state governments to strengthen and expand MPAs.	*www.oceanconservancy.org*
ReefGuardian International (formerly ReefKeeper International)	Working with a group of citizens from the Fort Lauderdale area, the Greater Fort Lauderdale MPA Committee, to establish two MPAs offshore of Broward County, Florida. Has made a formal request to the Florida Fish and Wildlife Conservation Commission for the establishment of these MPAs.	*www.reefguardian.org*

Table A2.1.—Continued.

Institution or organization	Involvement	Web site address
Pacific Coast Federation of Fishermen's Associations	MPAs page has links to policy statements, articles, books, organizations and other resources on MPAs.	www.pcffa.org/MPA.htm
Science and Management of Protected Areas Association (SAMPAA)	Encourages the use of science in protected areas management.	www.sampaa.org
Sustainable Ecosystems Institute (SEI)	Organization of scientists "committed to using their technical expertise to help solve ecological issues." Involved in marine reserves in the Caribbean and is developing monitoring and conservation guidelines for MPAs in the USA.	http://sei.org/ocean-protect.html
World Resources Institute (WRI)	Discusses MPAs as a promising tool for protecting marine biodiversity and marine habitats.	www.wri.org/wri/wr-96-97/ bi_txt6.html
International		
The World Conservation Union (IUCN), World Commission on Protected Areas	MPAs are one of the theme programs. Its goals include the establishment of a global network of marine protected areas, demonstrating the effectiveness of MPAs, and encouraging sustainable tourism.	http://iucn.org/themes/wcpa/
United Nations Environment Programme, Caribbean Environment Programme	The Special Protected Areas and Wildlife Program (SPAWP) supports MPA managers, meetings, training, and publications.	www.cep.unep.org/who/spaw.htm
Australia		
Australian Marine Conservation Society (AMCS)	Web site provides links to society policy and papers, description of Australia's involvement with MPAs, links to Australian government departments involved in MPAs, links to Australian MPA sites, and more.	www.amcs.org.au/links/mpa.htm
Coast and Marine Section, Environment Protection Agency, Department for Environment and Heritage, Government of South Australia	South Australia has set 2003 as the target date for establishment of a representative system of MPAs.	www.environment.sa.gov.au/coasts
Environment Australia's Marine Protected Areas, Commonwealth of Australia	Australia has several MPA programs, including a National Representative System of MPAs and the MPAs Program.	www.deh.gov.au/coasts/mpa
State of Victoria, Australia	Established a system of marine national parks and marine sanctuaries to protect representative examples of Victoria's marine environment. The system will cover approximately 54 000 hectares or 5.3 percent of Victoria's marine waters and will complement its system of terrestrial national parks and reserves. The web site also provides links to the recent legislation, minister's statement on the legislation, fact sheets, maps, and more.	www.dse.vic.gov.au/dse

Table A2.1.—Continued.

Institution or organization	Involvement	Web site address
	Canada	
Department of Fisheries and Oceans, Pacific Region, Canada	MPAs program includes a joint strategy document with British Columbia, four pilot MPAs, and the Central Coast Land and Coastal Resource Management Planning (CCLCRMP) process.	*www.pac.dfo-mpo.gc.ca/oceans/mpa*
Environment Canada, Canadian Wildlife Service	"Marine protected areas" is one of several programs utilized by the to further wildlife habitat conservation.	*www.cws-scf.ec.gc.ca*
National Marine Conservation Areas - Parks Canada charged with setting up a national system of MPAs.	Parks Canada National Marine Conservation Areas program is	*http://parkscanada.pch.gc.ca*
North American Commission for Environmental Cooperation	Through its North American Marine Protected Areas Network, promotes opportunities for collaboration and exchange among MPA practitioners and managers from Canada, USA, and Mexico.	*www.cec.org/programs_projects; www.cec.org/files/PDF/ BIODIVERSITY/NA-MPA-Network.pdf; www.cec.org/files/ PDF/BIODIVERSITY/ 215_e_EN.pdf*
Wildlife Conservation Society	Actively works to establish new marine reserves and conducts research to understand their effectiveness. Its new center in Papua New Guinea will study existing marine reserves from a socio-economic/artisanal fisheries perspective and reserves' effects on fish diversity and benthic invertebrate diversity. Working to establish new marine reserves in Antongil Bay, Madagascar, Belize, and Kenya.	*http://wcs.org/home/wild/marine/ 1733*
World Wildlife Fund Global	The Endangered Seas campaign works to establish MPAs as management tools to safeguard fisheries and marine biodiversity and has released a guide and toolkit providing examples of research and fully-protected marine reserves a round the world. WWF's Marine Ecosystems' South Pacific Programme works with Marine Protected Areas in Ono Island, Kadavu, Fiji, the Cook Islands, and Papua New Guinea.	*www.panda.org/endangeredseas/mpa; www.wwfpacific.org.fj/ono1.htm; www.wwfpacific.org.fj/cooks.htm; www.panda.org*

Appendix 3: Aquatic Protected Areas Web Sites

This collection includes web sites that provide information directly related to the management and science of marine protected areas (MPAs). Most of the sites in this list are portions of larger sites and provide offerings specific to MPAs. Web sites included in this list focus on national systems and describe MPA programs, research, and policies. Sites are listed in alphabetical order by name of site. Web site addresses are current as of May 2004; keep in mind that they may change.

Table A3.1.—Web sites with information directly related to the management and science of marine protected areas.

Name of site	Institution	URL address	Description
Bibliography of Marine Protected Areas	Department of Fisheries and Oceans, Canada	*www.dfo-mpo.gc.ca*	A bibliography of journal articles, conference proceedings, and reports from 1978–1995.
Bibliography of Marine Reserves	Florida Keys National Marine Sanctuary	*www.fknms.nos.noaa.gov/ tortugas/benefits/biblio/s.html*	Important papers on the theory and design of marine reserves and marine protected areas.
California Ocean Resources Management Program	California Resources Agency	*http://ceres.ca.gov/cra/ocean/*	Includes the report Improving California's System of Marine Managed Areas: Final Report of the State Interagency Marine Managed Areas Work Group.
California's Marine Life Protection Act	California Department of Fish and Game	*www.leginfo.ca.gov/bilinfo.html* (1999–2000 session, bill AB993)	Purpose of the Act is to improve the array of marine protected areas existing in California waters through the adoption of a Marine Life Protection Program and a comprehensive master plan.
Channel Islands MPA Network Recommendation	California Department of Fish and Game and the Channel Islands National Marine Sanctuary	*www.dfg.ca.gov/fg_comm/*	California Department of Fish and Game and the Channel Islands National Marine Sanctuary presented its report, A Recommendation for Marine Protected Areas in the Channel Islands National Marine Sanctuary to the California Fish and Game Commission on August 24, 2001.
Coral Parks Program, Coral Reef Alliance (CORAL)	Coral Reef Alliance (CORAL)	*www.coralreefalliance.org/parks/*	CORAL, a nonprofit, member-supported organization, will help to establish new marine parks to protect coral reefs by building partnerships around the world.
Course on Advanced Topics in Environmental Policy: Marine Protected Areas, ESM 297	Donald Bren School of Environmental Science and Management, University of California, Santa Barbara	*www.esm.ucsb.edu*	The course reviewed the role of MPAs as marine management tools and examined theoretical and applied issues surrounding marine reserves policy. Taught by Dr. Matthew Cahn and Dr. Bruce Kendall.
Course on Marine Protected Areas	Dalhousie University	*www.dal.ca*	The goal of the course was to provide marine professionals with the "state of the art in MPA theory, design and operation."
The Curtis and Edith Munson Distinguished Lecture Series 2001: Marine Protected Areas, Translating Science into Practice	Sponsored by the Center for Coastal and Watershed Systems, Yale School of Forestry and Environmental Studies	*www.yale.edu/ccws*	A series of 12 lectures that addressed critical biological, ecological, social, and economic issues pertaining to marine protected areas.

Table A3.1.—Continued.

Name of site	Institution	URL address	Description
Empty Oceans, Empty Nets	Habitat Media	*www.habitatmedia.org/index/html, www.pbs.org/emptyoceans, www.pbs.org/emptyoceans/cod/ index.html*	A two-part PBS series produced by Habitat Media examines the state of global fisheries and the efforts being made to restore them. The documentary, which aired on many public television stations on or near April 22, 2002, includes footage of marine environments around the world and interviews with fishermen, scientists, fish farmers, and citizens.
Fish or Famine	The Christian Science Monitor	*www.csmonitor.com/2001/0906/ p11s1-sten.html*	Peter N. Spotts, staff writer for the Christian Science Monitor, reviews the marine protected areas symposium held at Woods Hole Oceanographic Institution on August 27–29, 2001.
Geographic information system (GIS)— Marine and Coastal Protected Areas Database	Center for Marine Conservation, funded by U. S. Environmental Protection Agency Office of Wetlands, Oceans, and Watersheds	*www.epa.gov/owow/oceans/ maps/#mcpa*	GIS database of Federal Marine and Coastal Protected Areas in the United States and its territories, including Puerto Rico, U.S. Virgin Islands, Guam, American Samoa, and other possessions.
Protected Areas Geographic Information System (PAGIS) homepage	PAGIS Team, Coastal Services Center, National Ocean Service, National Oceanic and Atmospheric Administration	*www.csc.noaa.gov/pagis/index.htm*	The PAGIS project is designed to build GIS, spatial data management, and Internet capabilities at all National Estuarine Research Reserves and National Marine Sanctuaries. CD-ROMS with base data layers for each National Estuarine Research Reserve are available, with development of data for National Marine Sanctuaries in progress. Base layers include digital raster graphics, digital orthophoto quadrangles, nautical charts, and a digital boundary.
Giving Wild Fish a Break	The Christian Science Monitor	*www.csmonitor.com/durable/2001/ 05/03/fp18s2-csm.shtml*	Robert C. Cowen interviews authors of the Scientific Consensus Statement on Marine Reserves and Marine Protected Areas released by the National Center for Ecological Analysis and Synthesis at the University of California in Santa in Barbara during the American Association for the Advancement of Science Annual Meeting in February 2001.

Table A3.1.—Continued.

Name of site	Institution	URL address	Description
Global Directory of Marine (and Freshwater) Professionals (GLODIR)	Intergovernmental Oceanographic Commission of United Nations Educational, Scientific, and Cultural Organization	*www.nova.edu/ocean/glodir.html*	GLODIR is a directory of professionals worldwide engaged in marine and/or freshwater research and management. GLODIR's information includes: name, job title, job type, organization, address, description of activities by keyword, environment (marine, freshwater, brackish), and citations of papers for those included in the directory. To locate someone involved in MPA research and management, search on the activities keyword "marine protected areas." Users can also add their information to the directory, allowing others to locate them and their organizations.
Great Barrier Reef Marine Park Authority	Commonwealth of Australia	*www.gbrmpa.gov.au*	The Great Barrier Reef Marine Park Authority is the lead agency for Great Barrier Reef World Heritage Area issues. The Authority is the principal adviser to the Commonwealth Government on the care and development of the Great Barrier Reef Marine Park.
Gulf of Maine Protected Areas Project	National Oceanic and Atmospheric Administration, Ocean and Coastal Resource Management, Gulf of Maine Council on the Marine Environment, Woods Hole Marine Policy Center, Center for Marine Conservation, New England Aquarium, Center for Economic Enterprise	*www.gulfofmaine.org/library/ mpas/mpa.htm*	The project seeks to establish a network of MPAs in the Gulf of Maine region. A system of linked sites would "help to build a framework for an ecosystem approach to the management of marine resources" in the Gulf of Maine. The web site is an information clearinghouse for the project. Included are a definition of MPAs, background information on the project, a collection of full-text documents and reports, and a bibliography of articles, reports, and books on MPA topics.

Table A3.1.—Continued.

Name of site	Institution	URL address	Description
The Hawaii Coral Reef Assessment and Monitoring Program (CRAMP)	CRAMP	*http://cramp.wcc.hawaii.edu*	CRAMP is administered under the University of Hawaii in collaboration with the State of Hawaii Department of Land and Natural Resources, Division of Aquatic Resources, and includes scientists and managers from the Bishop Museum, Oceanic Institute, and Waikiki Aquarium. CRAMP is a research program designed to identify the controlling factors, both natural and anthropogenic, contributing to the stability, decline, or recovery of Hawaiian reefs.
The Hawaii Coral Reef Network	Washington State University	*www.coralreefnetwork.com/mpa/ default.htm, www.coralreefnetwork.com/ kona/default.htm*	Sponsors a comprehensive web site with information on the location, regulations, and research on MPAs in Hawaii and a detailed description of scientific research in a network of nine MPAs in Hawaii conducted during the West Hawaii Aquarium Project.
International Conference on the Economics of Marine Protected Areas	Fisheries Centre, University of British Columbia	*www.fisheries.ubc.ca*	A conference held for MPA practitioners to discuss the economics of marine protected areas, particularly the use of MPAs as
International MPA Site - Red Sea Marine Peace Park	Red Sea Marine Peace Park	*http://130.94.155.6/page2.html*	The Red Sea Marine Peace Park (RSMPP), coral reserves in the Gulf of Aqaba, is run cooperatively by Israel and Jordan.
California Marine Protected Areas Network Listserv	California Sea Grant Extension Program	*http://aquanic.org/infosrcs/marine.htm*	The listserve promotes information exchange among those interested in MPAs.
Fishfolk Listserv	MIT SeaGrant Advisory Program, Center for Marine Social Sciences	*http://web.mit.edu/seagrant/advisory/ cmssprojects.html#fishfolk*	Fishfolk is a listserv devoted to fisheries issues of relevance to social scientists, ishermen, biologists, managers, government officials, conservation group members, attorneys, and academia. Discussion of MPAs occurs on a regular basis. Excellent opportunity for communication among a wide variety of stakeholders.
Gulf of Maine Protected Areas Listserver	Stellwagen Bank National Marine Sanctuary	*http://stellwagen.nos.noaa.gov/ management*	Moderated email discussion forum hosted by the Stellwagen Bank National Marine Sanctuary and endorsed by the Gulf of Maine Council on the Marine Environment. Purpose is to facilitate an open discussion about the role of MPAs in the Gulf of Maine.

Table A3.1.—Continued.

Name of site	Institution	URL address	Description
WHALEWATCH Listserv	North American Commission for Environmental Cooperation and the Baja and California to Bering Sea Marine Conservation Initiative	*www.responsiblewhalewatching.org/ subscribe.html*	Dedicated to promoting sustainable tourism and marine protected areas on the West Coast of North America from the Baja to the Bering.
Marine Protected Areas and Marine Reserves	Pacific Coast Federation of Fishermen's Associations	*www.pcffa.org/MPA.htm*	Links to MPA web sites.
Marine Protected Areas of the United States	National Oceanic and Atmospheric Administration	*www.mpa.gov*	Information on MPAs as required by Executive Order 13158. It is jointly managed by the U.S. Department of Commerce and the U.S. Department of the Interior.
MPA Corner	United Nations Environment Programme, Caribbean Environment Programme	*www.cep.unep.org/pubs/cepnews/ v16n1mpa%20corner.htm*	News and links about MPAs.
MPA News: International News and Analysis on Marine Protected Areas	University of Washington, School of Marine Affairs	*www.mpanews.org*	Analyses of current activities in MPAs around the world, reports from MPA scientists, planners and managers, news from conference proceedings, book reviews, useful web sites; published monthly.
National Fisheries Conservation Center	National Fisheries Conservation Center	*www.nfcc-fisheries.org/mr_io.html*	Hosts a moderated online discussion on the use of marine reserves, provides opinion pieces written by leaders in the field with a diverse range of perspectives and interests, and provides links to web sites covering marine reserves issues.
NOAA Photo and Image Collection, National Marine Sanctuaries Database	National Oceanic and Atmospheric Administration Central Library and the Office of High Performance Computing and Communications	*www.photolib.noaa.gov/sanctuary/ index.html*	Contains more than 16,000 public-domain photos and images covering the oceans, atmosphere, and history of the pioneers who began the study of the environment in the USA. The National Marine Sanctuaries Database contains images of natural and cultural treasures from all of the U.S. national marine sanctuaries.
North American Marine Protected Areas Network	Commission for Environmental Conservation	*www.cec.org/programs_projects*	The goal of the project is to establish a permanent network of North American MPAs and promote information exchange and coordination among Canada, Mexico, and the United States.

Table A3.1.—Continued.

Name of site	Institution	URL address	Description
Northwestern Hawaiian Islands Coral Reef Ecosystem Reserve	National Ocean Service, National Oceanic and Atmospheric Administration	http://hawaiireef.noaa.gov	Created by President Clinton on December 4, 2000 by Executive Order 13178. Web site provides the public with information on the reserve and ongoing activities related to its management. NOAA has begun the process for designating the reserve as a national marine sanctuary.
Oceans Canada: Marine Protected Areas	Oceans Canada, Fisheries and Oceans Canada	www.dfo-mpo.gc.ca/oceanscanada/ newenglish/htmdocs/mpas/mpa.htm	Oceans Canada is a comprehensive web site designed to provide Canadians with a broad overview of oceans issues, activities and programs, from the federal government and beyond. The Marine Protected Areas section of the web site provides information on Canada's use of MPAs, links to background information, bibliographies, brochures, fact sheets, minister's statements, national framework processes, news releases, and policy statements.
PARKS: The International Journal for Protected Area Managers	The World Conservation Union, World Commission on Protected Areas	http://wcpa.iucn.org/pubs/ publications.html	Published to strengthen international collaboration among protected area professionals and to enhance their role, status and activities. Recent issues are available full-text on the web site. Volume 8, number 2 (June 1998) is devoted to MPAs. Volume 10, number 2 is devoted to non-material values of protected areas.
SeaWeb Ocean Citations: Selected Science Publications on Ocean Issues	SeaWeb	www.seaweb.org/background/ abstracts/	Citations to selected articles from scientific journals, conferences and proceedings, and other documents on MPAs and other ocean topics.
United States Coral Reef Task Force Documents	United States Coral Reef Task Force	http://coralreef.govdoc.cfm	Includes the documents "National Action Plan for Coral Reef Conservation," "Oversight of Agency Actions Affecting Coral Reef Protection," "Coral Reef Protected Areas: A Guide for Management," and "Building a National Network of MPAs for Coral Reefs."